T0374408

Bioconversion of Wastes to Value-added Products

Bioconversion of agricultural and industrial wastes into useful products plays an important role both in the economy and in the prevention of environmental pollution. This book presents technological approaches to the biotransformation of different wastes into valuable products and demonstrates developments in the field of organic waste disposal. Organized in four parts, *Bioconversion of Wastes to Value-added Products* addresses the bioconversion of wastes to (a) new food products, (b) energy, (c) biotechnological products, and (d) describes the construction of biosensors for food control.

Features:

- Covers the use of different food wastes to enrich meat, dairy, bakery and confectionery products
- Presents new technologies for utilization of wastes from the meat, dairy and wine industries, among others
- Promotes bioconversion of agricultural wastes into energy such as hydrogen or biogas
- Proposes the use of industrial wastes to produce exopolysaccharides using bacteria or macromycetes
- Describes design, construction and testing of biosensors for food control

The book is an aid to scientists and engineers contributing to manufacturing of useful products from non-recyclable wastes, as well as the creation of environmentally friendly technologies that protect the environment from potential contaminants.

Food Biotechnology and Engineering

Series Editor:
Octavio Paredes-López

Volatile Compounds Formation in Specialty Beverages
Edited by Felipe Richter Reis and Caroline Mongruel Eleutério dos Santos

Native Crops in Latin America
Biochemical, Processing, and Nutraceutical Aspects
Edited by Ritva Repo-Carrasco-Valencia and Mabel C. Tomás

Starch and Starchy Food Products
Improving Human Health
Edited by Luis Arturo Bello-Perez, Jose Alvarez-Ramirez, and Sushil Dhital

Bioenhancement and Fortification of Foods for a Healthy Diet
Edited by Octavio Paredes-López, Oleksandr Shevchenko, Viktor Stabnikov, and Volodymyr Ivanov

Latin-American Seeds
Agronomic, Processing and Health Aspects
Edited by Claudia Monika Haros, María Reguera, Norma Sammán, and Octavio Paredes-López

Advances in Plant Biotechnology
In Vitro Production of Secondary Metabolites of Industrial Interest
Edited by Alma Angélica Del Villar-Martínez, Juan Arturo Ragazzo-Sánchez, Pablo Emilio Vanegas-Espinoza, and Octavio Paredes-López

Bioconversion of Wastes to Value-added Products
Edited by Olena Stabnikova, Oleksandr Shevchenko, Viktor Stabnikov, and Octavio Paredes-López

For more information about this series, please visit: www.routledge.com/Food-Biotechnology-and-Engineering/book-series/CRCFOOBIOENG

Bioconversion of Wastes to Value-added Products

Edited by

Olena Stabnikova, Oleksandr Shevchenko, Viktor Stabnikov, and Octavio Paredes-López

CRC Press is an imprint of the
Taylor & Francis Group, an informa business

Designed cover image: © Shutterstock

First edition published 2024
by CRC Press
6000 Broken Sound Parkway NW, Suite 300, Boca Raton, FL 33487-2742

and by CRC Press
4 Park Square, Milton Park, Abingdon, Oxon, OX14 4RN

CRC Press is an imprint of Taylor & Francis Group, LLC

ISBN: 9781032348797 (hbk)
ISBN: 9781032359830 (pbk)
ISBN: 9781003329671 (ebk)

DOI: 10.1201/9781003329671

Typeset in Kepler
by codeMantra

Contents

Series Preface

BIOTECHNOLOGY—OUTSTANDING FACTS

The beginning of agriculture started about 12,000 years ago and ever since has played a key role in food production. We look to the farmers to provide the food we need but at the same time, now more than ever, to farm in a manner compatible with the preservation of the essential natural resources of the earth. Additionally, besides the remarkable positive aspects that farming has had throughout history, several undesirable consequences have been generated. The diversity of plants and animal species that inhabit the earth is decreasing. Intensified crop production has had undesirable effects on the environment (*i.e.*, chemical contamination of groundwater, soil erosion, exhaustion of water reserves). If we do not improve the efficiency of crop production in the short term, we are likely to destroy the very resource base on which this production relies. Thus, the role of the so-called sustainable agriculture in the developed and underdeveloped world, where farming practices are to be modified so that food production takes place in stable ecosystems, is expected to be of strategic importance in the future, but the future has already arrived.

Biotechnology of plants is a key player in these scenarios of the twenty-first century; nowadays, especially molecular biotechnology is receiving increasing attention because it has the tools of innovation for the agriculture, food, chemical and pharmaceutical industries. It provides the means to translate an understanding of, and ability to modify, plant development and reproduction into enhanced productivity of traditional and new products. Plant products, either seeds, fruits, and plant components and extracts, are being produced with better functional properties and longer shelf life, and they need to be assimilated into commercial agriculture to offer new options to small, and more than small, industries and finally to consumers. Within these strategies, it is imperative to select crops with larger proportions of edible parts as well, thus generating less waste; it is also imperative to consider the selection and development of a more environmentally friendly agriculture.

The development of research innovations for products is progressing, but the constraints of relatively long times to reach the marketplace, intellectual property rights, uncertain profitability of the products, consumer acceptance and even caution and fear with which the public may view biotechnology are tempering the momentum of all but the most determined efforts. Nevertheless, it appears uncontestable that food biotechnology of plants and microbials is and will emerge as a strategic component to provide food crops and other products required for human well-being.

FOOD BIOTECHNOLOGY AND ENGINEERING SERIES/ OCTAVIO PAREDES-LÓPEZ, PHD. SERIES EDITOR

The Food Biotechnology and Engineering Series aims at addressing a range of topics around the edge of food biotechnology and food microbial world. In the case of foods, it includes molecular biology, genetic engineering and metabolic aspects, science, chemistry, nutrition, medical foods and health ingredients, and processing and engineering with traditional- and innovative-based approaches. Environmental aspects to produce green foods cannot be left aside. At the world level, there are foods and beverages produced by different types of microbial technologies to which this series will give attention. It will also consider genetic modifications of microbial cells to produce nutraceutical ingredients, and advances of investigations on the human microbiome as related to diets.

TITLE: BIOCONVERSION OF WASTES TO VALUE-ADDED PRODUCTS

Editors: Olena Stabnikova, Oleksandr Shevchenko, Viktor Stabnikov, and Octavio Paredes-López

In the last century, problems associated with wastes from industries in general, agriculture and urban sites, received occasionally minor attention or no attention at all. It was estimated that there was enough space in the external environment to simply dump wastes and allow natural processes to transform or dispose of those which could be biodegradable.

Nowadays situation has changed dramatically. The present overpopulation, with societies which demand at the same time better ways of life including basic and selected food diets, with increasing water and energy consumption, among other intensive requirements, is producing notorious negative impacts on the mother nature. This is giving place to circumstances no longer sustainable in conjunction with the visible and recent effects produced by climatic changes.

There is today an increasing recognition of pollution problems generated by human groups which are, unconsciously up to certain extent, important generators of wastes; it is evident that high-income societies give place consequently to a higher level of wastes. Fortunately, the recycling of resources is becoming an important part of the solution to some of the most pressing problems.

As the book editors have clearly stated, the development of biotechnology has become a remarkable option for the bioconversion of wastes and by-products; thus, recycling is becoming a viable environmental protection and an economic activity. Some selected technological processes also show the potential to transform nonrecyclable wastes into useful products.

This book contains 11 chapters divided into four sections: Bioconversion of Wastes in New Food Products, Bioconversion of Waste into Energy, Bioconversion of Waste into Biotechnological Products and Biosensors for Food Control. They were written by 45 authors, 30 ladies and 15 gentlemen, most of them from Ukraine, plus two scientists from Pakistan and Poland. These well-known experts, including some young and intelligent researchers, are doing basic and applied research, and innovation in 15 different academic, scientific and technological institutions. During the development of this publication, editors and authors had in mind the purpose of inducing the reader in the exciting world of finding solutions to the challenges posed by the transformation of wastes and by-products because of the paramount importance of protecting the environment; challenges extended to the preservation of mother nature for the present and future generations of the twenty-first century.

It is a good opportunity to express full recognition and appreciation to all editors and authors for their excellent contribution, and to the editorial staff of CRC Press as well, especially to Ms. Laura Piedrahita, Mrs. Beth Hawkins and Mr. Stephen Zollo.

Preface

Organic wastes from the agriculture and food industry, as well as industrial and domestic wastewaters, contain valuable substances, some of which can be used as an additional source of important components to improve the nutritional properties of traditional foods; others can be converted into energy materials such as alcohol, bioethanol, biogas and hydrogen. Application of a variety of wastes from industry, agriculture and food processing to turn them into valuable products pursues two goals at the same time, namely, the systematic and economically feasible use of the existing potential of useful products from nonrecyclable wastes, as well as the creation of environmentally friendly technologies that protect the environment from potential contaminants. This publication covers application of organic wastes as additives to enhance nutritional value of food, bioconversion of solid and liquid wastes into energy, such as biogas and hydrogen, and biotransformation of industrial and food processing wastes into biotechnological products. At the end, in a chapter, design, construction and testing of biosensors for food control are described in detail.

Thus, this book represents technological approaches to biotransformation of different wastes into valuable products, demonstrates real developments in the field of organic waste disposal, and thus helps students at bachelor and graduate levels, lecturers, researchers and engineers of the food industry and environmental engineering by providing new knowledge and strategies on sustainable use of various wastes.

Olena Stabnikova, *National University of Food Technologies, Ukraine*

Oleksandr Shevchenko, *National Academy of Sciences, Ukraine*

Viktor Stabnikov, *National University of Food Technologies, Ukraine*

Octavio Paredes-López, *Instituto Politécnico Nacional, Mexico*

List of Contributors

Greta Adamczyk
Department of Food Technology and Human Nutrition
University of Rzeszow, Institute of Food Technology and Nutrition
Rzeszow, Poland

Zubair Ahmed
US-Pakistan Center for Advanced Studies in Water
Mehran University of Engineering and Technology
Jamshoro, Pakistan

Aleksandr Alexandrov
Department of Food Technology, Light Industry and Design
Ukrainian Engineering-Pedagogics Academy
Kharkiv, Ukraine

Valentyna Arkhypova
Department of Biomolecular Electronics
Institute of Molecular Biology and Genetics of the National Academy of Sciences of Ukraine
Kyiv, Ukraine

Olena Bilyk
Department of Bakery and Confectionary Goods Technology
National University of Food Technologies
Kyiv, Ukraine

Nina Bisko
Department of Mycology
Kholodny Institute of Botany, NAS of Ukraine
Kyiv, Ukraine

Olga Blahyi
Department of Food Technology, Light Industry and Design
Ukrainian Engineering-Pedagogics Academy
Kharkiv, Ukraine

Tatiana Brykova
Department of Technology and Organization of Hotel and Restaurant Business
Institute of Trade and Economics of National University of Trade and Economics
Chernivtsi, Ukraine

Vera Drobot
Department of Bakery and Confectionery Goods Technology
National University of Food Technologies
Kyiv, Ukraine

Anastsia Dubivko
Department of Milk and Dairy Technology
National University of Food Technologies
Kyiv, Ukraine

Sergei Dzyadevych
Department of Biomolecular Electronics
Institute of Molecular Biology and Genetics of the National Academy of Sciences of Ukraine
Kyiv, Ukraine

Larysa Dzyhun
Department of Industrial Biotechnology and Biopharmacy
National Technical University of Ukraine "Igor Sikorsky Kyiv Polytechnic Institute"
Kyiv, Ukraine

Oksana Fursik
Department of Technology of Meat and Meat Products
National University of Food Technologies
Kyiv, Ukraine

Nataliia Golub
Department of Bioenergy, Bioinformatics and Environmental Biotechnology
National Technical University of Ukraine "Igor Sikorsky Kyiv Polytechnic Institute"
Kyiv, Ukraine

Olena Gorodyska
Department of Chemistry, Technologies and Pharmacy
T.H. Shevchenko National University "Chernihiv Colehium"
Chernigiv, Ukraine

Nataliya Grevtseva
Department of International e-Commerce and Hotel and Restaurant Business
V.N. Karazin Kharkiv National University
Kharkiv, Ukraine

Sergey Gubsky
Department of Chemistry, Biochemistry, Microbiology and Hygiene of Nutrition
State Biotechnology University
Kharkiv, Ukraine

Tetiana Ivanova
Laboratory of Biofuel Biotechnology and Innovations in Green Energy
Institute of Food Biotechnology and Genomics, NAS of Ukraine
Kyiv, Ukraine

Inna Klechak
Department of Industrial Biotechnology and Biopharmacy
National Technical University of Ukraine "Igor Sikorsky Kyiv Polytechnic Institute"
Kyiv, Ukraine

Viktoriia Krasinko
Department of Biotechnology and Microbiology
National University of Food Technologies
Kyiv, Ukraine

Oksana Kochubei-Lytvynenko
Department of Milk and Dairy Technology
National University of Food Technologies
Kyiv, Ukraine

Tetiana Lazarieva
Department of Food Technology, Light Industry and Design
Ukrainian Engineering-Pedagogics Academy
Kharkiv, Ukraine

Andrii Marynin
Advanced Research Laboratory
National University of Food Technologies
Kyiv, Ukraine

Octavio Paredes-López
Instituto Politécnico Nacional
Mexico

Vasyl Pasichnyi
Department of Technology of Meat and Meat Products
National University of Food Technologies
Kyiv, Ukraine

Viktoriya Peshkova
Department of Biomolecular Electronics
Institute of Molecular Biology and Genetics of the National Academy of Sciences of Ukraine
Kyiv, Ukraine

Tetyana Pirog
Department of Biotechnology and Microbiology
National University of Food Technologies
and
Institute of Microbiology and Virology National Academy of Sciences
Kyiv, Ukraine

Olga Saiapina
Department of Biomolecular Electronics
Institute of Molecular Biology and Genetics of the National Academy of Sciences of Ukraine
Kyiv, Ukraine

Anatoliy Salyuk
Department of Food Chemistry
National University of Food Technologies
Kyiv, Ukraine

Tetyana Sergeyeva
Department of Biomolecular Electronics
Institute of Molecular Biology and Genetics of the National Academy of Sciences of Ukraine
Kyiv, Ukraine

Yevhenii Shapovalov
Department of Educational Knowledge System Creation
National Center "Junior Academy of Sciences"
Kyiv, Ukraine

Anastasiia Shevchenko
Department of Bakery and Confectionery Goods Technology
National University of Food Technologies
Kyiv, Ukraine

Oleksandr Shevchenko
Rector
National University of Food Technologies
Kyiv, Ukraine

Lyudmyla Shkotova
Department of Biomolecular Electronics
Institute of Molecular Biology and Genetics of the National Academy of Sciences of Ukraine
Kyiv, Ukraine

Alexei Soldatkin
Department of Biomolecular Electronics
Institute of Molecular Biology and Genetics of the National Academy of Sciences of Ukraine
Kyiv, Ukraine

Oleksandr Soldatkin
Department of Biomolecular Electronics
Institute of Molecular Biology and Genetics of the National Academy of Sciences of Ukraine
Kyiv, Ukraine

Viktor Stabnikov
Department of Biotechnology and Microbiology
National University of Food Technologies
Kyiv, Ukraine

Olena Stabnikova
Department of Biotechnology and Microbiology
Advanced Research Laboratory
National University of Food Technologies
Kyiv, Ukraine

Larysa Titova
Department of Industrial Biotechnology and Biopharmacy
National Technical University of Ukraine "Igor Sikorsky Kyiv Polytechnic Institute"
Kyiv, Ukraine

Oksana Topchii
Department of Technology of Meat and Meat Products
National University of Food Technologies
Kyiv, Ukraine

Lidiya Tovma
Department of Logistics
National Academy of the National Guard of Ukraine, Maidan Zakhysnykiv Ukrainy
Kharkiv, Ukraine

Sergey Tsygankov
Laboratory of Biofuel Biotechnology and Innovations in Green Energy
Institute of Food Biotechnology and Genomics, NAS of Ukraine
Kyiv, Ukraine

Iryna Tsykhanovska
Department of Food Technology, Light Industry and Design
Ukrainian Engineering-Pedagogics Academy
Kharkiv, Ukraine

Stanislav Usenko
Department of Environmental Safety and Occupational Health
National University of Food Technologies
Kyiv, Ukraine

Viktoria Yevlash
Department of Chemistry, Biochemistry, Microbiology and Hygiene of Nutrition
State Biotechnology University
Kharkiv, Ukraine

Sergey Zhadan
LLC "H2Holland Ukraine"
Kyiv, Ukraine

About the Editors

Olena Stabnikova achieved the degree of Engineer-Biotechnologist at the National University of Food Technologies, Kyiv, Ukraine in 1972; later, she received a Ph.D. in Technical Sciences in the National University of Food Technologies, Kyiv, Ukraine in 1978. She has 40 years of teaching and research experience in Biotechnology and Microbiology at the universities of Ukraine and Singapore, and now she is working as an Associate Professor at the Department of Microbiology and Biotechnology, National University of Food Technologies, Kyiv, Ukraine. Her research experience includes the supervision and participation in international and national projects on the increasing nutritional and health properties of food products, as well as utilization and management of organic wastes with the production of different value-added products such as compost, microbial biomass, biogas or hydrogen. She is the Editor-in-Chief of *Ukrainian Food Journal* and is a member of the Section "Scientific Problems of Food Technologies and Industrial Biotechnology", of the Scientific Council of the Ministry of Education and Science of Ukraine. Olena Stabnikova has published 150 research papers, 5 book chapters and 2 books. She has the h-index of 20 (Scopus) and 26 (Google Scholar).

Oleksandr Shevchenko, Doctor of Technical Sciences, Professor, is a Rector of the National University of Food Technologies (NUFT), Kyiv, Ukraine. He received a degree in Mechanical Engineering for food production from Kyiv Technological Institute of Food Industry, Ukraine. Later he received a Ph.D. in "Dynamics of transport systems of food enterprises" from Kyiv Technological Institute of Food Industry and then Doctor of Technical Sciences degree in "Development of the theory of physicochemical effects on storage of food products" from the National University of Food Technologies. He was the Head of the Department of Processes and Equipment of Food Production, Dean of the Faculty of Bakery and Confectionery, and Vice-Rector of Research at the National University of Food Technologies, Kyiv, Ukraine. He worked in the expert commission of the Higher Attestation Commission of Ukraine. He is the chairman and member of the specialized scientific councils and chairman of the Section "Scientific Problems of Food Technologies and Industrial Biotechnology", of the Scientific Council of the Ministry of Education and Science of Ukraine, and is also Secretary of the Presidium of the Council of Vice-Rectors for Research.

Dr. Shevchenko is Laureate of the State Prize of Ukraine in the field of science and technology (2012), Honored Worker of Education of Ukraine (2014), Laureate of the Prize of the Cabinet of Ministers of Ukraine for development and implementation of innovative technologies (2018) and Honored Worker of Science and Technology of Ukraine (2018). He is a scientific supervisor for graduate students and doctoral students and a member of the editorial board of three scientific journals of categories A and B. Since the early stages of Dr. Shevchenko's career, he has mainly been engaged in research in the field of mechanical engineering, machinery, processes and equipment for food production. The major theme in his research is the development of technology for long-term storage of food products using physical methods. Dr. Shevchenko has published 365 scientific and educational-methodical works, including 30 monographs, textbooks and manuals, 110 author's certificates, patents for inventions and utility models.

Viktor Stabnikov, Doctor of Technical Sciences, is a Professor and Head of the Department of Biotechnology and Microbiology, National University of Food Technologies, Kyiv, Ukraine. Viktor Stabnikov received a Bachelor's degree in Biotechnology, and then a M.Sc. degree in Biotechnology from the Ukrainian State University of Food Technologies, Kyiv, Ukraine. Later he received a Ph.D. in Biotechnology and Doctor of Technical Sciences in Biotechnology from the National University of Food Technologies, Kyiv, Ukraine. He also worked as Researcher-Intern at Nanyang Technological University, Singapore; Chief of the Department of Environmental Protection at Maxwell Scientific and Production Center of Oncology and Cardiology, Ukraine; Associate Professor at the Department of Biotechnology in the Institute of Municipal Activity at the National Aviation University, Ukraine; Visiting Research Fellow, Nanyang Technological University, Singapore and Associate Professor, Department of Microbiology and Biotechnology, National University of Food Technologies, Ukraine. Since the early stages of Dr. Stabnikov career, he has been engaged in production of dietary food, especially bakery selenium-enriched yeasts and selenium-enriched sprouts. His research also concerned prolongation of shelf life of food products due to application of microbial polysaccharides and prevention of bakery from microbial spoilage using coating with lactic acid bacteria. Dr. Stabnikov has published 61 peer-reviewed research papers, 13 book chapters, 1 book and many other reviews and abstracts. He has the h-index of 21 (Scopus) and 24 (Google Scholar). He is the Editor-in-Chief of *Ukrainian Journal of Food Science*, Editor of *Ukrainian Food Journal*, and a member of the Section "Scientific Problems of Food Technologies and Industrial Biotechnology", of the Scientific Council of the Ministry of Education and Science of Ukraine".

Octavio Paredes-López is a biochemical engineer and food scientist who obtained his Bachelor's and Master's degrees in Biochemical Engineering and Food Science, respectively, from the National Polytechnic Institute in Mexico City, and later earned a Master's degree in Biochemical Engineering from the Czechoslovak Academy of Sciences in Prague. He earned a Ph.D. (1980) in Plants Science from the University of Manitoba, Canada, which awarded him an honorary Doctor of Science in 2005. He is a past president of the Mexican Academy of Sciences, a founding member of the International Academy of Food Science and Technology, and an ex-member of the Governing Board of the National Autonomous University of Mexico. He received the National Prize for Arts and Sciences in 1991, the highest scientific recognition in Mexico and the Academy of Sciences of the Developing World Award (previously the Third World Academy of Sciences) in 1998, Trieste, Italy. Paredes-López has conducted research and postdoctoral stays in the USA, Canada, Britain, France, Germany, Switzerland and Brazil. He has served as a general editor of the journal *Plant Foods for Human Nutrition*, and an editorial board member of *Critical Reviews in Food Science and Nutrition, Frontiers in Food Science and Technology* and ten other international journals. He has authored over 300 publications in indexed journals, book chapters, and books and over 150 articles in newspapers in Mexico, USA and France. His h-index is 64.

Chapter 1

The Use of Wastes from the Flour Mills and Vegetable Processing for the Enrichment of Food Products

Anastasiia Shevchenko, Oksana Fursik, Vera Drobot, and Oleksandr Shevchenko
National University of Food Technologies

CONTENTS

DOI: 10.1201/9781003329671-1

1.1 INTRODUCTION

1.1.1 Oat Bran, Wastes from the Flour Mills, and Pumpkin Cellulose, Wastes from Vegetable Processing, as the Sources of Dietary Fiber for Food Products Manufacturing

Secondary processing products of various industries are a valuable source of dietary fiber for food enrichment. Finding rational approaches to the reuse of such raw materials is an important problem for the modern food industry. It is most acutely for the grain and vegetable industry, because the obtained waste still remains underutilized due to the lack of sufficiently studied and substantiated principles of their processing and application (Elik et al., 2019). Dietary fibers are polysaccharide components of plants which are resistant to digestion and partial adsorption in the human intestine. These are food additives which regulate important technological characteristics of food products, and affect their safety, sensory properties and quality (Shevchenko, 2023; Strashynskiy et al., 2021a). Vegetable raw materials, namely grains, legumes, vegetables and fruits, are the main sources of dietary fiber. Fibers are usually obtained from the by-products of cereal processing. It is known that cereals constitute 50% of the amount of fiber consumed in Western European countries. The amount of fiber in these crops ranges from 1.1% to 17.3% and depends on the type of raw material. The largest amounts of insoluble dietary fibers are in barley and rye seeds, and fewer amounts are in oat, wheat, corn and millet. The lowest fiber contents are in sorghum, rice and fonio seeds. However, these indicators are relative and depend on the variety or cultivar of the selected crop (Esteban et al., 2017). For legumes, the dietary fiber content is in the range from 15.3% to 29.7% and depends on the type of crop. In general, the highest dietary fiber content is in beans, chickpeas and peas, and the lowest is in lentil seeds (Esteban et al., 2017). Special attention should be paid to the use of dietary fibers from grains and vegetables due to the high content of soluble and insoluble dietary fibers, lower content of phytic acid in their composition, lower caloric content, increased capacity for fermentation and improved functionality in relation to water and fat molecules (Garcia-Amezquita et al., 2018).

Oat and pumpkin on their own contain biologically active substances and are recommended to be used for maintaining human homeostasis (Delgado-Vargas

& Paredes-López, 2002); meanwhile, oat and pumpkin processing by-products are widely used in preparation of different foods such as bakery, confectionery, meat and dairy products. There is a lack of sufficient information in the field of obtaining dietary fibers from by-products, as well as regarding their characteristics and recommendations for use. Accordingly, there are opportunities for studying this topic, and especially waste from the flour milling and vegetable industries (Hussain et al., 2020).

Pumpkin and oat processing waste are used in dairy production. It was shown that the extract obtained from pumpkin seeds has the ability to coagulate milk because the peptidase activity of this extract is observed in a wide pH range, which is due to the presence of active catalytic peptides (Abhiraman & Soumya, 2020). The formed gel had a strong structure with constant sensory properties. The maximum value of the modulus of springiness and complex viscosity was 75.2 Pa and 7.78 Pa·s, respectively. The value of the loss coefficient was in the range from 0.54 to 0.20, which indicates constant gel formation. Dry pumpkin powder obtained from shredded peel and pulp in a ratio of 70:30 was used as a carrier for probiotic lactic acid bacteria *Lactobacillus casei* ATCC-393 in the composition of diary supplement, which was used in production of chocolate or soy milk and apple juice-based beverage (Genevois et al., 2016). The use of pumpkin and its waste in the food industry is often associated with the need to include carotene to the diet. A study of its quality showed that addition of 15% pumpkin pulp and 10% brown sugar to buffalo milk led to an increase in the carotene content to 1.2 mg/100 g (Patel et al., 2020). Enrichment of yogurt with oat proteins in the amount of 12.3%, 13.8% and 15.3% to dry matter will allow for replacing defatted milk powder while maintaining good sensory properties (Brückner-Gühmann et al., 2019).

Pumpkin processing waste as a source of fiber is also used in the technologies of meat and fish products as a valuable functional and technological additive rich with irreplaceable dietary fibers, which are absent in meat and fish raw materials. Pumpkin pulp in the amount of 10%, 15%, 20% and 25% was added to minced fish to form semifinished products—burgers. Adding 15%–20% of pumpkin puree reduced the yield of the product by 7.99%–12.31%. The indicators of fat retention and hardness increased by 4.48%–5.34% and 19.83%–26.61%, respectively, in comparison with the sample without adding pumpkin puree. Sensory properties (taste, flavor, consistency) of fish burgers with pumpkin puree were highly evaluated (Ali et al., 2017). Extruded product obtained from pumpkin seed flour was used in the composition of meat–vegetable burgers in the amount of 30% for replacing meat raw material. It was found that fiber content increased to 2–2.2 g per 100 g of the product, which is 5.5 times higher compared to a control sample; meanwhile, high sensory properties of the product were preserved (Baune et al., 2021).

A new type of sausage product with increased content of dietary fiber includes products based on minced fish with the addition of oat fiber in the amount of 2.5%, 5%, 7.5% and 10% for replacing fish raw material. The addition of oat fibers led to a decrease in the moisture content of all samples and significantly affected the strength of the gel and water-holding capacity. The formation of compact gel structure in the presence of oat fibers, which provided an excellent texture of the finished product, was confirmed by using scanning electron microscopy. Increasing the amount of oat fiber led to a slight smell of oat. It is recommended to enrich minced sausage with oat fibers in the amount of 2.5% (Gore et al., 2022). The use of oat fiber concentrate in meat substitutes based on extruded faba bean protein concentrates affected the mechanical, physicochemical and rheological properties of minced meat and the finished product. A constant protein system was formed when the ratio of oat fibers and faba bean proteins was 25:75 (Ramos-Diaz et al., 2022).

Pumpkin and oat processing wastes are also used in production of bakery and confertionary. For example, pumpkin paste was added to the recipe of semifinished biscuit dough for flour confectionery products. The paste added to the egg–sugar mixture for biscuit in the amount of 10%–20% to the mass of flour contributes to the improvement of foaming and stability of the whipped mass (Ashurova & Sulaymanova, 2021). Pumpkin flour added to the cookie recipe in amounts of 10%, 15%, and 20% increased the content of beta-carotene in cookies to 8.457, 9.796 and 12.712 μg/100 g, whereas cookies made of wheat flour without additives did not contain beta-carotene. Cookies with 20% pumpkin flour had a bitter aftertaste (Fathonah et al., 2018). Oat fiber was added to the biscuit dough in the amount of 5%, 10%, 15%, 20% and 30% to the weight of flour. An increase in oat fiber content increased dough density, decreased springiness and ensured more viscous consistency. The crust and pulp became darker. It is possible to obtain a product with high sensory characteristics if no more than 20% of oat fiber is added (Majzoobi et al., 2015).

Pumpkin puree of the "Bambino" variety (early, medium ripe) was used in the production of bakery products from wheat flour using various technologies. The improvement of sensory properties of bread was established in case of the addition of no more than 10%. Increasing the content of pumpkin pulp to 20% in wheat dough led to a decrease in bread volume and deterioration of porosity. Pumpkin puree has a high biological potential, especially inhibitory and antioxidant activities (Różyło et al., 2014). A mixture of pumpkin and spelt flour was added to wheat flour in bread production. It was established that with the increase in the dosage of the mixture of flours, the volume of bread decreased. The total content of carotene in wheat bread increased with the addition of pumpkin products (Kampuse et al., 2015; Rakcejeva et al., 2011). Pumpkin flour, 5%–20%, to replace wheat flour, causes a decrease in the gas-forming capacity of the dough by 1.9%–7.4%. The accumulation of sugars in the dough during fermentation also decreases by 7.6%–16.2%, but the rate of sugars fermentation increases by 16.9%–20.3% (Shevchenko et al., 2022). The replacement of wheat flour by oat flour in the recipe of bread in the amount of 10%, 30% and 50% led to a change in the volume of bread. It was 550, 450 and 388 mL, respectively, compared to the volume of wheat bread 838 mL. Bread with 20% and 30% oat flour had a higher hardness compared to other samples. The cohesiveness of bread with wheat flour was higher than the cohesiveness of bread with oat flour, and it increased with increasing storage time (Lee & Ha, 2011).

The problem of using food processing waste as a source of valuable nutrients to enrich products and increase their nutritional value is relevant (Stabnikova et al., 2021). It is necessary to use nontraditional raw materials in the recipes, which can change not only sensory properties of the products but also enrich them with necessary essential nutrients increasing their functional properties. Moreover, sources of dietary fiber are recommended to add to the diet with low content of FODMAP (fermentable oligosaccharides, disaccharides, monosaccharides and polyols) (Gibson et al., 2015) for people who suffer from diseases of gastrointestinal tract, especially irritable bowel syndrome. Different food components can provoke the appearance of symptoms of the gastrocolonic reflex or aggravate them. These components include some carbohydrates, which are split in the intestines, but poorly absorbed or not absorbed at all, namely lactose, fructose, fructans, galactans, and polyols that were combined into the FODMAP group. A lack of plant fibers not only limits the amount of vitamins and minerals which enter the body with food, but also makes people overeat, since fiber is responsible for the feeling of satiety (Hill et al., 2017). According to the low-FODMAP diet, oat, pumpkin and their by-products are recommended for consumption for people with irritable bowel syndrome, in particular as sources of dietary fiber (Nanayakkara et al., 2016). However, addition of these ingredients can change the properties of the products. It is advisable to add pumpkin cellulose and oat bran into bread recipes, but there are insufficient data on their impact on the technological process of manufacturing bakery products. That is why a relevant task is to develop approaches to enrich bakery products with food processing waste, in particular pumpkin cellulose and oat bran.

1.2 CHEMICAL COMPOSITION AND STRUCTURAL PROPERTIES OF OAT BRAN AND PUMPKIN CELLULOSE

Oat bran obtained as waste after processing oat into flour (Ukrainian producer LLC "Organic-Eco-Produkt") and pumpkin cellulose obtained from the kernels of pumpkin seeds (Ukrainian producer TM "Only oil") were used as sources of fibers to be added to bakery products. Oat bran is produced by grinding clean oat groats into flour and separating the coarser particles of bran from the flour by sieving and bolting so that the oat bran fraction is not more than 50% of

TABLE 1.1 CHEMICAL COMPOSITION OF RAW MATERIALS*

Content, %	High-Grade Wheat Flour	Oat Bran	Pumpkin Cellulose
Water	11.5±0.1	17.3±0.1	6.4±0.1
Protein	10.3±0.1	17.0±0.1	42.0±0.1
Fat	1.1±0.1	7.0±0.1	6.0±0.1
Fiber	3.5±0.1	15.4±0.1	32.0±0.1
β-glucan	-	5.1±0.1	-
Essential Amino Acids (EAA), g/100 g of Raw Material			
Valine	0.42±0.01	0.96±0.02	1.64±0.02
Isoleucine	0.36±0.01	0.67±0.02	1.36±0.02
Leucine	0.71±0.02	1.37±0.02	2.57±0.02
Lysine	0.23±0.01	0.76±0.02	1.35±0.02
Methionine	0.40±0.01	0.34±0.01	0.78±0.02
Threonine	0.28±0.01	0.50±0.01	1.12±0.02
Tryptophan	0.13±0.01	0.33±0.01	0.72±0.02
Phenylalanine	0.52±0.01	0.91±0.02	1.86±0.02

* Results given as: M±SD (mean±standard deviation) of triplicate trials.

the starting material (Burrows, 2011; Guo, 2009). Pumpkin cellulose is a valuable by-product obtained after the extraction of oil and obtaining pumpkin flour from seeds (Dhiman et al., 2018). Ripe pumpkin was cut into halves, and seeds were extracted from the fluffy portion of fibrous strands. The seeds were washed, oven-dried, manually decorticated, crushed by using a mechanical grinder and defatted by soaking in n-hexane for 36 h with change of solvent every 8 h (Atuonwu & Akobundu, 2010). Dietary cellulose was extracted using the modified enzymatic–gravimetric method, and then dried at 55°C in an air oven and then cooled (Kim et al., 2016). Chemical compositions of these products as well as high-grade wheat flour are shown in Table 1.1.

Oat bran and pumpkin cellulose contain 1.6 and 4.0 times more protein and 4.4 and 9.1 times more dietary fiber than high-grade wheat flour, respectively. They also contain a significant amount of the amino acid lysine, which is limited in wheat flour: 0.76 g/100 g of oats and 1.35 g/100 g of pumpkin cellulose in comparison with 0.23 g/100 g of wheat flour.

Oat bran has creamy color and consists of small particles with a weighted average particle size of 0.44 mm (Figure 1.1a). Pumpkin cellulose is greenish in color and consists of rather large particles with the weighted average particle size of 0.67 mm (Figure 1.1b).

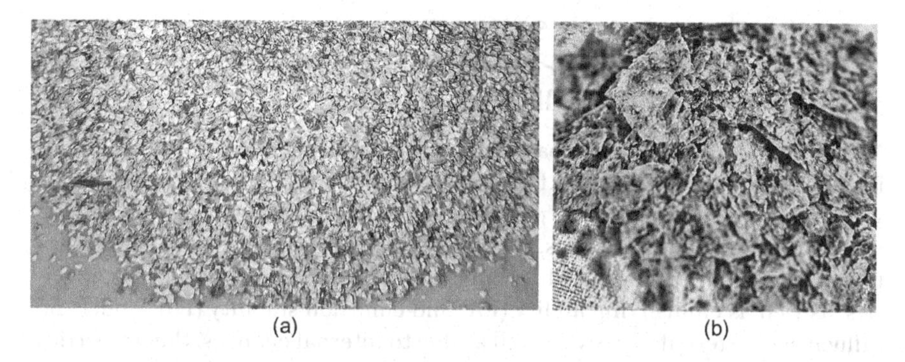

Figure 1.1 Oat bran (a) and pumpkin cellulose (b). 1, suspension of oat bran and water; 2, suspension of oat bran and water after heat treatment; 3, suspension of pumpkin cellulose and water; 4, suspension of pumpkin cellulose and water after heat treatment.

TABLE 1.2 PARTICLE SIZE OF OAT BRAN AND PUMPKIN CELLULOSE COMPARED TO WHEAT FLOUR

Size Indicators, No. of Sieve	Hole Size (μm)	Wheat Flour, Variety			Oat Bran	Pumpkin Cellulose
		First	Second	Wholemeal		
The residue on sieve %, no more:						
No. 33/36 (35)	220	2	-	-	-	-
No. 27	260	-	2	-	55.36	
No. 067	670	-	-	2	44.64	100
Passage through a sieve, % no less:						
No. 49/52 PA (43)	132	80	-	-	-	-
No. 41/43 (38)	160	-	65	35	-	-

The process of making bread depends on the particle size of the recipe ingredients. Raw material, which differs in granulometric composition from wheat flour, can significantly affect the structural and mechanical properties of the dough and the quality of the finished product. The evaluation of the granulometric composition of raw materials showed that both oat bran and pumpkin cellulose had significantly larger particle sizes than wheat flour, as the residue on sieve N 067 with hole size 670 μm exceeded the limit normalized value for wheat flour by 22 and 50 times, respectively (Table 1.2).

1.3 FUNCTIONAL AND STRUCTURAL-MECHANICAL PROPERTIES OF OAT BRAN AND PUMPKIN CELLULOSE

Main functional properties of oat bran and pumpkin cellulose, namely the ability to interact with water and fat molecules and stabilize dispersed systems, were studied to determine the possibility of their use as food ingredients. The most important characteristics, which allow to assess their functional properties, are water-absorbing, fat-absorbing, water-holding and fat-holding capacities, as well as emulsifying ability (EA) and emulsion stability (ES). Under the influence of external factors, as well as due to internal changes, the properties of dietary fibers are modified, which will ultimately affect the ability to interact with water and fat molecules. Therefore, to determine the ability of oat bran and pumpkin cellulose to bind and hold water and fat, which are present in most food products, water-absorbing capacity (WAC), water-holding capacity (WHC), fat-absorbing capacity (FAC) and fat-holding capacity (FHC) were determined. Test samples were mixed with water/fat and centrifugated. The amount of liquid separated by centrifugation was determined. When determining WHC and FHC, samples with water/fat were preheated before centrifugation (Raikos et al., 2014). Setting limits for these parameters allow to determine the required amount of water and fat, which is an important factor that must be taken into account when developing product recipes (Figure 1.2).

The WAC of oat bran was 21% higher compared to pumpkin cellulose. This is explained by the difference in the granulometric composition of the experimental samples. Oat bran has smaller particle size than pumpkin cellulose, which makes the system more porous and leads to increase in the water absorption

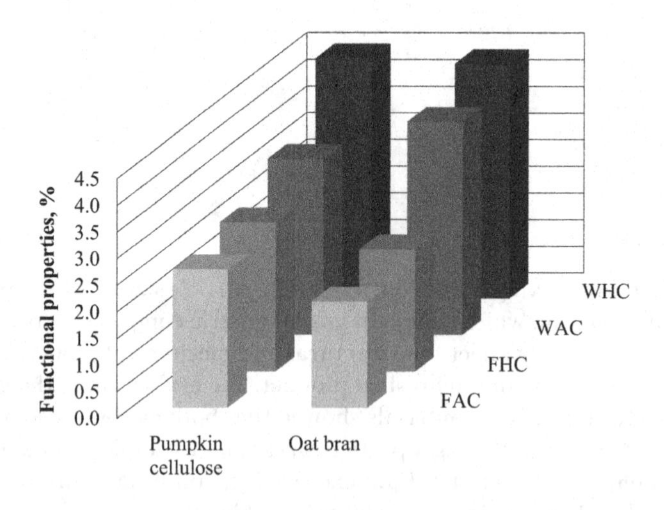

Figure 1.2 Functional properties of oat bran and pumpkin cellulose.

rate. In addition, chemical composition indicates the presence of a significant amount of water-soluble carbohydrates in the composition of oat bran (Zhang et al., 2021). The study of the WHC indicator, which characterizes the ability of a system to retain moisture in the formed structure after heat treatment, showed the redistribution in the interaction of used sources of dietary fibers with water. For pumpkin cellulose, this indicator is at the level of oat bran and increases by 36% compared to the WAC indicator (for oat bran the increase was 10%). The observed changes in the absorption of water molecules may be related to the fact that the polymer chains of pumpkin cellulose are not mobile, which limits the water-absorbing properties of the system without the additional influence of external factors (Yu et al., 2021). Heat treatment contributed to the release of side polar groups of protein, soluble molecules and areas of fibers which have hydrophilic properties, which increase the amount of retained moisture. There is radical redistribution for the indicators of FAC and FHC. These indicators are higher for pumpkin cellulose than for oat bran; FAC by 30%, FHC by 22%, which may be due to the higher content of hydrophobic polysaccharides contained in pumpkin seeds (Wang et al., 2017b). The increase in the ability to hold fat was at the level of 7.7%–15% for both additives. The different distribution of the studied properties for the used dietary fibers in relation to water and fat molecules is explained by the difference in the ratio of hydrophilic and hydrophobic particles in the composition of the samples of fibers (Wang et al., 2017b).

The EA and ES indicators characterize the ability of multicomponent systems to form constant dispersions of water and fat and to resist their stratification over time and under the influence of external factors. These indicators were determined by measuring the amount of separated oil from prepared emulsions by centrifugation before and after heat treatment. The ability of the formed emulsion to maintain constant structure depends on the nature of the used additives, which act as stabilizers and thickeners. In this regard, the ability of selected samples of dietary fibers to form and stabilize water/fat systems was studied (Figure 1.3).

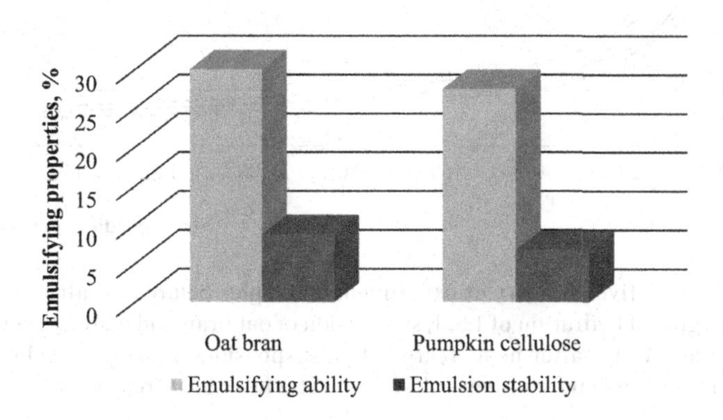

Figure 1.3 Emulsifying properties of oat bran and pumpkin cellulose.

The advantages of oat bran over pumpkin cellulose in terms of the ability to form constant emulsions were shown: the EA index of oat bran was higher by 9%, and the ES index was higher by 29% than that of pumpkin cellulose. This may be due to the content of β-glucan, which increases the viscosity of the water system, thereby contributing to a more effective surface coating of fat molecules (Anttila et al., 2004). The increase in the difference in the manifestation of emulsifying properties after heat treatment between oat bran and pumpkin cellulose is determined by the ability of pumpkin lignin carbohydrate complexes to promote the formation of interpolymeric bonds between polysaccharides, which bind protein molecules and create more tough and impermeable system (Aminzadeh et al., 2017). Hydrated dietary fibers are dispersed systems, the properties of which depend on the volume ratio of the dispersed phase and the dispersion medium, the nature and strength of the bonds of the components. Study of the structural and mechanical properties, namely the effective viscosity of the hydrated

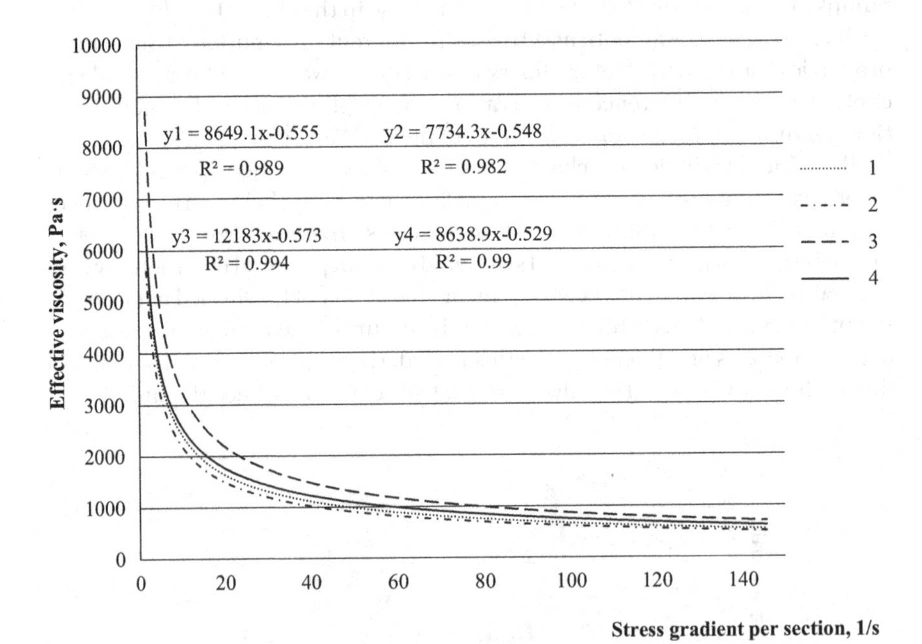

Figure 1.4 Effective viscosity of experimental samples before and after heat treatment at a degree of hydration of 1:3. 1, suspension of oat bran and water; 2, suspension of oat bran and water after heat treatment; 3, suspension of pumpkin cellulose and water; 4, suspension of pumpkin cellulose and water after heat treatment.

suspensions of oat bran and pumpkin cellulose, was conducted for more complete characterization of the samples. The effective viscosity index at three degrees of hydration 1:3, 1:6 and 1:9 with water at a temperature of 15±3°C was determined. The effect of heat treatment leading to a temperature of 95±3°C and boiling for 30 min and the ability to thicken aqueous dispersions were studied using the Rheotest-2 device (Strashynskiy et al., 2021b) (Figures 1.4–1.6).

Comparison of effective viscosity of oat bran and pumpkin cellulose at different degrees of hydration showed a decrease in the ability to resist applied shear forces with an increase in the amount of dispersion medium in suspension. The highest decrease in effective viscosity was for pumpkin cellulose before and after heat treatment by an average of 50%–55% with each step of increasing the amount of added water. The obtained data indicate that oat bran has higher by 11%–90% effective viscosity compared to pumpkin cellulose and the difference increases with the increase in the degree of hydration. This may be due to the presence of β-glucan in oat bran, which is

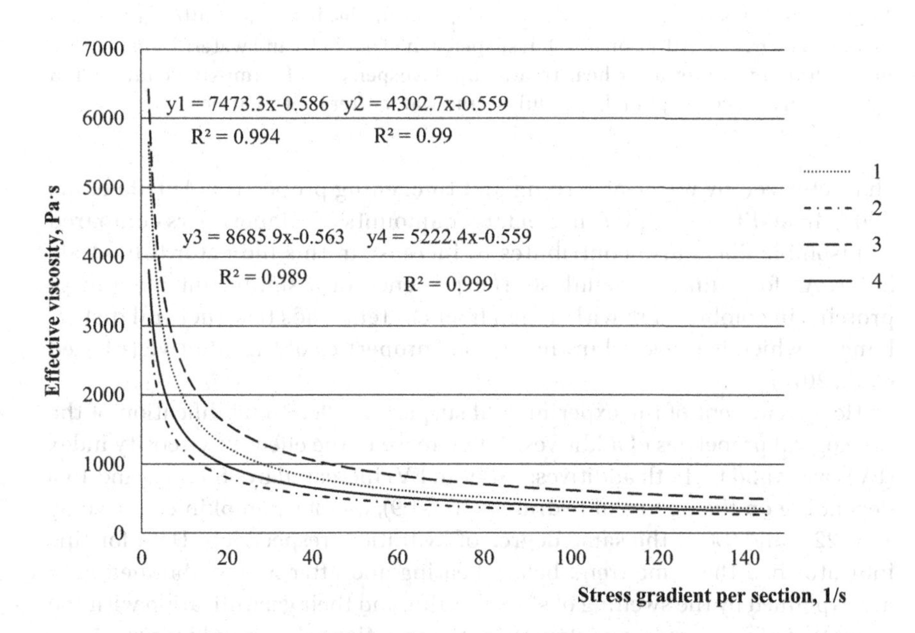

Figure 1.5 Effective viscosity of experimental samples before and after heat treatment at a degree of hydration of 1:6. 1, suspension of oat bran and water; 2, suspension of oat bran and water after heat treatment; 3, suspension of pumpkin cellulose and water; 4, suspension of pumpkin cellulose and water after heat treatment.

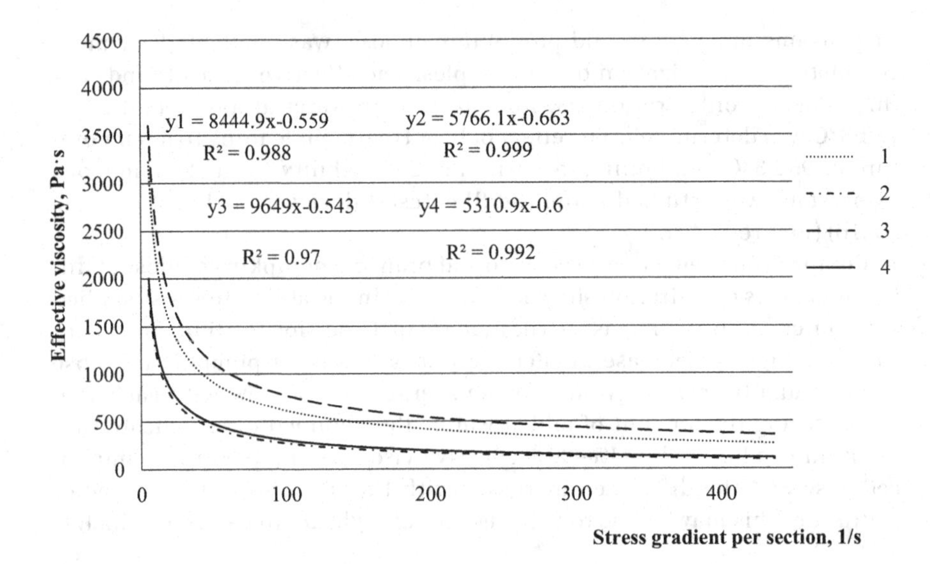

Figure 1.6 Effective viscosity of experimental samples before and after heat treatment at a degree of hydration of 1:9. 1, suspension of oat bran and water; 2, suspension of oat bran and water after heat treatment; 3, suspension of pumpkin cellulose and water; 4, suspension of pumpkin cellulose and water after heat treatment.

characterized by water-absorbing and thickening properties (Anttila et al., 2004). In addition, the presence of larger amounts of soluble fibers compared to insoluble fibers also contributes to increase in this indicator (Cui et al., 2013). As for pumpkin cellulose, the presence of a significant amount of proteins in combination with a high fiber content leads to structural disturbances, which is reflected in the viscous properties of the additive (Ahmed et al., 2014).

Heat treatment of the experimental suspensions led to modification of the rheological properties of additives. An increase in the effective viscosity index (EVI) was valid for both additives: oat bran EVI increased by 38%, 21% and 16% depending on the degree of hydration (1:3–1:9), and for pumpkin cellulose by 16%, 22% and 9% at the same degree of hydration, respectively. Data for this indicator had the same trend before heating and after it. The obtained data are explained by the swelling of starch grains and their gelatinization with the formation of a more viscous structure. The equations shown in Figures 1.4–1.6 describe the dependence of the effective viscosity on the shear stress gradient. Their approximation coefficients (R^2) testify to the high reliability of the equations, which characterize the structural and mechanical properties of samples of food fibers.

1.4 INFLUENCE OF OAT BRAN ADDITION ON THE TECHNOLOGICAL PROCESS OF BAKERY PRODUCT MANUFACTURING AND QUALITY OF BREAD

1.4.1 Microbiological and Biochemical Processes in the Dough and Structural and Mechanical Properties of the Dough with Oat Bran

Microbiological processes in dough play an important role in bread making. During fermentation of dough, biotransformation of its polymers takes place, the intensity of which depends on the chemical composition of the recipe components. Intensity of microbiological processes in the dough with oat bran was evaluated by the amount of carbon dioxide released during the fermentation and keeping of the dough and the dynamics of its release. These indicators were determined by the volumetric method, namely by the volume of CO_2 released at constant temperature and pressure (Munteanu et al., 2019; Verheyen et al., 2015). The dough samples were prepared from high-grade wheat flour added with pressed bakery yeast, 3%; salt, 1.5%; sunflower lecithin as an emulsifier and a source of phosphatidylcholine, 3%, to the mass of flour (Partridge et al., 2019), with the replacement of 5.0%; 7.0%; and 10.0% and 15.0% of wheat flour by oat bran. Dough without additives was used as control (Figure 1.7).

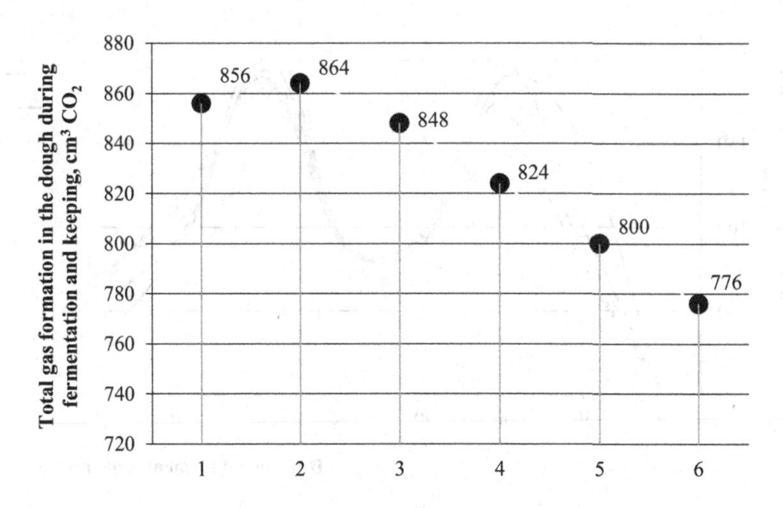

Figure 1.7 Total gas formation in the dough during fermentation and keeping. 1, control sample; 2, sample with lecithin; 3, sample with lecithin and 5% oat bran to replace wheat flour; 4, sample with lecithin and 7% oat bran to replace wheat flour; 5, sample with lecithin and 10% oat bran to replace wheat flour; 6, sample with lecithin and 15% oat bran to replace wheat flour.

Gas-forming capacity of the dough increased by 1% in samples with lecithin. This indicates an improvement in the enzymatic capacity of yeast (Dotsenko et al., 2019). With an increase in the dosage of oat bran from 5% to 15%, the gas-forming capacity decreased by 0.9%–9.3% compared to the sample with lecithin and by 1.8%–10.2% compared to the control sample. This can be explained by changes with the addition of fiber-rich ingredients such as oat β-glucan (Wang et al., 2017a). Protein length and branching rate significantly increased in the presence of oat bran due to the presence of β-glucan (De Bondt et al., 2021), and protein complexes with wheat flour starch were formed. A decrease in the fermentation activity of yeast in the presence of oat bran will affect the dynamics of carbon dioxide released during dough fermentation and keeping (Figure 1.8).

It was established that gas formation occurs less intensively in the dough with oat bran, because fermentation is delayed due to a decrease in the availability of nutrients. The graph of the dynamics of the release of carbon dioxide shows that the first peak of gas formation in the dough with the replacement of wheat flour with oat bran is observed after 60 min, while in the control sample after 50 min. This is explained by the fact that the amylolytic enzymes of oat bran are less active than those of wheat flour. With an increase in the dosage of oat bran, the amount of released carbon dioxide in the dough decreased. The reason for

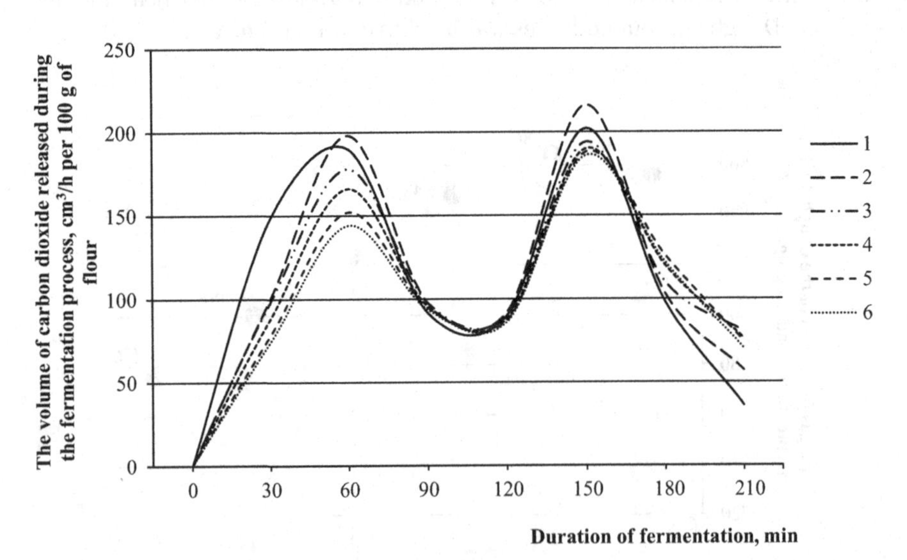

Figure 1.8 Dynamics of gas formation in dough with different dosage of oat bran: 1, control sample; 2, sample with lecithin; 3, sample with lecithin and 5% oat bran to replace wheat flour; 4, sample with lecithin and 7% oat bran to replace wheat flour; 5, sample with lecithin and 10% oat bran to replace wheat flour; 6, sample with lecithin and 15% oat bran to replace wheat flour.

this is a decrease in the availability of starch for amylolysis. The second peak of gas formation for all samples is observed after 150 min. The biochemical processes, which take place in the dough, mean the breakdown of the components of flour, mainly proteins and starch, under the action of the flour own enzymes, as well as the enzymes of yeast and other microorganisms presented in flour. Sugars and nitrogenous substances are accumulated in the dough. The content of sugars depends on the ratio between their accumulation and usage during dough fermentation (Drobot et al., 2014). The amount of sugars produced during yeast-free dough fermentation was determined by the difference between their content in the dough after 180 min of fermentation and immediately after kneading. The amount of fermented sugars was determined by the difference between the sum of the amount of sugars at the beginning of fermentation of the yeast dough and the amount of sugars formed in the yeast-free dough and the amount of sugars contained in the yeast dough after 180 min of fermentation. The amount of accumulated sugars in the dough was determined by the accelerated iodometric method (Shevchenko et al, 2022). It was shown that in the sample with lecithin, the amount of formed and fermented sugars increased by 1.2% and 12.1%, respectively, compared to the control sample, which is associated with the action of the phospholipid component (Table 1.3).

The addition of increasing dosage of oat bran up to 15% contributed to a decrease in the amount of formed sugars by 15.5%–16.3%, and fermented sugars by 18.6%–28.9%. This can be explained by the fact that oat bran proteins form complexes with wheat flour starch and therefore impair the access of enzymes to starch grains. The components of the recipe have a significant impact on the structural and mechanical properties of the dough system, and subsequently on the quality of the bread. The basis of the formation of the structural and mechanical properties of the dough is the gluten framework formed by proteins. The elasticity and springiness of the dough, and the shape stability of the products at the stage of keeping and baking depend on the properties of the gluten framework. The structural and mechanical properties of the dough were characterized by the gas-holding capacity (the specific volume of the dough was used as an indicator) and the shape-holding capacity (spreading of the ball of dough during 180 min was used as an indicator). It was found that when using lecithin itself and in a mixture with oat bran, the gas-holding capacity of the dough decreased by 12.9%–29.6% because lecithin does not contain gluten proteins, while the gluten framework of the dough loses its elasticity and weakens (Figure 1.9).

At the same time, the viscosity of the dough changed. The addition of lecithin helps to reduce the thinning of the dough by 3.7% due to lower water absorption capacity. The diameter of the dough ball with oat bran during fermentation decreased by 22.2%–29.6%, depending on its dosage. The addition of oat bran improved the shape-holding capacity of all dough samples, which can be explained by the high sorption properties of oat components (Figure 1.10).

TABLE 1.3 ACCUMULATION AND FERMENTATION OF SUGARS (IN TERMS OF MALTOSE), % OF DRY MATTER OF DOUGH, DURING THE DOUGH FERMENTATION*

Stage of Bread Making	Control Sample	Sample with Lecithin	Oat Bran to Replace Wheat Flour (%)			
			5	7	10	15
Yeast-Free Dough						
After kneading	2.10±0.10	2.10±0.10	2.10±0.10	2.10±0.10	2.10±0.10	2.10±0.10
After 180 min of fermentation	3.15±0.13	3.39±0.17	3.19±0.13	3.19±0.13	3.19±0.13	3.18±0.13
Accumulated sugars	1.05±0.01	1.29±0.03	1.09±0.01	1.09±0.01	1.09±0.01	1.08±0.01
Yeast Dough						
After kneading	2.12±0.10	2.15±0.12	2.10±0.10	2.06±0.10	2.06±0.10	2.06±0.10
After 180 min of fermentation	1.69±0.06	1.78±0.08	1.84±0.08	1.86±0.08	1.92±0.08	1.96±0.09
Fermented sugars	1.48±0.05	1.66±0.06	1.35±0.04	1.29±0.03	1.23±0.03	1.18±0.03

* Results given as: M±SD (mean±standard deviation) of triplicate trials.

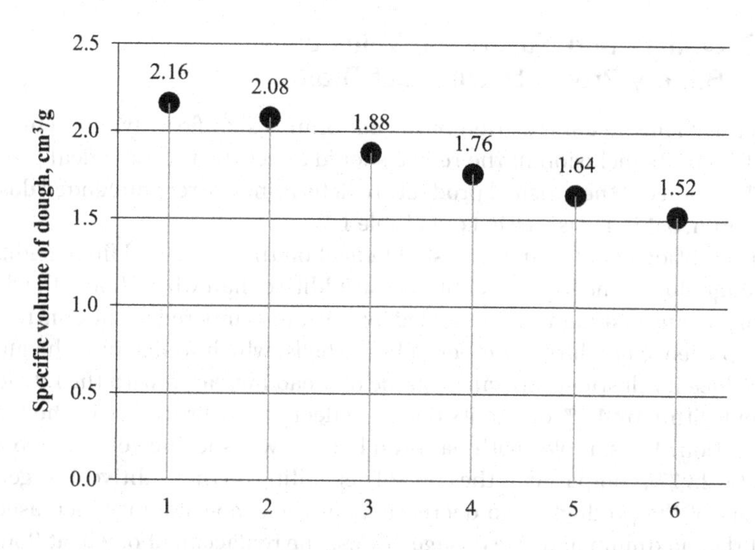

Figure 1.9 Specific volume of dough. 1, control sample; 2, sample with lecithin; 3, sample with lecithin and 5% oat bran to replace wheat flour; 4, sample with lecithin and 7% oat bran to replace wheat flour; 5, sample with lecithin and 10% oat bran to replace wheat flour; 6, sample with lecithin and 15% oat bran to replace wheat flour.

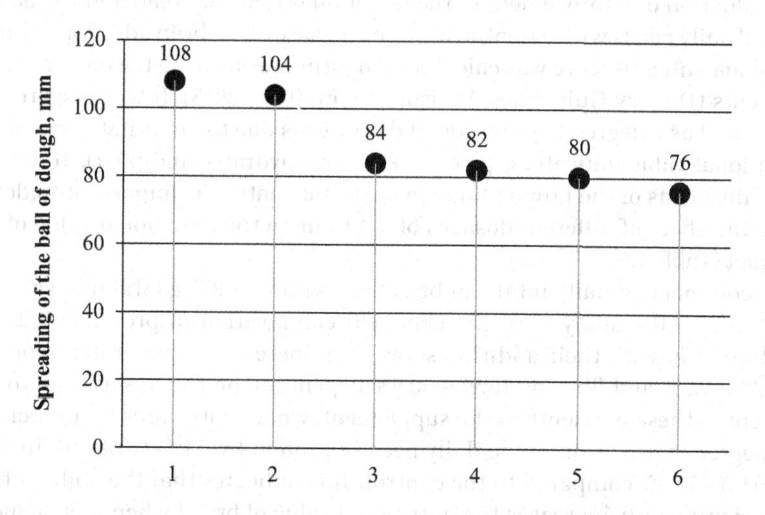

Figure 1.10 Spreading of dough ball. 1, control sample; 2, sample with lecithin; 3, sample with lecithin and 5% oat bran to replace wheat flour; 4, sample with lecithin and 7% oat bran to replace wheat flour; 5, sample with lecithin and 10% oat bran to replace wheat flour; 6, sample with lecithin and 15% oat bran to replace wheat flour.

1.4.2 Quality and Nutritional Value of Bakery Products with Oat Bran

Since the chemical composition of the oat bran is significantly different from wheat flour, its inclusion in the recipe should affect the technological process and the quality of the finished product. To determine the recommended dosage of oat bran, baking was carried out (Table 1.4).

The addition of oat bran had a slight effect on the initial and final acidity of the dough due to the higher acidity of the additive than wheat flour. The duration of keeping of dough was extended by 2–9 min compared to the control due to the presence of a large number of bran shells, which wedge into the gluten and reduce its elasticity. Specific volume of bread increased with the introduction of lecithin by 4.5% due to its positive effect on the processes in the dough preparation. The samples with oat bran had a lower specific volume of bread, by 3.0%–23.2%, compared to the control depending on the additive dosage. The porosity of the products also decreased, but the shape stability increased by 8% at the maximum additive dosage of 15%. The replacement of wheat flour by oat bran had a significant impact on the sensory parameters of bread. When replacing wheat flour by 10%–15% of oat bran, the presence of cracks on the surface of the products and large porosity was observed. There was also inclusion of bran on the surface, but this did not impair its consumer properties. A pleasant oat smell and taste was felt.

Evaluation of improvement of the nutritional value of bread and the ensurance of daily needs with essential nutrients in case of oat bran addition in bread was done. Integral score was calculated to estimate ensuring the daily need for nutrients (Dietary Guidelines Advisory Committee, 2015). Integral score was calculated as a degree of provision of daily needs due to quantitative values of nutritional value indicators (protein, fat, carbohydrates and fiber). To ensure the daily needs of the human body in basic nutrients, it is important to determine the effect of different dosages of oat bran on the nutritional value of the product (Table 1.5).

Recommended daily intake of bread in Ukraine is 277 g (Stabnikova et al., 2019; 2023). The analysis of the chemical composition of products without oat bran and with their addition showed an increase in the protein content by 3.2%–9.3% and fiber by 16.9%–50.9% depending on the dosage due to the content of these nutrients in the supplement, what contributes to an increasing degree of provision of the daily need in protein by 3.1%–9.3% and in fiber by 16.9%–50.9% compared to the control. This indicates that the ability of oat bran significantly increases the nutritional value of bread when it is included in the recipe.

An expert evaluation of the products was carried out with the participation of tasters. Fifty experts evaluated bread quality indicators and established

TABLE 1.4 QUALITY INDICATORS OF BREAD WITH OAT BRAN

Indicators	Control Sample	Sample with Lecithin	Oat Bran to Replace Wheat Flour (%)			
			5	7	10	15
Dough						
Moisture, %	41.8	41.7	41.8	42.0	41.8	41.7
Acidity, degrees • initial • final	1.8 2.4	1.8 2.4	1.8 2.4	1.8 2.5	1.8 2.6	1.9 2.6
Duration of fermentation, min	150					
Duration of keeping, min	44	44	46	48	51	53
Bread						
Specific volume of bread, $cm^3/100\,g$	224	234	227	209	204	172
Shape stability, H/D	0.56	0.61	0.61	0.65	0.65	0.66
Porosity of the crumb, %	73	75	70	68	67	60
Final acidity, degrees	2.0	2.0	2.1	2.2	2.2	2.3
Surface condition	Smooth, without cracks			With small cracks	With significant cracks	With large cracks
Color of the crumb	Light					
Color of the crust	Light yellow		Light yellow, with minor inclusion of bran		Light yellow, with inclusion of bran	Light yellow, with inclusion of bran
Porosity structure	Uniform, small, thin-walled					Uneven, large, thick-walled
Taste	Inherent in the product			With oat taste		

TABLE 1.5 NUTRITIONAL VALUE OF BREAD WITH PARTIAL REPLACEMENT OF WHEAT FLOUR WITH OAT BRAN

Content, %	Control Sample	Sample with Lecithin	Oat Bran to Replace Wheat Flour (%)			
			5	7	10	15
Proteins	8.10	8.10	8.36	8.46	8.61	8.86
Fats	1.09	3.06	3.28	3.37	3.51	3.73
Carbohydrates	52.97	52.97	52.22	51.92	51.47	50.72
Fiber	2.65	2.65	3.10	3.28	3.55	4.00
Ensuring the Daily Need for Nutrients Due to the Consumption of 277 g of Bread, %						
Proteins	22.45	22.45	23.15	23.43	23.85	24.55
Fats	3.02	8.48	9.10	9.34	9.72	10.34
Carbohydrates	42.65	42.65	42.05	41.81	41.45	40.84
Fiber	29.38	29.38	34.37	36.37	39.37	44.36

the weighting coefficient of each indicator. The data were averaged and a total score was determined on a 100-point scale. The results were processed by statistical procedures (Table 1.6).

Therefore, to maximize the enrichment of products with important nutrients, in particular dietary fiber, while maintaining the quality of bakery products, it is recommended to replace no more than 10% of wheat flour with oat bran.

1.5 INFLUENCE OF PUMPKIN CELLULOSE ADDITION ON THE BAKERY MANUFACTURING PROCESS AND QUALITY OF BREAD

1.5.1 Microbiological and Biochemical Processes in the Dough and Structural and Mechanical Properties of the Dough with Pumpkin Cellulose

Microbiological processes in dough with pumpkin cellulose were evaluated by the amount of released carbon dioxide during fermentation and keeping dough and the dynamics of its release. Dough samples were prepared from high-grade wheat flour added with pressed bakery yeast, 3%; salt, 1.5%; sunflower lecithin as an emulsifier and source of phosphatidylcholine, 3%, to the mass of flour with the replacement of 5.0%; 7.0%; and 10.0% and 15.0% of wheat flour by pumpkin cellulose. Dough without additives was used as control (Figure 1.11).

TABLE 1.6 SENSORY ASSESSMENT OF THE BREAD PREPARED WITH THE PARTIAL REPLACEMENT OF WHEAT FLOUR WITH OAT BRAN ON A 100-POINT SCALE TAKING INTO ACCOUNT THE WEIGHTING FACTOR OF QUALITY INDICATORS*

Indicators	Weighting Factor	Control Sample	Sample with Lecithin	Oat Bran to Replace Wheat Flour (%)			
				5	7	10	15
Specific volume of bread, $cm^3/100\,g$	0.15	4.6±0.3	4.9±0.3	4.7±0.3	4.0±0.3	3.8±0.3	3.2±0.3
Shape stability of bread, baked without form	0.15	Convex upper crust	Convex upper crust	Convex upper crust	Convex upper crust	Upper crust with depressions	Upper crust with depressions
		5.0±0.3	5.0±0.3	5.0±0.3	5.0±0.3	4.0±0.3	3.5±0.3
Color of the crust	0.05	Light yellow	Light yellow	Light yellow, with minor inclusions of bran	Light yellow, with minor inclusions of bran	Light yellow, with inclusions of bran	Light yellow, with inclusions of bran
		5.0±0.3	5.0±0.3	5.0±0.3	5.0±0.3	5.0±0.3	5.0±0.3
Surface condition	0.05	Smooth, without cracks	Smooth, without cracks	Smooth, without cracks	With small cracks	With significant cracks	With large cracks
		5.0±0.1	5.0±0.1	5.0±0.1	4.6±0.1	4.0±0.1	3.0±0.1
Color of the crumb	0.05	Світлий	Світлий	Світлий	Світлий	Світлий	Світлий
		5.0±0.1	5.0±0.1	5.0±0.1	5.0±0.1	5.0±0.1	5.0±0.1

(*Continued*)

TABLE 1.6 (*Continued*) SENSORY ASSESSMENT OF THE BREAD PREPARED WITH THE PARTIAL REPLACEMENT OF WHEAT FLOUR WITH OAT BRAN ON A 100-POINT SCALE TAKING INTO ACCOUNT THE WEIGHTING FACTOR OF QUALITY INDICATORS*

Indicators	Weighting Factor	Control Sample	Sample with Lecithin	Oat Bran to Replace Wheat Flour (%)			
				5	7	10	15
Porosity structure	0.09	Even, small, thin-walled	Even, small, thin-walled	Even, small, thin-walled	Even, small, thin-walled	Even, small, thin-walled	Uneven, large, thick-walled
		4.8±0.3	4.8±0.3	4.8±0.3	4.8±0.3	4.8±0.3	3.5±0.3
Elasticity of the crumb	0.12	Elastic	Elastic	Elastic	Elastic	Inelastic	Inelastic
		4.8±0.1	5.0±0.1	4.8±0.1	4.8±0.1	4.0±0.1	3.8±0.1
Aroma	0.11	Inherent in the product	Inherent in the product	Inherent in the product	Expressed, with an oat aroma	Brightly expressed, with an oat aroma	Brightly expressed, with an oat aroma
		4.6±0.3	4.6±0.3	4.6±0.3	5.0±0.3	5.0±0.3	5.0±0.3
Taste	0.13	Inherent in the product			With oat flavor	With oat flavor	With oat flavor
		4.8±0.3	4.8±0.3	4.8±0.3	4.9±0.3	5.0±0.3	5.0±0.3
Chewiness of the crumb	0.10	Good chewiness	Good chewiness	Good chewiness	Good chewiness	Good chewiness	Good chewiness
		5.0±0.1	5.0±0.1	5.0±0.1	5.0±0.1	5.0±0.1	4.0±0.1
Score (out of 100 points)		97.2±0.2	98.2±0.2	97.4±0.2	96.2±0.2	91.2±0.2	82.0±0.2

* Results given as: M±SD (mean±standard deviation) of triplicate trials.

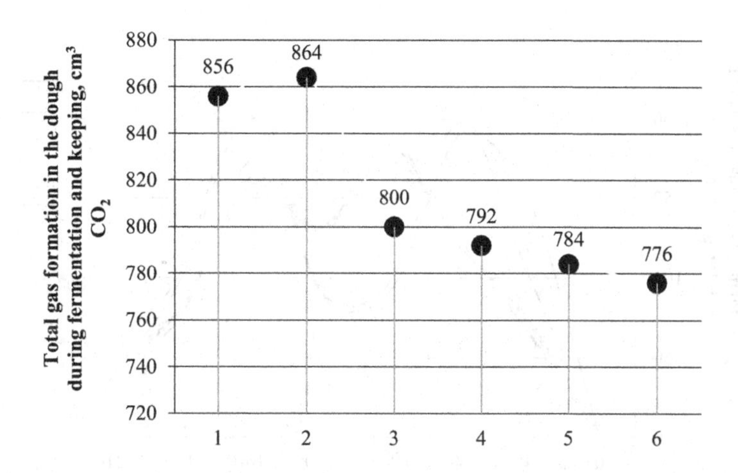

Figure 1.11 Total gas formation in the dough during fermentation and keeping. 1, control sample; 2, sample with lecithin; 3, sample with lecithin and 5% pumpkin cellulose to replace wheat flour; 4, sample with lecithin and 7% pumpkin cellulose to replace wheat flour; 5, sample with lecithin and 10% pumpkin cellulose to replace wheat flour; 6, sample with lecithin and 15% pumpkin cellulose to replace wheat flour.

With an increase in the dosage of pumpkin cellulose, the gas-forming capacity of the dough decreased by 0.9%–9.3% compared to the control, because complexes of proteins of pumpkin cellulose and wheat flour starch are formed, which causes a decrease in the intensity of carbon dioxide release due to the poorer availability of starch for the action of amylolytic enzymes (Wang et al., 2022). The first peak of gas formation in the dough with the replacement of wheat flour with pumpkin cellulose is observed after 90 min, while in the control after 50 min (Figure 1.12).

It is explained by the fact that the amylolytic enzymes of pumpkin cellulose are less active than those of wheat flour (Shevchenko et al., 2022). The amount of released carbon dioxide in the dough decreased with an increase in the dosage of the additive due to decrease in the availability of starch for amylolysis. The second peak of gas formation for all samples is observed after 150 min.

Addition of pumpkin cellulose to the dough can affect the biochemical processes in it, and the action of enzymes and microorganisms. It was shown that the replacement of a part of wheat flour with pumpkin cellulose increased the amount of formed sugars by 1.5%–5.4%, and fermented sugars by 4.2%–10.2% with an increase in its dosage. It indicates a positive effect on the enzyme complex of yeast and leads to an increase in sugar-forming capacity due to the depolymerization of carbohydrates of the additive (Medina-López et al., 2022) (Table 1.7).

Since pumpkin cellulose has significantly larger particles than wheat flour, and differs in protein composition and structural properties, it will influence the

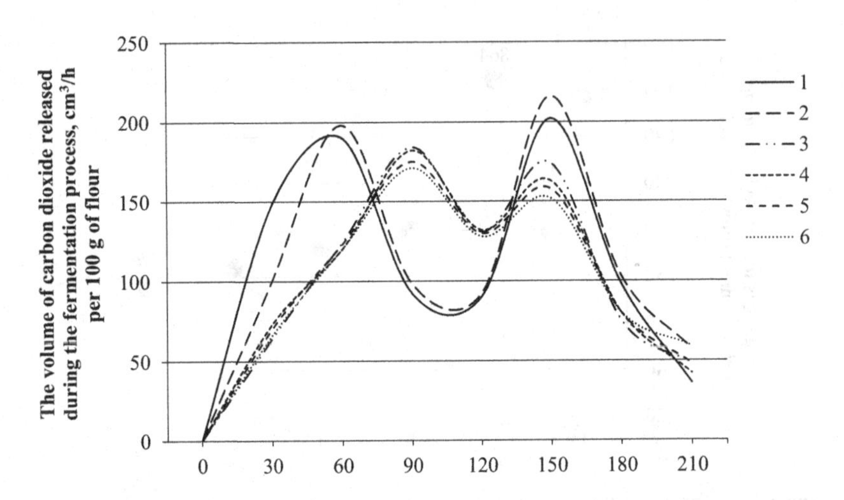

Figure 1.12 Dynamics of gas formation in dough with different dosage of pumpkin cellulose. 1, control sample; 2, sample with lecithin; 3, sample with lecithin and 5% pumpkin cellulose to replace wheat flour; 4, sample with lecithin and 7% pumpkin cellulose to replace wheat flour; 5, sample with lecithin and 10% pumpkin cellulose to replace wheat flour; 6, sample with lecithin and 15% pumpkin cellulose to replace wheat flour.

TABLE 1.7 ACCUMULATION AND FERMENTATION OF SUGARS DURING THE FERMENTATION OF THE DOUGH (IN TERMS OF MALTOSE),% TO DRY MATTER*

Indicators	Control Sample	Sample With Lecithin	Pumpkin Cellulose to Replace Wheat Flour (%)			
			5	7	10	15
Yeast-Free Dough						
After kneading	2.10±0.10	2.10±0.10	2.06±0.10	2.03±0.10	1.99±0.09	1.95±0.09
After 3 h of fermentation	3.15±0.13	3.39±0.17	3.37±0.17	3.36±0.17	3.34±0.16	3.31±0.16
Accumulated sugars	1.05±0.01	1.29±0.03	1.31±0.03	1.33±0.03	1.35±0.04	1.36±0.04
Yeast Dough						
After kneading	2.12±0.10	2.15±0.12	2.14±0.12	2.10±0.10	2.05±0.10	2.02±0.10
After 3 h of fermentation	1.69±0.06	1.78±0.08	1.72±0.07	1.68±0.07	1.58±0.06	1.55±0.06
Fermented sugars	1.48±0.05	1.66±0.06	1.73±0.07	1.75±0.07	1.82±0.08	1.83±0.08

* Results given as: M±SD (mean±standard deviation) of triplicate trials.

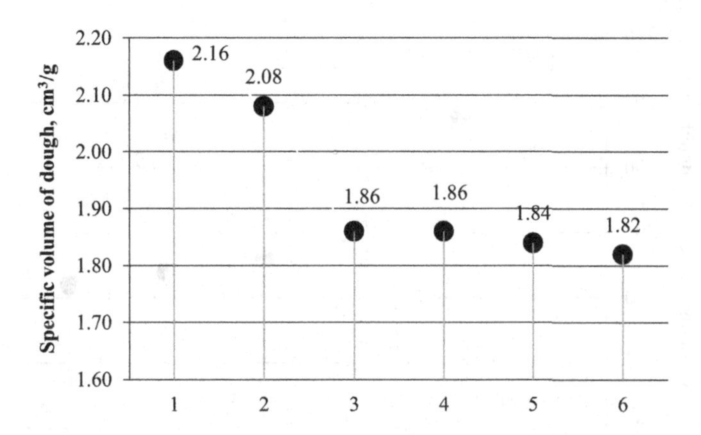

Figure 1.13 Specific volume of dough. 1, control sample; 2, sample with lecithin; 3, sample with lecithin and 5% pumpkin cellulose to replace wheat flour; 4, sample with lecithin and 7% pumpkin cellulose to replace wheat flour; 5, sample with lecithin and 10% pumpkin cellulose to replace wheat flour; 6, sample with lecithin and 15% pumpkin cellulose to replace wheat flour.

structural and mechanical properties of the dough system, and subsequently the quality of bread. Thus, the gas-holding capacity of the dough decreased by 10.5%–12.5% when pumpkin cellulose is added. This is due to the addition of pectin substances to the dough along with pumpkin cellulose, which increases dough viscosity and weakens gluten (Figure 1.13).

The viscosity of the dough also changes. The diameter of the ball of dough with the additive during fermentation was smaller by 10.5%–12.5%, depending on its dosage (Figure 1.14).

The shape stability of the dough improved due to the introduction of an increased amount of dietary fibers and pentosans with pumpkin cellulose, which have high water absorption and water-holding capacity (Table 1.1 and Figure 1.2). The redistribution of moisture and, accordingly, the decrease in the rate of diffusion during the osmotic swelling of flour proteins are associated with the attraction of water molecules by hydrophilic compounds, which include pumpkin dietary cellulose with high water-binding capacity.

1.5.2 Quality and Nutritional Value of Bakery Products with Pumpkin Cellulose

To obtain bread of high quality, it is important to establish a dosage of the components in the recipe. Baking of bread was conducted to determine the recommended dosage of pumpkin cellulose to replace wheat flour (Table 1.8).

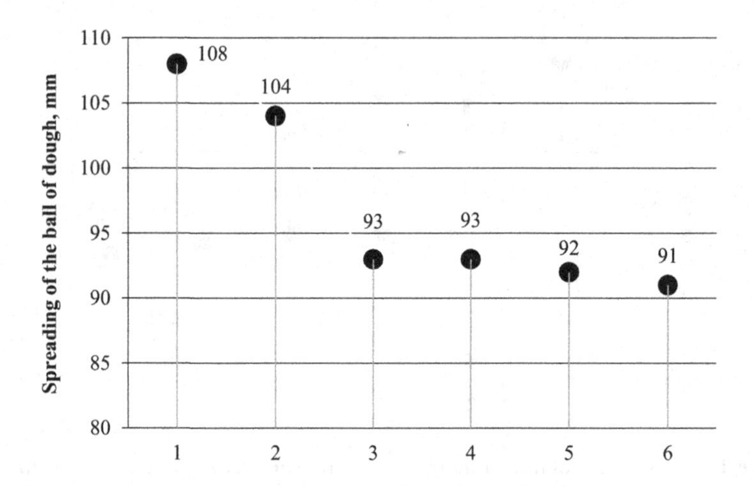

Figure 1.14 Spreading of dough ball. 1, control sample; 2, sample with lecithin; 3, sample with lecithin and 5% pumpkin cellulose to replace wheat flour; 4, sample with lecithin and 7% pumpkin cellulose to replace wheat flour; 5, sample with lecithin and 10% pumpkin cellulose to replace wheat flour; 6, sample with lecithin and 15% pumpkin cellulose to replace wheat flour.

Increase of the amount of wheat flour replaced with pumpkin cellulose leads to a rise of the initial acidity due to the higher acidity of pumpkin cellulose than flour, and the final acidity, which is associated with increased acid accumulation during fermentation of dough with fiber. It was shown that the specific volume of bread was reduced by 7.7%–41.0% depending on the dosage of this additive. Porosity and shape stability of the products also decreased. Lower indicators of specific volume and porosity of bread can be explained by the swelling specificity of pumpkin cellulose, which does not lose its shape when swollen and is in the form of inclusions in the dough. In the process of fermentation and keeping, the swollen pumpkin cellulose is embedded in the gluten frame, thereby destroying its integrity. Its addition had a significant impact on the sensory parameters of bread. With an increase in the dosage, the presence of greenish shade of the crumb and crust of the products, inclusions of fiber particles, pumpkin taste, and aroma were observed.

It is important to determine the effect of different dosages of pumpkin cellulose on the nutritional value of the product to ensure the needs of the body in basic nutrients (Table 1.9).

Ensuring the daily need for nutrients was calculated according to the daily consumption of bakery products by Ukrainian people, which is estimated as 277 g. The analysis of the chemical composition of the products without pumpkin cellulose and with its addition showed an increase in the protein content

TABLE 1.8 QUALITY INDICATORS OF BREAD WITH PUMPKIN CELLULOSE

Indicators	Control Sample	Sample With Lecithin	Pumpkin Cellulose to Replace Wheat Flour (%)			
			5	7	10	15
Dough						
Moisture, %	41.8	41.7	41.8	42.0	41.8	41.7
Acidity, degrees • initial • final	1.8 2.4	1.8 2.4	1.8 3.1	1.9 3.2	2.1 3.5	2.3 3.7
Duration of fermentation, min	150	150	150	150	150	150
Duration of keeping, min	44	44	43	41	40	38
Bread						
Specific volume of bread, $cm^3/100\,g$	224	234	216	214	161	138
Shape stability, H/D	0.56	0.61	0.59	0.57	0.57	0.57
Porosity of the crumb, %	73	75	72	72	70	70
Final acidity, degrees	2.0	2.0	2.2	2.5	2.9	3.1
Surface condition	Smooth, without cracks					Smooth, without cracks, uneven
Color of the crumb	Light			With a barely noticeable darkening shade	With a greenish shade	With a noticeable greenish shade
Color of the crust	Light yellow		Light yellow, with inclusions of fiber		With a greenish shade, with inclusions of fiber	
Porosity structure	Uniform, small, thin-walled				Uniform, small, thick-walled	
Taste	Inherent in the product		With pumpkin taste		With a pronounced pumpkin taste	

TABLE 1.9 NUTRITIONAL VALUE OF BREAD WITH PARTIAL REPLACEMENT OF WHEAT FLOUR WITH PUMPKIN CELLULOSE

Content, %	Control Sample	Sample With Lecithin	Pumpkin Cellulose to Replace Wheat Flour (%)			
			5	7	10	15
Proteins	8.10	8.10	9.30	9.78	10.50	11.71
Fats	1.09	3.06	3.25	3.32	3.43	3.62
Carbohydrates	52.97	52.97	50.93	51.92	51.47	50.72
Fiber	2.65	2.65	3.73	4.16	4.81	5.89
Ensuring the Daily Need for Nutrients due to the Consumption of 277 g of Bread, %						
Proteins	22.45	22.45	25.77	27.10	29.10	32.42
Fats	3.02	8.48	8.99	9.20	9.51	10.02
Carbohydrates	42.65	42.65	41.01	41.81	41.45	40.84
Fiber	29.38	29.38	41.34	46.12	53.30	65.26

by 1.1–1.4 times and fiber by 1.4–2.2 times depending on the dosage. This contributes to an increase in providing the daily need for protein by 14.8%–44.4%, and in fiber by 40.7%–122.1% compared to the control. An expert evaluation of the products was carried out with the participation of sensory tests. The results were processed by statistical analysis (Table 1.10).

Therefore, for the maximum enrichment of products with important nutrients, in particular dietary fibers, and to obtain high-quality bakery products, it is worth replacing no more than 7% of wheat flour with pumpkin cellulose.

1.6 CONCLUSION

Bakery products enriched with fibers are useful to be included to the human diet for people with diseases of the gastrointestinal tract, especially irritable bowel syndrome. Studies of the chemical composition of oat bran and pumpkin cellulose showed higher by 1.6 and 4 times amounts of protein and by 4.4 and 9.1 times amounts of fiber compared to wheat flour, respectively. The ability of oat bran to absorb water was higher by 10%–21% and form stable emulsions by 29% compared to pumpkin cellulose. Pumpkin cellulose has a higher capability to interact with fat molecules. The higher amount of soluble dietary fibers and β-glucan in the composition of oat bran led to an increase of effective viscosity of aqueous solutions by 11%–90% compared with pumpkin cellulose.

TABLE 1.10 SENSORY ASSESSMENT OF THE BREAD PREPARED WITH THE PARTIAL REPLACEMENT OF WHEAT FLOUR WITH PUMPKIN CELLULOSE ON A 100-POINT SCALE TAKING INTO ACCOUNT THE WEIGHTING FACTOR OF QUALITY INDICATORS*

Indicators	Weighting Factor	Control Sample	Sample With Lecithin	Pumpkin Cellulose to Replace Wheat Flour (%)			
				5	7	10	15
Specific volume of bread, $cm^3/100\,g$	0.15	4.6±0.3	4.9±0.3	4.7±0.3	3.9±0.3	3.2±0.3	2.9±0.3
Shape stability of bread, baked without form	0.15	Convex upper crust	Convex upper crust	Convex upper crust	Convex upper crust	Convex upper crust	Convex upper crust
		5.0±0.3	5.0±0.3	5.0±0.3	5.0±0.3	5.0±0.3	5.0±0.3
Color of the crust	0.05	Light yellow	Light yellow	Light yellow, with inclusions of fiber	Light yellow, with inclusions of fiber	With a greenish tint, with inclusions of fiber	With a greenish tint, with inclusions of fiber
		5.0±0.3	5.0±0.3	5.0±0.3	5.0±0.3	5.0±0.3	5.0±0.3
Surface condition	0.05	Smooth, without cracks	Smooth, without cracks	Smooth, without cracks	Smooth, without cracks	Smooth, without cracks	Smooth, without cracks, uneven
		5.0±0.1	5.0±0.1	5.0±0.1	5.0±0.1	5.0±0.1	4.0±0.1
Color of the crumb	0.05	Light	Light	Light	With a barely noticeable darkening shade	With a greenish tint	With a noticeable greenish tint
		5.0±0.1	5.0±0.1	5.0±0.1	4.7±0.1	4.5±0.1	4.0±0.1

(*Continued*)

TABLE 1.10 (*Continued*) SENSORY ASSESSMENT OF THE BREAD PREPARED WITH THE PARTIAL REPLACEMENT OF WHEAT FLOUR WITH PUMPKIN CELLULOSE ON A 100-POINT SCALE TAKING INTO ACCOUNT THE WEIGHTING FACTOR OF QUALITY INDICATORS*

Indicators	Weighting Factor	Control Sample	Sample With Lecithin	Pumpkin Cellulose to Replace Wheat Flour (%)			
				5	7	10	15
Porosity structure	0.09	Even, small thin-walled	Even, small thin-walled	Even, small thin-walled	Even, small thin-walled	Even, small thick-walled	Even, small thick-walled
		4.8±0.3	4.8±0.3	4.8±0.3	4.8±0.3	4.0±0.3	4.0±0.3
Elasticity of the crumb	0.12	Elastic	Elastic	Elastic	Elastic	Elastic	Elastic
		4.8±0.1	5.0±0.1	5.0±0.1	5.0±0.1	5.0±0.1	5.0±0.1
Aroma	0.11	Inherent in the product	Inherent in the product	Inherent in the product	Inherent in the product	Expressed, with pumpkin aroma	Brightly expressed, with pumpkin aroma
		4.6±0.3	4.6±0.3	4.6±0.3	4.8±0.3	5.0±0.3	5.0±0.3
Taste	0.13	Inherent in the product	Inherent in the product	With pumpkin taste	With pumpkin taste	With brightly expressed pumpkin taste	With brightly expressed pumpkin taste
		4.8±0.3	4.8±0.3	5.0±0.3	5.0±0.3	5.0±0.3	5.0±0.3
Chewiness of the crumb	0.10	Good chewiness	Good chewiness	Good chewiness	Good chewiness	Good chewiness	Good chewiness
		5.0±0.1	5.0±0.1	5.0±0.1	5.0±0.1	5.0±0.1	5.0±0.1
Score (out of 100 points)		97.2±0.2	98.2±0.2	98.2±0.2	96.4±0.2	93.4±0.2	89.8±0.2

* Results given as: M±SD (mean±standard deviation) of triplicate trials.

Replacement of 5%–15% of wheat flour by oat bran or pumpkin cellulose led to a decrease in the gas-forming capacity of dough compared to the control sample: by 1.8%–10.2% in the case of the addition of oat bran and by 0.9%–9.3% with pumpkin cellulose. Oat bran caused a decrease in the amount of accumulated sugars by 15.5%–16.3% and fermented sugars by 18.6%–28.9%. On the contrary, in the dough with pumpkin cellulose the amount of accumulated sugars increased by 1.5%–5.4%, and fermented sugars by 4.2%–10.2%. Addition of oat bran or pumpkin cellulose increased the nutritional value of bakery products, in particular the content of protein and dietary fiber. However, to obtain bread of high quality, it is recommended to replace no more than 5%–7% of wheat flour by oat bran or pumpkin cellulose.

REFERENCES

Abhiraman, K., & Soumya, S. (2020). Rheological and physico-chemical properties of milk gel using isolate of pumpkin (Cucurbita moschata) seeds: A new source of milk clotting peptidase, *Food Hydrocolloids, 106*, 1058–1066. https://doi.org/10.1016/j.foodhyd.2020.105866

Ahmed, J., Al-Foudari, M., Al-Salman, F., & Almusallam, A.S. (2014). Effect of particle size and temperature on rheological, thermal, and structural properties of pumpkin flour dispersion, *Journal of Food Engineering, 124*, 43–53. https://doi.org/10.1016/j.jfoodeng.2013.09.030

Ali, H.A., Mansour, E.H., ElBedawey, A.E.F., & Osheba, A.S. (2017). Evaluation of tilapia fish burgers as affected by different replacement levels of mashed pumpkin or mashed potato, *Journal of the Saudi Society of Agricultural Sciences, 18*(2), 127–132. https://doi.org/10.1016/j.jssas.2017.01.003

Aminzadeh, S., Zhang, L., & Henriksson, G. (2017). A possible explanation for the structural inhomogeneity of lignin in LCC networks, *Wood Science and Technology, 51*(2), 1365–1376. https://doi.org/10.1007/s00226-017-0941-6

Anttila, H., Sontag-Strohm, T., & Salovaara, H. (2004). Viscosity of beta-glucan in oat products, *Agricultural and Food Science, 13*(1), 80–87. https://doi.org/10.2137/1239099041838012

Ashurova, M.Z., & Sulaymanova, G.X. (2021). Development of technology for functional confectionery products, *International Journal on Orange Technologies, 3*(11), 9–15. https://doi.org/10.31149/ijot.v3i11.2369

Atuonwu, A.C., & Akobundu, E.N.T. (2010). Nutritional and sensory quality of cookies supplemented with defatted pumpkin (Cucurbita pepo) seed flour, *Pakistan Journal of Nutrition, 9*(7), 672–677. https://doi.org/10.3923/pjn.2010.672.677

Baune, M.C., Jeske, A.L., Profeta, A., Smetana, S., Broucke, K., Van Royen, G., Gibis, M., Weiss, J., & Terjung, N. (2021). Effect of plant protein extrudates on hybrid meatballs – changes in nutritional composition and sustainability, *Future Foods, 4*, 100081. https://doi.org/10.1016/j.fufo.2021.100081

Brückner-Gühmann, M., Benthin, A., & Drusch, S. (2019). Enrichment of yoghurt with oat protein fractions: Structure formation, textural properties and sensory evaluation, *Food Hydrocolloids, 86*, 146–153. https://doi.org/10.1016/j.foodhyd.2018.03.019

Burrows, V.D. (2011). Hulless oat development, applications, and opportunities. In Webster, F.H., Wood, P.J. (Eds.), *Oats: Chemistry and technology*, 2nd Edition, American Association of Cereal Chemists, pp. 31–50. https://doi.org/10.1016/B978-1-891127-64-9.50008-7

Cui, S.W., Wu, Y., & Ding, H. (2013). The range of dietary fibre ingredients and a comparison of their technical functionality. In Delcour, J.A., Poutanen, K. (Eds.), *Fibre-rich and wholegrain foods*, Woodhead Publishing, Sawston, pp. 96–119. https://doi.org/10.1533/9780857095787.1.96

De Bondt, Y., Hermans, W., Moldenaers, P., & Courtin, C.M. (2021). Selective modification of wheat bran affects its impact on gluten-starch dough rheology, microstructure and bread volume, *Food Hydrocolloids, 113*, 106348. https://doi.org/10.1016/j.foodhyd.2020.106348

Delgado-Vargas, F., & Paredes-López, O. (2002). *Natural colorants for food and nutraceutical uses*. CRC Press, Boca Raton, London, New York, Washington, DC.

Dhiman, A., Bavita, K., Attri, S., & Ramachandran P. (2018). Preparation of pumpkin powder and pumpkin seed kernel powder for supplementation in weaning mix and cookies, *International Journal of Chemical Studies, 6*(5), 167–175.

Dietary Guidelines Advisory Committee. (2015). *Scientific report of the 2015 dietary guidelines advisory committee: Advisory report to the secretary of health and human services and the secretary of agriculture*. U.S. Department of agriculture, Agricultural research service, Washington, DC.

Dotsenko, V., Medvid, I., Shydlovska, O., & Ishchenko, T. (2019). Studying the possibility of using enzymes, lecithin, and albumen in the technology of gluten-free bread, *Eastern-European Journal of Enterprise Technologies, 1*(11(97)), 42–51. https://doi.org/10.15587/1729-4061.2019.154957

Drobot, V., Semenova, A., Smirnova, J., & Mykhonik, L. (2014). Effect of buckwheat processing products on dough and bread quality made from whole-wheat flour, *International Journal of Food Studies, 3*(1), 1–12. https://doi.org/10.7455/ijfs/3.1.2014.a1

Elik, A., Yanik, D.K., Istanbullu, Y., Guzelsoy, N.A., Yavuz, A., & Gogus, F. (2019). Strategies to reduce post-harvest losses for fruits and vegetables, *Strategies, 5*, 29–39. https://doi.org/10.7176/JSTR/5-3-04

Esteban, R.M., Mollá, E., & Beníte, V. (2017). Sources of fiber. In Samaan, R. (Eds.), *Dietary fiber for the prevention of cardiovascular disease*, Academic Press, Los Angeles, CA, pp. 121–146. https://doi.org/10.1016/B978-0-12-805130-6.00007-0

Fathonah, S., Setyaningsih, R., Paramita, D., Istighfarin, O., & Litazkiyati, N. (2018). The sensory quality and acceptability of pumpkin flour cookies, In Proceedings of the 7th Engineering International Conference on Education, Concept and Application on Green Technology, 439–445. https://doi.org/10.5220/0009012804390445

Garcia-Amezquita, L.E., Tejada-Ortigoza, V., Serna-Saldivar, S.O., & Welti-Chanes, J. (2018). Dietary fiber concentrates from fruit and vegetable by-products: Processing, modification, and application as functional ingredients, *Food Bioprocess Technology, 11*, 1439–1463. https://doi.org/10.1007/s11947-018-2117-2

Genevois, C., Flores, S., & de Escalada Pla, M. (2016). Byproduct from pumpkin (Cucurbita moschata Duchesne ex poiret) as a substrate and vegetable matrix to contain Lactobacillus casei, *Journal of Functional Foods, 23*, 210–219. https://doi.org/10.1016/j.jff.2016.02.030

Gibson, P.R., Varney, J., Malakar, S., & Muir, J.G. (2015). Food components and irritable bowel syndrome, *Gastroenterology, 148*, 1158–1174. https://doi.org/10.1053/j.gastro.2015.02.005

Gore, S.B., Xavier, K.A.M., Nayak, B.B., Tandale, A.T., & Balange, A.K. (2022). Technological effect of dietary oat fiber on the quality of minced sausages prepared from Indian major carp (Labeo rohita), *Bioactive Carbohydrates and Dietary Fibre, 27*, 1003–1005. https://doi.org/10.1016/j.bcdf.2021.100305

Guo, M. (2009). *Dietary fiber and dietary fiber rich foods, functional foods: Principles and technology*. CRC Press, Boca Roca, FL, pp. 63–111. https://doi.org/10.1533/9781845696078.63.

Hill, P., Muir, J., & Gibson, P. (2017). Controversies and recent developments of the low-FODMAP, *Diet, Gastroenterology & Hepatology, 13*(1), 36–45. https://pubmed.ncbi.nlm.nih.gov/28420945/

Hussain, S., Jõudu, I., & Bhat, R. (2020). Dietary fiber from underutilized plant resources—A positive approach for valorization of fruit and vegetable wastes, *Sustainability, 12*(13), 5401. https://doi.org/10.3390/su12135401

Kampuse, S., Ozola, L., Straumite, E., & Galoburda, R. (2015). Quality parameters of wheat bread enriched with pumpkin (Cucurbita Moschata) by-products, *Acta Universitatis Cibiniensis Series E Food Technology, 19*(2), 3–14. https://doi.org/10.1515/aucft-2015-0010

Kim, C.J., Kim, H.W., Hwang, K.E., Song, D.H., Ham, Y.K., Choi, J.H., Kim, Y.B., & Choi, Y.S. (2016). Effects of dietary fiber extracted from pumpkin (Cucurbita maxima Duch.) on the physico-chemical and sensory characteristics of reduced-fat frankfurters, *Food Science of Animal Resources, 36*(3), 309–318. https://doi.org/10.5851/kosfa.2016.36.3.309

Lee, N., & Ha, K. (2011). Quality characteristics of bread added with oat flours, *Korean Journal of Crop Science, 56*(2), 107–112. https://doi.org/10.7740/kjcs.2011.56.2.107

Majzoobi, M., Habibi, M., Hedayati, S., Ghiasi, F., & Farahnaky, A. (2015). Effects of commercial oat fiber on characteristics of batter and sponge cake, *Journal of Agricultural Science and Technology, 17*(1), 99–107. http://jast.modares.ac.ir/article-23-3754-en.html

Medina-López, S., Zuluaga-Domínguez, C., Fernández-Trujillo, J., & Hernández-Gómez, M. (2022). Nonconventional hydrocolloids' technological and functional potential for food applications, *Foods, 11*(3), 401. https://doi.org/10.3390/foods11030401

Munteanu, G.M., Voicu, G., Ferdeş, M., Ştefan, E.M., Constantin, G.A., & Tudor, P. (2019). Dynamics of fermentation process of bread dough prepared with different types of yeast, *Scientific Study and Research: Chemistry and Chemical Engineering, Biotechnology, Food Industry, 20*(4), 575–584.

Nanayakkara, W.S., Skidmore, P.M., O'Brien, L., Wilkinson, T.J., & Gearry, R.B. (2016). Efficacy of the low FODMAP diet for treating irritable bowel syndrome: The evidence to date, *Clinical and Experimental Gastroenterology, 9*, 131–142. https://doi.org/10.2147/CEG.S86798

Partridge, D., Lloyd, K.A., Rhodes, J.M., Walker, A.W., Johnstone, A.M., & Campbell, B.J. (2019). Food additives: Assessing the impact of exposure to permitted emulsifiers on bowel and metabolic health – Introducing the FADiets study, *Nutrition Bulletin, 44*(4), 329–349. https://doi.org/10.1111/nbu.12408

Patel, A.S., Bariya, A.R., Ghodasara, S.N., Chavda, J.A., & Patil, S.S. (2020). Total carotene content and quality characteristics of pumpkin flavoured buffalo milk, *Heliyon, 6*, e04509. https://doi.org/10.1016/j.heliyon.2020.e04509

Raikos, V., Neacsu, M., Russell, W., & Duthie, G. (2014). Comparative study of the functional properties of lupin, green pea, fava bean, hemp, and buckwheat flours as affected by pH, *Food Science & Nutrition, 2*(6), 802–810. https://doi.org/10.1002/fsn3.143

Rakcejeva, T., Galoburda, R., Cude, L., & Strautniece, E. (2011). Use of dried pumpkins in wheat bread production, *Procedia Food Science, 1*, 441–447. https://doi.org/10.1016/j.profoo.2011.09.068

Ramos-Diaz, J.M., Kantanen, K., Edelmann, J.M., Jouppila, K., Sontag-Strohm, T., & Piironen, V. (2022). Functionality of oat fiber concentrate and faba bean protein concentrate in plant-based substitutes for minced meat, *Current Research in Food Science, 5*, 858–867. https://doi.org/10.1016/j.crfs.2022.04.010

Różyło, R., Gawlik-Dziki, U., Dziki, D., Jakubczyk A., Karaś, M., & Różyło, K. (2014). Wheat bread with pumpkin (Cucurbita maxima L.) pulp as a functional food product, *Food Technology and Biotechnology, 52*(4), 430–438. https://doi.org/10.17113/b.52.04.14.3587

Shevchenko, A. (2023). Artichoke powder and buckwheat bran in diabetic bakery products. In Paredes-López, O., Shevchenko, O., Stabnikov, V., Ivanov, V. (Eds.), *Bioenhancement and fortification of foods for a healthy diet*, CRC Press, Boca Raton, London, pp. 115–134. https://doi.org/10.1201/9781003225287

Shevchenko, A., Drobot, V., & Galenko, O. (2022). Use of pumpkin seed flour in preparation of bakery products, *Ukrainian Food Journal, 11*(1), 90–101. https://doi.org/10.24263/2304-974X-2022-11-1-10

Stabnikova, O., Antonuk, M., Stabnikov, V., & Arsen'eva, L. (2019). Ukrainian dietary bread with selenium-enriched malt. *Plant Foods for Human Nutrition, 74*(2), 157–163. https://doi.org/10.1007/s11130-019-00731-z

Stabnikova, O., Marinin, A., & Stabnikov, V. (2021). Main trends in application of novel natural additives for food production, *Ukrainian Food Journal, 10*(3), 524–551. https://doi.org/10.24263/2304-974X-2021-10-3-8

Stabnikova, O., Stabnikov, V., Antoniuk, M., Arsenieva, L., & Ivanov, V. (2023). Bakery products enriched with organoselenium compounds. In Paredes-López, O., Shevchenko, O., Stabnikov, V., Ivanov, V. (Eds.), *Bioenhancement and fortification of foods for a healthy diet*, CRC Press, Boca Raton, London, pp. 89–111. https://doi.org/10.1201/9781003225287-6

Strashynskiy, I., Grechko, V., Fursik, O., Pasichnyi, V., & Marynin, A. (2021a). Determining the properties of chia seed meal gel, *Eastern-European Journal of Enterprise Technologies, 6*(11 (114)), 90–98. https://doi.org/10.15587/1729-4061.2021.245505

Strashynskiy, I., Grechko, V., Fursik, O., Pasichnyi, V., & Marynin, A. (2021b). Influence of temperature modes of processing chia seed meal gel on its rheological and functional-technological properties, *ScienceRise, 6*, 11–17. https://doi.org/10.21303/2313-8416.2021.002226

Verheyen, C., Albrecht, A., Elgeti, D., Jekle, M., & Becker, T. (2015). Impact of gas formation kinetics on dough development and bread quality, *Food Research International, 76*(3), 860–866. https://doi.org/10.1016/j.foodres.2015.08.013

Wang, L., Liu, F., Wang, A., Yu, Z., Xu, Y., & Yang, Y. (2017a). Purification, characterization and bioactivity determination of a novel polysaccharide from pumpkin *(Cucurbita moschata)* seeds, *Food Hydrocolloids, 66,* 357–364. https://doi.org/10.1016/j.foodhyd.2016.12.003

Wang, L., Ye, F., Li, S., Wei, F., Chen, J., & Zhao, G. (2017b). Wheat flour enriched with oat b-glucan: A study of hydration, rheological and fermentation properties of dough, *Journal of Cereal Science, 75,* 143–150. https://doi.org/10.1016/j.jcs.2017.03.004

Wang, Y., Ral, J., Saulnier, L., & Kansou, K. (2022). How does starch structure impact amylolysis? Review of current strategies for starch digestibility study, *Foods, 11*(9), 1223. https://doi.org/10.3390/foods11091223

Yu, G., Zhao, J., Wei, Y., Huang, L., Li, F., Zhang, Y., & Li, Q. (2021). Physicochemical properties and antioxidant activity of pumpkin polysaccharide (Cucurbita moschata Duchesne ex Poiret) modified by subcritical water, *Foods, 10*(1), 197. https://doi.org/10.3390/foods10010197

Zhang, K., Dong, R., Hu, X., Ren, C., & Li, Y. (2021). Oat-based foods: Chemical constituents, glycemic index, and the effect of processing, *Foods, 10*(6), 1304. https://doi.org/10.3390/foods10061304

Chapter 2

Biotransformation of Collagen-Containing Meat Materials into Valuable Product

Oksana Topchii, Vasyl Pasichnyi,
Andrii Marynin, and Olena Stabnikova
National University of Food Technologies

CONTENTS

2.1 INTRODUCTION

2.1.1 Meat Processing Co-Products for Human Consumption

Currently, many countries have a significant problem connected with the limited resources of animal-based protein. At the same time, it is expected that by 2050 the global demand for protein will increase by another 50% (Henchion et al., 2017; Westhoek et al., 2011). Meanwhile, protein is an important part of the human diet supplying essential amino acids needed for maintaining a

DOI: 10.1201/9781003329671-2

healthy body (Adesogan et al., 2020). According to the Food and Agriculture Organization of the United Nations, the number of undernourished people in the world is continually increasing. In 2020, it was estimated between 720 and 811 million, and one billion people worldwide could suffer from an acute deficiency of animal protein. Thus, the processing of waste products from the meat industry for their further use in the manufacturing of food products is undoubtedly of considerable interest to provide the world population with complete diets.

A huge quantity of valuable by-products, approximately 150 million tons annually, are produced during meat processing. The main supplier of these by-products of the meat processing industry is slaughterhouses. The live weight of the animals, the weight taken immediately before slaughter, consists of edible, inedible and discardable by-products. The discardable part of the live weight of cattle, lamb and pigs is estimated as 66%, 52% and 80%, respectively, and consists of bones, skin, blood and internal organs (Limeneh et al., 2022). According to Russ & Pittroff (2004), the specific waste index (SWI)—mass of accumulated waste divided by the mass of saleable product—consists of cow, sheep, pig and calf 0.56, 0.1, 0.2 and 0.87, respectively. Only minor share of those wastes are converted into valuable by-products such as animal feed, fertilizer, glue and gelatin (Adhikari et al., 2018; Almeida et al., 2013; Hidaka & Liu, 2013; Jayathilakan et al., 2012; Limeneh et al., 2022). Animal blood consists of 2.4% to 8.0% of the animal's live weight and is considered to be a source of potentially valuable materials, such as protein and heme iron, which can be used in the food industry (Jayathilakan et al., 2012). There is an attempt to use bovine blood as an emulsifier or fat replacer in food preparation (Silva & Silvestre, 2003).

The edible meat by-products include internal organs such as spleen, heart, lungs, liver, kidney and others and constitute a significant part of the animal's live weight. For example, it consists of 10%–30% for pork and beef (Nollet & Toldra, 2011). Proteins contained in these by-products, offal of the second category, have a complete set of essential amino acids. The protein content in bovine lung was 19.2±0.4% or 71.6±0.3% in dried lung (Lafarga & Hayes, 2017a). Protein content in the offal of Hanwoo cattle was, %: 18.6±0.5 in the heart; 18.6±0.9 in the liver; 16.0±0.9 in the kidney; 17.6±0.7 in the lung; 22.2±2.3 in the bladder; 21.3±1.8 in the duodenum; and 6.8±0.2 in the reproductive organ (Seong et al., 2014). Protein content in tripe, ears, lips, lungs and heart of cattle was, %: 14.3; 24.6; 24.9; 15.2 and 16.8, respectively (Zinina et al., 2019). With the exception of beef heads and spleen, offal of the second category contains a complete set of essential amino acids. The ratio of total essential amino acids to amino acids varied from 38.37 to 47.41% for different edible bovine by-products (Seong et al., 2014). Most of these by-products have a low fat content of up to 10%, which allows using them in the manufacturing of meat products as protein raw materials. Bioactive hydrolysates and peptides could be obtained from

TABLE 2.1 THE CONTENT OF ESSENTIAL AMINO ACIDS IN THE PROTEINS OF BEEF OFFALS

Offals	Essential Amino Acids[a], g/100 g Protein								
	Trp	Lys	Thr	Val	Ile	Leu	Phe+Tyr	Met+Cys	Total
Lungs	0.78	6.23	3.53	5.78	3.65	8.5	9.15	2.08	39.67
Meat trimmings	0.92	8.41	4.94	6.05	4.22	8.02	3.46	3.80	39.82
Ears	0.52	4.18	2.17	3.47	2.07	4.14	3.85	1.67	22.06
Tripe	0.90	5.80	3.61	3.90	3.41	6.10	5.85	1.57	31.13
Heart	1.36	7.98	4.94	6.13	4.88	8.87	8.27	4.44	46.86

[a] Trp, tryptophan; Lys, lysine; Thr, threonine; Val, valine; Ile, isoleucine; Leu, leucine; Phe, phenylalanine; Tyr, tyrosine; Met, methionine; Cys, cys.

meat by-products to be used as functional food ingredients (Lafarga & Hayes, 2017a, b). The content of essential amino acids (tryptophan, lysine, threonine, valine, isoleucine, leucine, phenylalanine, and methionine) in the proteins of the beef offal is shown in Table 2.1 (adapted from Jayawardena et al., 2022; Zinina et al., 2019).

Slaughterhouse by-products contain such resource as collagen, oligopeptide by its chemical nature, the principal structural element of the connective tissues such as animal skin, bone, cartilage, tendon and blood vessels, which is widely used in the food and beverage industries as additive, edible film and coating, and carrier (Hashim et al., 2015; Tang et al., 2022). Collagen is a fibrous protein, which is present in almost all animal tissues in various forms and accounts for about 30% of the total protein (Hashim et al., 2015; Owczarzy et al., 2020). For example, veins, tendons and cartilages of slaughtered animals, separated during meat degreasing in the sausage and canning industry, contain 88.5% protein, consisting of 61%–78% of collagens. Being composed of protein, collagen can be converted into a valuable by-product for further use in food preparations (Hashim et al., 2015; Jayawardena, et al., 2022; Lafarga & Hayes, 2017a, b; Zinina et al., 2019). The development of food products using low-grade collagen-containing raw materials is becoming increasingly important, especially when combined with muscle proteins. The main sources of collagen are shown in Figure 2.1.

Thus, collagen-containing meat raw materials are an affordable and promising source of essential amino acids, organic iron, and other macro- and micronutrient; their application as additives to food products could increase their nutritional and consumable values.

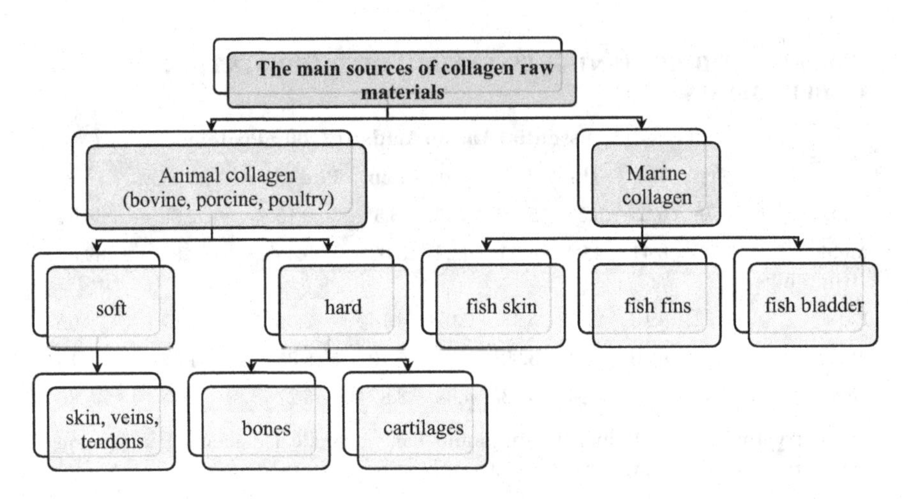

Figure 2.1 The main sources of collagen raw materials.

2.2 COLLAGEN AND ITS PRACTICAL APPLICATIONS

The collagens are a family of fibrous proteins found in almost all animal tissues (Owczarzy et al., 2020). They are synthesized mainly in the cells of fibroblasts, specialized cells, which are the most common cell type of connective tissues and contribute a lot to their formation. Collagens serve as a building material for most tissues and organs and provide a basic framework for them (Wu et al., 2022). There are 29 various types of collagen proteins. All collagens have a stable triple-helix structure consisting of three collagen polypeptide chains, named α-chains, wound together (Shoulders & Raines, 2009). Collagen molecules are stabilized by electrostatic, hydrophobic interactions, as well as hydrogen and covalent cross-links between amino acids of protein α-chains. Every third residue of amino acid in peptide chain in all types of collagen is glycine (Gly), resulting in a repeating X-Y-Gly sequence, where X and Y can be any amino acid. The most abandoned amino acids in collagens are glycine, proline, and hydroxyproline and glycine is every third amino acid in collagen protein. Glycine is the smallest of all the amino acids that allow it to form a tight configuration of peptide chains in the molecule of collagen helping it to withstand stress. Proline accounts for about 10% of the total amino acid of collagen, and collagen serves as a good source of this amino acid in the human diet because its content in other animal proteins is very low. Collagen proteins have broad functional properties: strong swelling in electrolyte solutions, low solubility in water, special microstructural properties and dense arrangement of molecules, which leads to high strength. In particular, the properties of native collagen determine its practical application: the ability to maintain the molecular

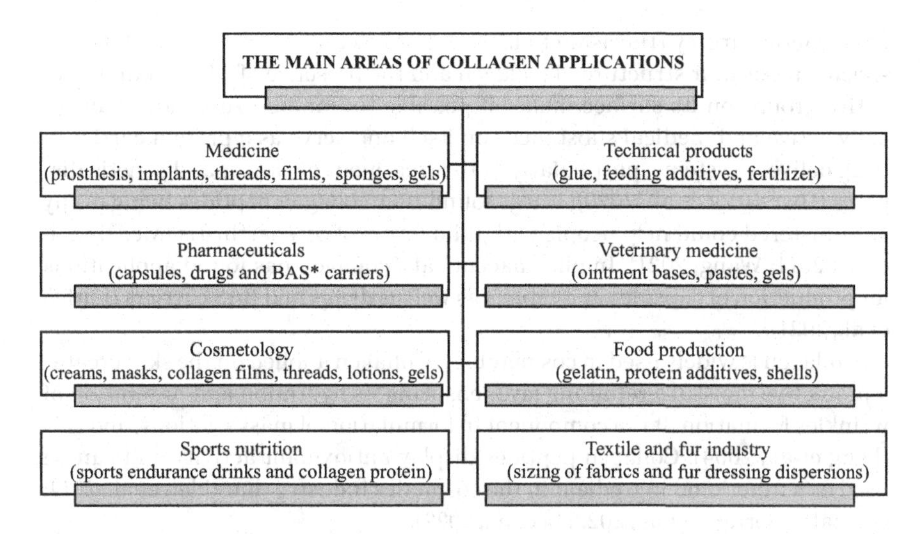

Figure 2.2 The main areas of collagen applications. *BAS—biological active substances.

structure when isolated from tissues; the possibility of transferring into the solution to obtain artificial materials and modifications. Native collagen is poorly soluble in water and has low thermostability (Alexandretti et al., 2019). It cannot be dissolved in a pH buffer at neutral pH and would be denatured at a temperature above 38°C. This limits its use in many fields, for example, the production of injectable biomaterials and cosmetics. However, collagen is a relatively cheap material having broad functional properties, so it finds diverse applications (Figure 2.2).

Collagen is one of the most promising biomaterials widely used in the world for medical, pharmaceutical, veterinarian and cosmetological practices. More than 50,000 tons of collagen and gelatin are used annually for medicinal purposes (Olsen et al., 2003). The advent of soluble collagen products has expanded the scope of its application for medical goals. It is used as a substrate for a protective agent, as well as a transporter of drugs and biologically active substances (BAS) (Lee et al., 2001). Collagen films are used in ophthalmology; collagen itself is used for preparation of sponges for covering wounds and burns; for production of collagen gels, nanoparticles/microspheres for enzyme immobilization, derivatives for transgenic engineering; and tolerance inducers used in the treatment of rheumatoid arthritis (Tenorová et al., 2022). Collagen is widely used in regenerative medicine for tissue regeneration (Kirti & Khora, 2020), in tissue engineering for temporary replacement of skin and bones, as a component of artificial blood vessels and valves, in dental practice and as an implant in cosmetic surgery (Hayashi et al., 2014; Lutfee et al., 2021; Wang, 2021). The specific molecular structure of collagen and the presence of a large number of active groups on its surface makes it possible to immobilize various biologically active and medical substances on itself and serve as a protein carrier in drug delivery systems, particularly in the treatment of cancer and genetic diseases (Owczarzy et al., 2020). It was found that collagen peptides being orally administered could help people with mild forms of osteoarthritis (McAlindon et al., 2011; Wang, 2021). In pharmaceuticals, collagen has found applications for production of capsules and tablets as well as drugs and BAS carriers (Lutfee et al., 2021).

Collagen is widely used in cosmetology. Collagen applied on the skin creates a protective moisture-retaining layer ensuring its hydration and prevention of wrinkles formation. It is a component in formulation of masks, lotions, and gels (Peng et al., 2004). Collagen peptides display antioxidant activity and can be used as a functional ingredient in the cosmetics industry (Barzideh et al., 2014; González-Serrano et al., 2022; Li et al., 2022).

It was predicted that the production and consumption of sports food, which provides enhanced consumption of special nutrients needed for sports activity will be increased in 2021–2030 (Ivanov et al., 2021). Recently, the use of hydrolyzed collagen preparations as an element of sports food, which primarily serves to correct the nutrition of athletes, with high loads on the joints and ligaments has become more widespread (König et al., 2021). Consumption of collagen peptides leads to an increase in muscle strength and to an increase in fat-free mass (Oertzen-Hagemann et al., 2019). Collagen is involved in the construction of protein and allows to restore the cartilaginous surface of the menisci, and intervertebral discs, as well as strengthen the articular-ligamentous apparatus of the spine (Clifford et al., 2019; Jendricke et al., 2020).

Collagen found its application in the manufacturing of technical products. Animal adhesives are made from collagen and are widely used in restorations as adhesives and coatings (Ntasi et al., 2022). In the production of compound feed, collagen is used in the manufacturing of granular extruded and plate feeds to increase their biological value, strength and solubility in water. Especially important is the use of collagen hydrolysis products in the production of starter feed for young valuable fish and aquatic species such as sturgeon, trout, salmon, and sea urchin, which contribute to an increase in their survival rate. Collagen has a high content of methionine (6%) and lysine (19%), so it is recommended to be used as a feed supplement in the poultry industry (Park et al., 2021). Collagen hydrolyzate is rich in nitrogen, phosphorus and potassium and when encapsulated can be used as a biodegradable biofertilizer to improve poor soil (Stefan et al., 2020).

Veterinary medicine uses collagen for manufacturing ointment bases, pastes, and gels. It was proposed to treat osteoarthritis, a common disease in dogs (Schunck et al., 2017).

Collagen hydrolyzate could be used for the development of nutraceuticals—food or food ingredients with certain physiological effects (Lupu et al., 2020). An effective method for biotreatment of collagen-containing raw materials is enzymatic hydrolysis which significantly increases its bioavailability in the human body (Cao et al., 2021). It has been reported that addition of collagen could improve the rheological properties and quality indicators of meat products (Hashim et al., 2015). It is also used for producing food films and coatings, as a structure former in fillings for canned food and minced fish, in the production of artificial caviar, broths, jelly, sauce products, various health drinks and cocktails and as additives in the manufacturing of bakery and confectionery goods (Hashim et al., 2015; Nogueira et al., 2019). Gelatin, which is widely used in food preparation as a gelling and emulsifying agent, is derived by partial hydrolysis of collagen by breaking the amino acid chains of the peptides (Mokrejs et al., 2009).

In textile collagen is used for sizing of fabrics, and in the fur industry collagen dispersions are used for fur dressing dispersions (Rafikov et al., 2020).

In recent years, a lot of research has been carried out to find ways to modify collagen-containing raw materials to be used as emulsifier, stabilizer, source of dietary fiber and antioxidants in the production of a wide range of food products (Meyer, 2019; Vidal et al., 2022). Processing collagenous tissues includes many operations and methods, such as mechanical treatment (fleshing and splitting to purify skins, degumming of small intestine submucosa and mincing and homogenization); physical treatments (extraction, temperature treatment and radiation); chemical treatment to obtain purified connective tissue (acid and alkaline treatment, hypo- and hypertonic treatment using solutions of different salts, treatment with organic solvents, detergents, chelating agents, reductive and oxidative treatment), and biological treatment using enzymes (mainly proteases to break the peptide bonds in collagen, as well as lipases to cleave ester bonds of triglycerides and cholesterolesters, nucleases to remove cellular remnants of RNA and DNA), and transglutaminases that catalyzes the formation of isopeptide bonds between proteins (Matinong et al., 2022; Meyer, 2019).

Collagen has great potential for application as an extender, moisture emulsifier, and texture improver in the food industry, particularly it was shown that its addition to meat raw material improved its technological and rheological properties, as well as nutritional value by increasing protein content in meat products (Cao et al., 2021). Thus, a source of structural collagen protein, tripe, treated with dioxide, was introduced into sausage mince to replace the lard. Obtained semismoked sausage showed higher stability during storage due to antioxidant abilities of additives and prevention of the oxidation of lipids (Shukurlu et al., 2022). Collagen could be used as a binding agent, mainly due to its high water-retention capacity. Addition of pork collagen in boneless cured pork increased water-holding capacity (WHC) and improved the protein functionality characteristics of the product (Schilling et al., 2003). The increase of WHC and improvement of the texture of deli rolls was observed when collagen was included in the recipe (Daigle et al., 2005). Collagen fiber obtained from the inner layers of bovine hide was proposed to be used in the recipe of chicken ham to replace vegetable proteins without changing its acceptability (Prestes et al., 2013). Replacement of 50% fat with collagen gel extracted from chicken feet in the recipe of chicken sausages ensured maintaining higher quality of the product during storage at 4°C for 42 days: highest WHC (81.05%), the lowest thiobarbituric acid reactive substances value (0.38 mg MDA/kg) and the highest antioxidant activity in comparison with control (Aráujo et al., 2021). Addition of fish collagen hydrolyzate in the meatballs in amounts 1%, 2% and 3% (w/w of meat) resulted in enhancement of their resistance to fat oxidation; meanwhile, the sensorial assessment showed no differences between quality of control meatballs and all experimental samples (Palamutoglu, 2019). Addition of collagen peptide to replace 50% fat-produced frankfurter low-fat sausage with increased protein content without reducing the quality of sausages (Samara et al., 2017). Incorporation of collagen superfine powder obtained by heat treatment from pig skin used as a filler and textural modifier for preparation of Harbin red sausage led to reduction of cooking loss, improving rheological properties, particularly, an increase in elasticity, compatibility and homogeneity, with no change of the sausages sensory quality (Wang et al., 2018). Thus, due to the low production cost of collagen and its ability to improve physicochemical and sensory characteristics of meat products, its use for partial replacement of raw meat has great potential. The use of animal proteins from collagen-containing raw materials makes it possible to enrich meat products with dietary fibers, significantly improving the rheological parameters of food products and, above all, the consistency. Collagen has a number of positive biological and functional properties (high water-retaining and texture-forming capacity), which allows it to be used in various food systems. To improve the functional characteristics of the collagen-containing materials, the application of enzymatic hydrolysis could be an effective method for its treatment.

2.3 APPLICATION OF ENZYMES TO HYDROLYZE MEAT COLLAGEN-CONTAINING MATERIALS

The growing interest in the use of enzyme preparations in the manufacturing of meat products with a high content of connective tissue is associated with the structure and properties of the collagen present in it. The low functional and technological properties and the value of collagen-containing raw materials justify the expediency of using enzyme preparations to increase the nutritional and technological value of raw materials with high protein content. Application of enzymes in the meat industry allows to intensify the technological process, use nontraditional, lower-grade raw materials to maximize yields of marketable products, and has a positive effect on functional–technological, rheological and sensory characteristics of final products (Marques et al., 2010). The main areas of enzymes application in the meat industry include (a) meat tenderization; (b) cross-linking meats; (c) flavor development, and (d) nutrition improvement (Dong et al., 2017). Proteolytic enzyme transglutaminase is used to improve the technological, sensory, and nutritional potential of meat products (Kaić et al., 2021; Marques et al., 2010). Transglutaminases cross-link glutamine of one protein molecule and lysine of another protein molecule by the formation of amide bond and finally resulting in protein polymerization. This ability of transglutaminases to catalyze cross-linking of protein molecules, especially collagen, is used in the food industry to improve meat product quality (Ivanov et al., 2021).

Proteases, which are used to treat collagen-containing materials, are related to class hydrolases, which split chemical bonds using water. Proteolytic enzymes are capable of hydrolyzing peptide bonds (-CO=N-) in protein molecules breaking them down into polypeptides with lower molecular mass and amino acids. Proteases may be from animal sources (trypsin, pepsin, chymotrypsin, pancreatic elastase, and renin), plants (bromelain from pineapples, papain from papaya, ficin from figs) or from microbial origin (collagenase, proteinase) origin. There are some commercially available, single or mixed, proteases such as Alcalase® (Novozymes, Bagsværd, Denmark), Nutrase® (Nutrex, Hoogbuul, Belgium), Flavourzyme® (Novozymes, Bagsværd, Denmark) and Protamex® (Novozymes, Bagsværd, Denmark). Pepsin is most widely used for collagen treatment, which is followed by collagen extraction.

Enzymatic treatment of collagen-containing materials has such advantages as (a) direction action, i.e., does not destroy, but splits proteins with the formation of low molecular weight peptides and amino acids; (b) it occurs at moderate conditions preventing loss of amino acids; (c) higher yield and purity of hydrolyzate; (d) concentration of enzymes is low, so there is no need to remove them from the product; (e) not long processing time; (f) low salt content in the final product; (g) consumes less energy; (h) produces less waste; and (j) allows more accurately control of the process for an adequate final product accumulation (Matinong et al., 2022; Schmidt et al., 2016). Even though the cost of the enzyme is quite high, the advantages, of using enzymatic processing of collagen-containing raw materials, are obvious. Some examples of enzyme application to treat collagen-containing meat tissues are shown in Table 2.2.

Enzymatic hydrolysis may be used in combination with some traditional methods. Thus, the combination of ultrasonic and pepsin treatment of cattle

TABLE 2.2 APPLICATION OF ENZYMES TO TREAT COLLAGEN-CONTAINING MATERIALS

Collagen-Containing Material	Enzymatic Treatment	Aim	References
Cattle tendon	Ultrasonic+Pepsin treatment	Purification for biomedical applications	Ran and Wang (2014)
Chicken wooden breast	0.2% papain and 20 min on ultrasound	Improving the tenderness of meat	Lima et al. (2022)
Beef muscles and collagen	Papain and bromelin	Improvement of the beef meat tenderness	Ionescu et al. (2008)
Collagen-containing tissues	Actinidin	Hydrolyze collage	Mostafaie et al. (2008)
Pork and rabbit muscle	Actinidin	Meat tenderization	Zhang et al. (2017)
Beef connective tissue	Papain, bromelain, actinidin and zingibain	Hydrolyze proteins of beef connective tissue	Ha et al. (2012)
Camel meat burger patties	Ginger extract, 7%, and papain, 0.01%	Increased collagen solubility and burger sensory properties	Abdel-Naeem and Mohamed (2016)
Spent hen chicken	Kiwifruit protease, 10%	Meat tenderization due to hydrolysis of muscle fibers	Sharma and Vaidya (2018)
Bovine Achilles tendon	Protease, producer *B. subtilis*	Meat tenderizers to hydrolyze collagen or connective tissue protein	Sorapukdee et al. (2020)
Collagen	Cold-adapted collagenase, producer *Pseudoalteromonas* sp. SM9913	Reduction of meat shear force and keeping the fresh color and moisture of meat during storage	Zhao et al. (2012)

tendon to collagen extraction included tendon pre-swollen for 12 h in 0.5 mol/L acetic acid at 4°C; then, treatment with pepsin 50 U/mg of sample for 18 h, and thereafter ultrasonic–pepsin treatment for 30 h was proposed (Ran & Wang, 2014). Wooden breast is a myopathy caused by deteriorating nutritional and technological properties due to an increase in the amount of connective tissue and the presence of collagen fibers. Treatment of chicken wooden breast with 0.2% papain and 20 min on ultrasound improved the tenderness of meat (Lima et al., 2022).

Proteases have found their applications in many industrial processes in various fields, including pharmaceuticals, medicine, production of detergents, and food technology (Rastogi & Bhatia, 2019; Scanlon et al., 2018). In the meat industry, plant proteases found application for meat tenderization due to their ability to hydrolyze muscle proteins, collagen and elastin. As meat tenderizers, proteolytic enzymes are best suited to degrade collagen in connective tissues at relatively low pH and low temperature (Singh et al., 2018; Sun et al., 2016; Tantamacharik et al., 2018). The most important proteases of plant origin, which are commonly used in the meat industry, are papain, bromelain, and ficin. Among them, papain, which has high thermal and pressure stability, can be regarded as the most popular enzyme used in the processing of all types of meat raw materials (Marques et al., 2010). Papain is inactivated only under extreme conditions treatment such as 900 mPa, 80°C for 22 min (Marques et al., 2010). However, the use of papain has certain disadvantages such as insufficient penetration into the thickness of the muscle tissue, which leads to uneven softening of the finished product, as well as the relatively high cost of enzymatic preparations based on papain and the need to introduce a high concentration of the enzyme to achieve the desired consistency (Scanlon et al., 2018). It was shown that the simultaneous use of papain (EC 3.4.4.10) and bromelain (EC 3.4.4.24) to treat beef muscles and collagen proteins led to their limited hydrolysis and an improvement of the beef meat tenderness (Ionescu et al., 2008). At the same time, it was noted that the collagenase activity of bromelain was approximately twice higher than papain activity. Enzyme ficin (EC 3.4.22.3) is an endoproteolytic enzyme from trees of the genus *Ficus* that can be also used as a meat tenderizer and for collagen hydrolysis (Troncoso et al., 2022). Among the enzymes of plant origin used in meat production, one can note actinidin, abundant in Kiwifruit and commercially applied as a meat tenderizer (Husain, 2018). The efficiency of actinidin application for collagen hydrolysis by Mostafaie et al. (2008) was shown. Plant proteases can be combined to achieve higher yields as was done in the study, where a combination of four plant enzymes—papain, bromelain, actinidin, and zingibain—was used to hydrolyze proteins of beef connective tissue at 55°C and pH 6.0 (Ha et al., 2012). Application of a combined treatment of camel burger patties with ginger extract, 7%, and papain, 0.01%, increased collagen solubility and burger sensory properties such as juiciness, tenderness and overall acceptability; at the same time, the shear force values tended to reduce significantly and lipid stability to improve during storage of burgers. Interestingly, the destructive effect on the connective tissue occurred mainly due to the action of papain extract (Abdel-Naeem & Mohamed, 2016). It was demonstrated that ginger proteases have collagenase activity and can be applied as a meat tenderizer (Choi & Laursen, 2000; Hashimoto et al., 1991; Kim et al., 2007). Thus, proteases of plant origin have a great application potential for the bioconversion of collagen-containing meat materials into products with high-added value through low-cost processes. Nearly 95% of the enzymes used in the meat industry in most European countries and the United States are plant proteases such as papain and bromelain, while the use of proteases of microbiological origin remains a less explored category and enzymes of this type are not widely used at present time (Bhat et al., 2018; Lafarga & Hayes, 2017b).

However, proteases of microbial origin produced by nonpathogenic bacterial and fungal strains have been found to be useful in meat tenderization (Arshad et al., 2016; Ryder et al., 2015). The main producers of bacterial collagenolytic proteases are related to *Bacillus* genus and are representatives of different species such as *Bacillus subtilis, B. amyloliquefaciens, B. pumilus, B. licheniformis* and *B. polyfermenticus*. Proteases produced by *B. subtilis* subsp. *subtilis* B13 (isolated from beef butchery) and *B. siamemsis* S6 (isolated from soil) showed a high collagenolytic activity to hydrolyze collagen from bovine Achilles tendon with multiple cleavage sites and could be used as meat tenderizers when collagen or connective tissue protein caused the meat toughness (Sorapukdee et al., 2020). A common producer of commercially available proteases, which could hydrolyze collagen and elastin and, to a limited extent, myofibrillar proteins, is the fungus *Aspergillus oryzae* (Ryder et al., 2015). In addition to the genus *Aspergillus*, collagenase producers were found among the fungi from genera *Cladosporium, Penicillium, Alternaria* and others (Lima et al., 2011; Wanderley et al., 2017).

Special interest in the meat industry are microbial producers of collagenases, which are proteases capable of cleaving native collagen under physiological conditions in *vivo* and in *vitro* (Holmbeck & Birkedal-Hansen, 2013; Pooja et al., 2022). Collagenases have numerous industrial applications, and in the past, they were isolated from animal tissues (Pal & Suresh, 2016). However, now microbial collagenases could be used instead of those from animal sources. Microbial collagenases (EC 3.4.24.3) digest native collagen in the triple-helical region at X–Gly bonds. The first discovered and studied microbial collagenase was produced by *Clostridium histolyticum* (Van Wart, 2013). This bacterial culture secretes a mixture of collagenases and other proteases, which have hydrolytic activity toward connective tissues. Microbial collagenases, purified from bacterial strain *Pseudoalteromonas* sp. SM9913, which are capable of degrading collagen, have been proposed to be an alternative to plant proteases for meat tenderness (Zhao et al., 2012). Treatment with these cold-adapted collagenolytic protease beef meat reduced beef meat shear force and preserved the fresh color and moisture during storage. Thus, the treatment of collagen-containing meat raw materials with microbial collagenases can be an advisable option.

2.4 ENZYMATIC TREATMENT OF COLLAGEN-CONTAINING MEAT RAW MATERIALS TO OBTAIN PROTEIN ENRICHER FOR MEAT PRODUCTS

To use connective tissue components in the preparation of chopped semifinished meat products, the possibility of enzyme hydrolysis of low-value collagen-containing meat raw materials was studied. As collagen-containing raw materials, meat trimmings and lungs of beef were used. The determination of the protein fractions in studied beef materials showed that the alkali-soluble fraction consisted of a significant portion of the total protein content and was 36.3% in the lungs and 86.3% in meat trimmings (Figure 2.3).

Amino acid content in collagen-containing beef raw materials was determined, and amino acid score (AAS) was calculated for each essential amino acid. Amino acid score, a widely used method for evaluation of protein quality, is the ratio of the amount of amino acid in the test protein to the amount of the same amino acid in a hypothetical protein with an ideal amino acid scale; and it was calculated for essential amino acids by the formula:

AAS = mg AC in 1 g of test protein/mg AC in 1 g of ideal protein and shown in Table 2.3.

When calculating the amino acid score, the FAO reference protein scale (FAO, 2013) was used. It is accepted that the first limiting amino acid in a test protein determines its biological value (Boye et al., 2012; Schaafsma, 2012). It was evident that lungs had a more balanced amino acid composition than meat trimmings. Methionine+Cysteine were the first limiting amino acid for both raw materials but AAS for lungs was equal to 88%, meanwhile, AAS for meat trimming was only 28% (Table 2.2). Meat trimmings with a large inclusion of connective tissue were characterized with unbalanced composition of amino acids: they had low content of essential amino acids, especially leucine, isoleucine, lysine, and threonine and high content of glycine, arginine and proline, which agrees with the reports of other authors studied the nutritive value of collagen (Miller, 1996; Mokrejs et al., 2009).

Enzymatic treatment of low-grade raw materials was carried out with a collagenase-containing preparation STABICOL SKIN, "food composition for meat products" (Priority International LLC, USA) to destroy the connective tissue,

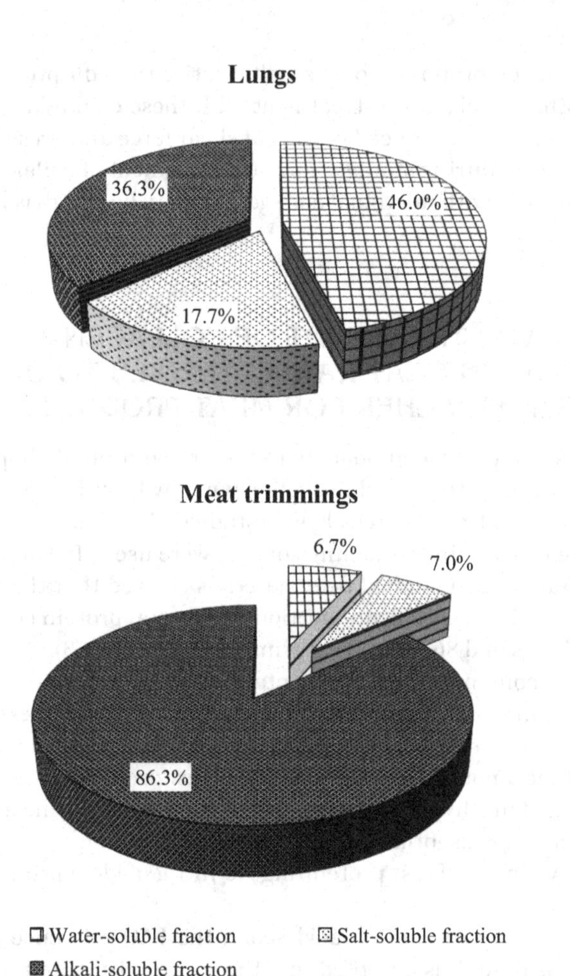

Figure 2.3 Distribution of protein fractions in studied beef materials.

change its fractional composition, and increase its technological functionality. Priority International LLC is an enterprise specialized in the production and distribution of food ingredients including enzymes for the meat industry. To obtain protein enrichers (PE), the washed collagen-containing material—lungs and meat trimmings—was crushed on a meat grinder equipped with a set of grinding plates with 6–8-mm-diameter holes, mixed in a ratio of 1:1, and this mixture was treated for 6 h with enzyme solutions (1:1.5 w/v) with concentration 0.010%; 0.025% and 0.050% (PE 1, 2 and 3, respectively); and for 12 h with enzyme solutions with a concentration of 0.010%; 0.025% and 0.050%

TABLE 2.3 THE CONTENT OF AMINO ACIDS IN THE COLLAGEN-CONTAINING BEEF RAW MATERIALS

Amino Acid (AA)	Lungs		Meat Trimmings	
	g/100 g protein	AAS[a], %	g/100 g protein	AAS, %
Glycine	9.61±0.81		24.73±2.09	
Alanine	6.53±0.45		9.09±0.71	
Valine	5.60±0.43	140	4.32±0.25	108
Leucine	8.57±0.67	141	4.58±0.34	75
Isoleucine	3.68±0.23	123	2.24±0.09	75
Serine	5.31±0.34		3.67 ±0.26	
Threonine	4.81±0.39	192	2.51±0.12	101
Methionine+Cysteine	2.03±0.10	88	0.63±0.04	28
Phenylalanine+Tyrosine	5.71±0.45	139	2.96±0.03	72
Lysine	6.32±0.56	132	2.15±0.09	45
Arginine	6.48±0.45		7.37±0.07	
Histidine	3.69±0.28		0.71±0.06	
Aspartic acid	5.41±0.41		6.04±0.45	
Glutamic acid	12.29±1.09		12.80±1.13	
Proline	7.62±0.67		12.94±1.09	

[a] AAS, amino acid score.

(PE 4, 5 and 6, respectively) at room temperature. Biotreated collagen-containing material was separated from the excessive moisture and finely ground passing through the holes of the grid with a diameter of 2–3 mm and protein enricher was obtained. Changes in protein and moisture content in protein enrichers obtained after enzymatic treatment of collagen-containing material are shown in Table 2.4.

Protein losses were minimal regardless of processing parameters. The water-binding capacity (WBC) of protein enrichers is shown in Figure 2.4.

Water-binding capacity of protein enrichers increased significantly, and this new property is essential in case of their use to improve texture characteristics and increase the ability to bind water in low-fat meat products (Argel et al., 2020). Water-binding capacity was higher in samples of protein enrichers obtained by enzymatical treatment of collagen-containing material for 6 h compared with those treated for 12 h. Analysis of amino acid composition of the

TABLE 2.4 CHANGE OF THE CONTENT OF PROTEIN AND MOISTURE IN PROTEIN ENRICHERS OBTAINED FROM ENZYMATICALLY TREATED COLLAGEN-CONTAINING MATERIAL BY DIFFERENT MODES

Indicator, %	Control	Protein Enricher (PE)					
	0[a]	1	2	3	4	5	6
Protein	16.8±0.5	13.9±0.4	13.9±0.4	14.0±0.4	13.6±0.5	13.8±0.5	13.9±0. 5
Moisture	69.3±2.0	69.6±2.1	71.2±2.1	72.9±2.1	73.3±2.2	74.1±2.2	75.0±2.2
Protein loss	0	2.9±0.1	2.9±0.1	2.8±0.1	3.1±0.1	3.0±0.1	2.9±0.1

[a] 0, control; protein enrichers treated for 6 h with enzymatic solution with concentration 0.010% (1); 0.025% (2); 0.05% (3); and for 12 h with enzymatic solution with concentration 0.010% (4); 0.025% (5); 0.05% (6).

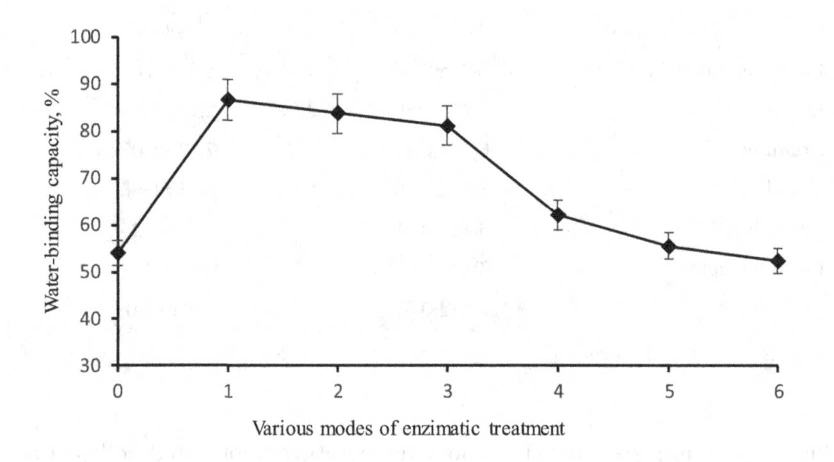

Figure 2.4 Water-binding capacity (WBC) of protein enrichers (PE) obtained from collagen-containing material treated by different modes: 0 (control); 1, 2 and 3—treatment with enzymatic solutions with concentration 0.01%, 0.025% and 0.05%, respectively, for 6 h; 4, 5 and 6—treatment with enzymatic solutions with concentration 0.01%, 0.025% and 0.05%, respectively, for 12 h.

PE obtained after treatment of collagen-containing material with enzymatic solution with a concentration of 0.025% for 6 h did not show any significant differences in comparison with control, and with protein enricher obtained after treatment of collagen-containing material with enzymatic solution with the same concentration of 0.025% but for 12 h (Table 2.5).

TABLE 2.5 AMINO ACID CONTENT, G/100 G OF PROTEIN ENRICHER OBTAINED AFTER TREATMENT WITH ENZYMATIC SOLUTION WITH CONCENTRATION 0.025% FOR 6 H (PE2) AND 12 H (PE5)

Amino Acids	Control		Treatment for 6 h		Treatment for 12 h	
	g/100 g	% of total	g/100 g PE2	% of total	g/100 g PE5	% of total
Lysine	0.87	8.30	0.80	8.25	0.74	7.77
Histidine	0.31	2.95	0.24	2.48	0.22	2.25
Arginine	0.78	7.44	0.73	7.58	0.81	8.45
Aspartic acid	0.75	7.14	0.74	7.64	0.74	7.70
Threonine	0.48	4.60	0.46	4.73	0.42	4.42
Serine	0.48	4.57	0.43	4.41	0.42	4.41
Glutamic acid	1.59	15.20	1.64	16.99	1.59	16.57
Proline	0.78	7.44	0.67	6.96	0.77	8.06
Glycine	0.84	7.98	0.80 •	8.26	0.86	8.93
Alanine	0.91	8.64	0.81	8.41	0.89	9.33
Cystine	0.08	0.75	0.08	0.80	0.09	0.97
Valine	0.37	3.55	0.33	3.46	0.30	3.11
Methionine	0.13	1.22	0.12	1.24	0.07	0.72
Isoleucine	0.32	3.06	0.34	3.53	0.30	3.08
Leucine	0.97	9.24	0.80	8.33	0.74	7.68
Tyrosine	0.34	3.27	0.27	2.81	0.25	2.59
Phenylalanine	0.49	4.66	0.40	4.14	0.38	3.97

Based on the presented results and taking into account protein losses and changes in the functional and technological properties of protein enrichers, the processing mode was determined as 6 h with an enzyme preparation concentration of 0.025%. The final product was used as an additive in the preparation of chopped semifinished meat products. PE was also introduced into minced meat in the form of a protein/fat emulsion (PFE), which was produced by adding ice water to the protein enricher in a ratio of 1:1, and sesame oil, 25% by weight of the protein enricher. This intermediate product was mixed thoroughly for 4 min and stored at 0–4°C before being used.

2.5 EFFECT OF THE PROTEIN ENRICHER ADDITION ON QUALITY OF CHOPPED SEMIFINISHED MEAT PRODUCTS

Protein additives–protein enricher—collagen-containing materials treated with enzyme solution 0.025% and protein enricher in the form of fat emulsion—were added to replace minced meat (semifat pork and chicken) in the chopped semifinished meat products. Replacement of minced meat with proposed additives was not higher than 20% according to recommendations that collagen cannot be substituted for more than 30% of the required daily supply of protein because of its nutritive imbalance (Miller, 1996). Recipes of the chopped semifinished meat products with partial replacement of minced meat with protein enricher and protein/fat emulsion are shown in Table 2.6.

To identify changes in nutritional value, physicochemical and functional-technological indicators, the properties of chopped semifinished meat products with the replacement of 10%–20% minced meat with protein additives were studied. Qualitative indicators of the chopped semifinished meat products are shown in Table 2.7.

Thus, the introduction of a protein enricher into the recipes of chopped semifinished meat products makes it possible to increase the water-binding capacity and moisture content of the produced meat products, while the water content in samples with PE from a collagen-containing material treated with enzymes for 6 h was higher than in similar samples but treated for 12 h.

The ability to swell the protein enricher is due to the significant area of macromolecules, as well as the presence of polar groups. An important functional property of PE is the ability to emulsify fat creating an oil/water dispersion system. When creating food enrichers, it is important to evaluate their functional properties, such as water-binding capacity (WBC) and water-holding capacity (WHC). The latter two parameters are important indicators, which determine the technological properties of a product. WHC and WBC are the weight of water retained by 1 g dry material, but WHC is the water content measured after an external force (e.g., thermal treatment during cooking) has been applied, whereas WBC is measured without stress (Tiwari & Cummins, 2021). The ability of the chopped semifinished meat products to retain its water was determined based on minced semifat pork and chicken with replacement from 10% to 20% with protein additives (protein enricher and protein/fat emulsion) (samples 1, 2 and 3 from Table 2.6) (Figure 2.5a and b).

WHC is one of the most important indicators of raw minced meat affecting product texture and its behavior during storage. WHC characterizes the moisture content in chopped semifinished meat products and the amount of moisture removed during the heat treatment. This indicator is closely related to the output of finished products. A graphical interpretation of the pattern of

TABLE 2.6 RECIPES OF THE CHOPPED SEMIFINISHED MEAT PRODUCTS WITH PROTEIN ENRICHER

Raw Materials	kg/100 kg						
	Control	S1[b]	S2	S3	S4	S5	S6
Pork	50	45	40	40	45	40	40
Chicken	15	25	25	20	25	25	20
Protein/fat emulsion	-	10	15	-	10	15	-
Protein enricher[a]	-	-	-	20	-	-	20
Raw offal	15	-	-	-	-	-	-
Eggs	3	3	3	3	3	3	3
Onion	5	5	5	5	5	5	5
Bread	7	7	7	7	7	7	7
Breadcrumbs	3	3	3	3	3	3	3
Food salt	1.7	1.7	1.7	1.7	1.7	1.7	1.7
Spices	0.3	0.3	0.3	0.3	0.3	0.3	0.3

[a] Protein enricher obtained from collagen-containing material treated for 6 h with enzymatic solution with concentration 0.025%.

[b] S, sample.

TABLE 2.7 QUALITATIVE INDICATORS OF THE CHOPPED SEMIFINISHED MEAT PRODUCTS

Indicators	Control	S1[a]	S2	S3	S4	S5	S6
Moisture, %	75.6	82.4	84.6	80.2	78.4	80.2	76.3
Protein, %	16.9	16.7	16.5	16.1	16.4	16.3	15.9
pH	6.45	6.42	6.4	6.4	6.35	6.33	6.28
WBC[a], %	61.0	68.1	71.5	69.7	62.5	61.3	58.4

[a] WBC, water-binding capacity; the chopped semifinished meat products prepared according to recipes from Table 2.6.

S, sample.

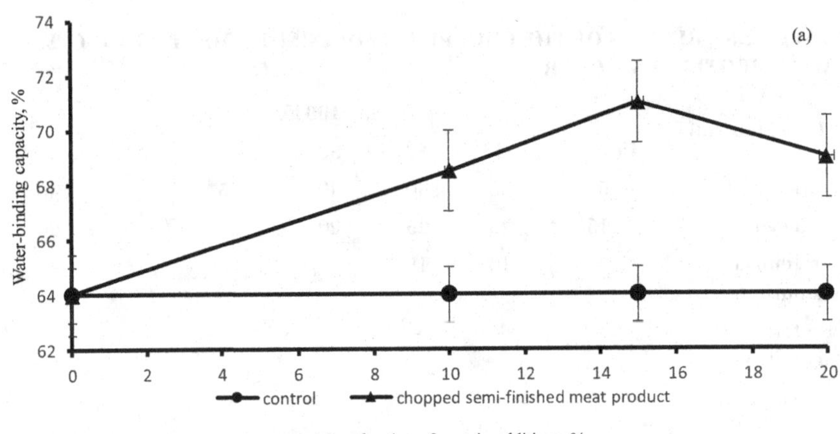

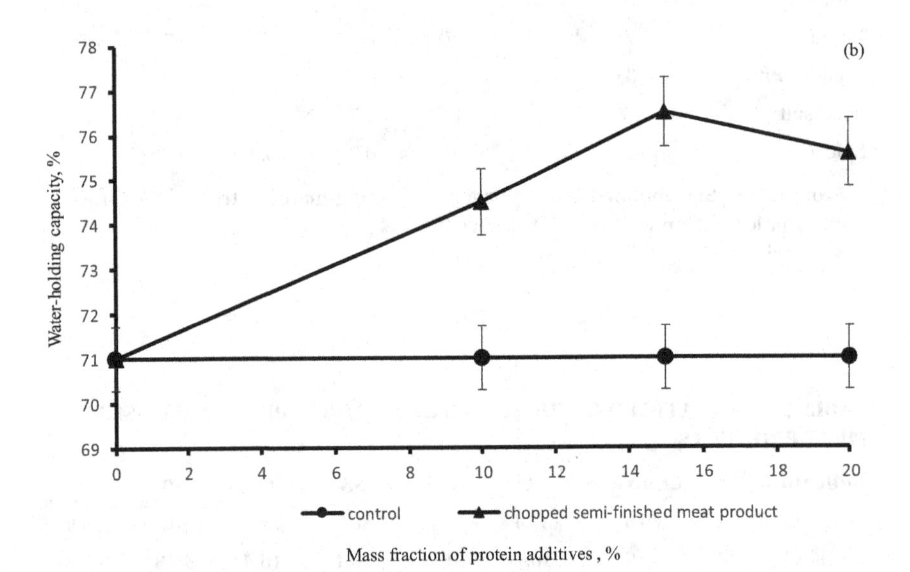

Figure 2.5 Change of water-binding capacity (WBC) (a) and water-holding capacity (WHC) (b) depending on the mass fraction of protein additives in the chopped semifinished meat products.

changes in WBC (Figure 2.5a) and WHC (Figure 2.5b) shows that the replacement of 15% of the minced meat with protein additives is the best solution to increase water-binding and water-holding abilities of the chopped semifinished meat products. The increase in the WHC of the chopped semifinished

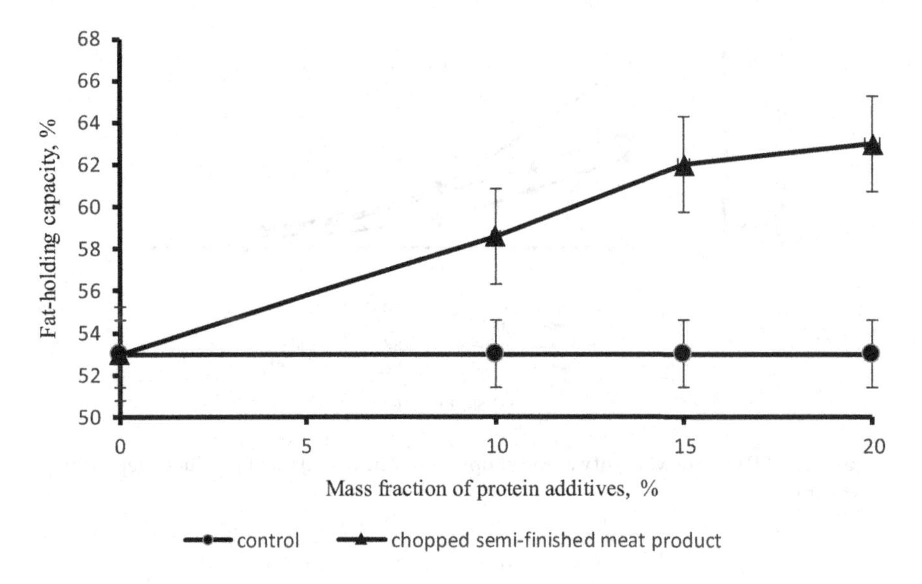

Figure 2.6 Change of fat-binding capacity (FBC) depending on the mass fraction of protein additives in the chopped semifinished meat products.

meat products with the addition of protein additives can be explained by the fact that gelatinization and swelling of these components occur during the heat treatment, and a colloidal system is formed.

Changes in fat-holding capacity (FHC) of chopped semifinished meat products are shown in Figure 2.6.

With an increase in the mass fraction of the protein additives in the chopped semifinished meat products, an increase in the FHC is observed. This can be useful for the technology of meat products containing a large mass fraction of fat. The rheological characteristics of the chopped semifinished meat products were studied using a viscometer Brookfield DV-II+Pro (Brookfield Engineering Laboratories, Inc., USA) to evaluate their strength by the magnitude of shear stress due to shear force in comparison with the control (Figure 2.7).

The finished products (cutlets) from the chopped semifinished meat products with 15% replacement of minced meat with protein enricher treated for 6 h with enzyme solution with concentration 0.01% (1), 0.025% (2) and 0.05% (3) were prepared. A sensory evaluation (appearance, consistency, flavor, color and taste) of the final products was carried out using a five-point scale by the profile method (Figure 2.8).

The score of the finished products, averaged over organoleptic indicators (consistency, color and taste), for the control was 3.9, for the products with

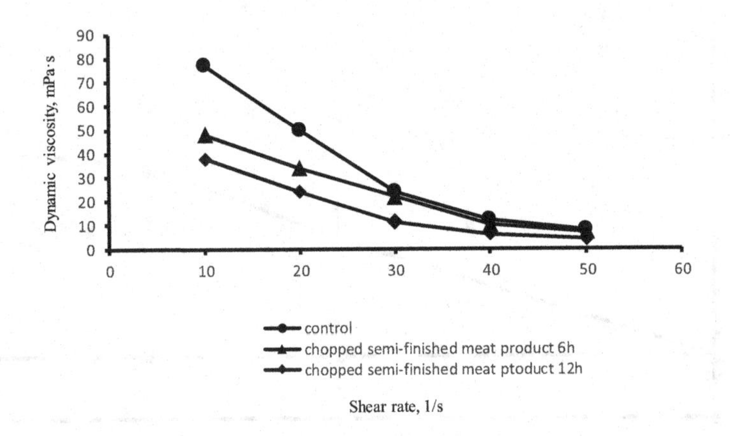

Figure 2.7 Dynamic viscosity of the chopped semifinished meat products depending on shear rate.

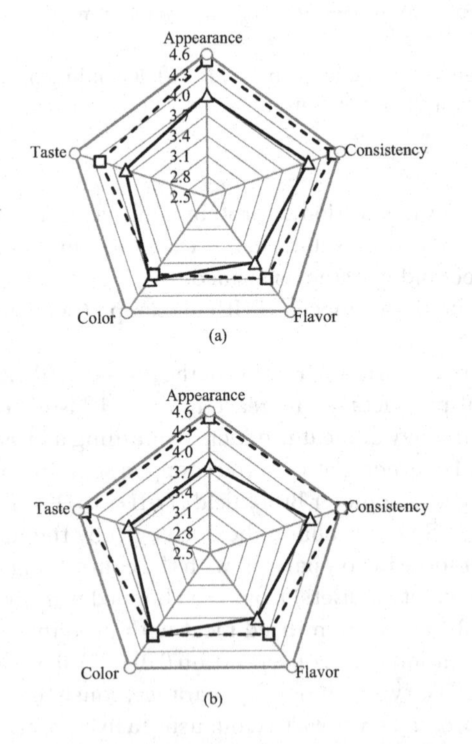

Figure 2.8 Sensorial parameters of the final products prepared from the chopped semifinished meat products with 15% replacement of minced meat with protein enrichers treated for 6h with enzyme solution with concentration 0.01% (a), 0.025% (b) and 0.05% (c): Δ—control; □—finished products.

protein additives (1 and 3) 4.0 and (2) 4.3. Characteristics of finished products are shown in Table 2.8.

Thus, the finished product prepared from the chopped semifinished meat product with 15% replacement of minced meat with protein additives from collagen-containing material treated for 6h with enzyme solution 0.025% (2) was chosen as the best one. The composition of amino acids of control and final product with protein additive was determined by ion-exchange liquid column chromatography using an AAA T-339 automatic amino acid analyzer (Table 2.9; Figure 2.9).

It was found that the final product with the protein additives had better performance than the control: water-binding capacity increased by 6.0%, the product yield increased by 2.7%, and amino acid score of first limiting amino acids methionine+cysteine increased from 53.7% in control to 61.1%.

TABLE 2.8 CHARACTERISTICS OF THE FINAL PRODUCTS

Parameters	Control	Sample 1	Sample 2	Sample 3
Water, %	56.4±0.14	58.5±0,21	59.2±0.16	57.5±0.34
pH	6.5±0.02	6.4±0.02	6.4±0.02	6.4±0.02
Protein, %	12.7±0.3	13.2±0.25	13.9±0.35	13.0±0.2
WBC, % of total water	83.24±0.62	86.19±0,49	89.61±0.32	85.15±0.42
Yield, %	81.1±0.43	82.9±0,51	83.8±0.47	80.1±0.59

TABLE 2.9 COMPOSITION OF AMINO ACIDS IN CONTROL AND FINAL PRODUCT WITH PROTEIN ADDITIVES (PE)

	Control		With PE	
Amino Acids	g/100g protein	AAS[a], %	g/100g protein	AAS, %
Valine	4.47	89.4	4.79	95.8
Isoleucine	3.92	98.0	3.29	82.2
Leucine	6.1q	87.1	5.11	73.0
Lysine	4.62	83.6	5.56	101.1
Methionine+Cysteine	1.88	53.7	2.14	61.1
Threonine	3.54	88.5	2.99	74.7
Tryptophan	3.29	54.8	4.94	82.3
Phenylalanine+Tyrosine	1.11	110.0	0.89	89.0

[a] AAS, amino acid score.

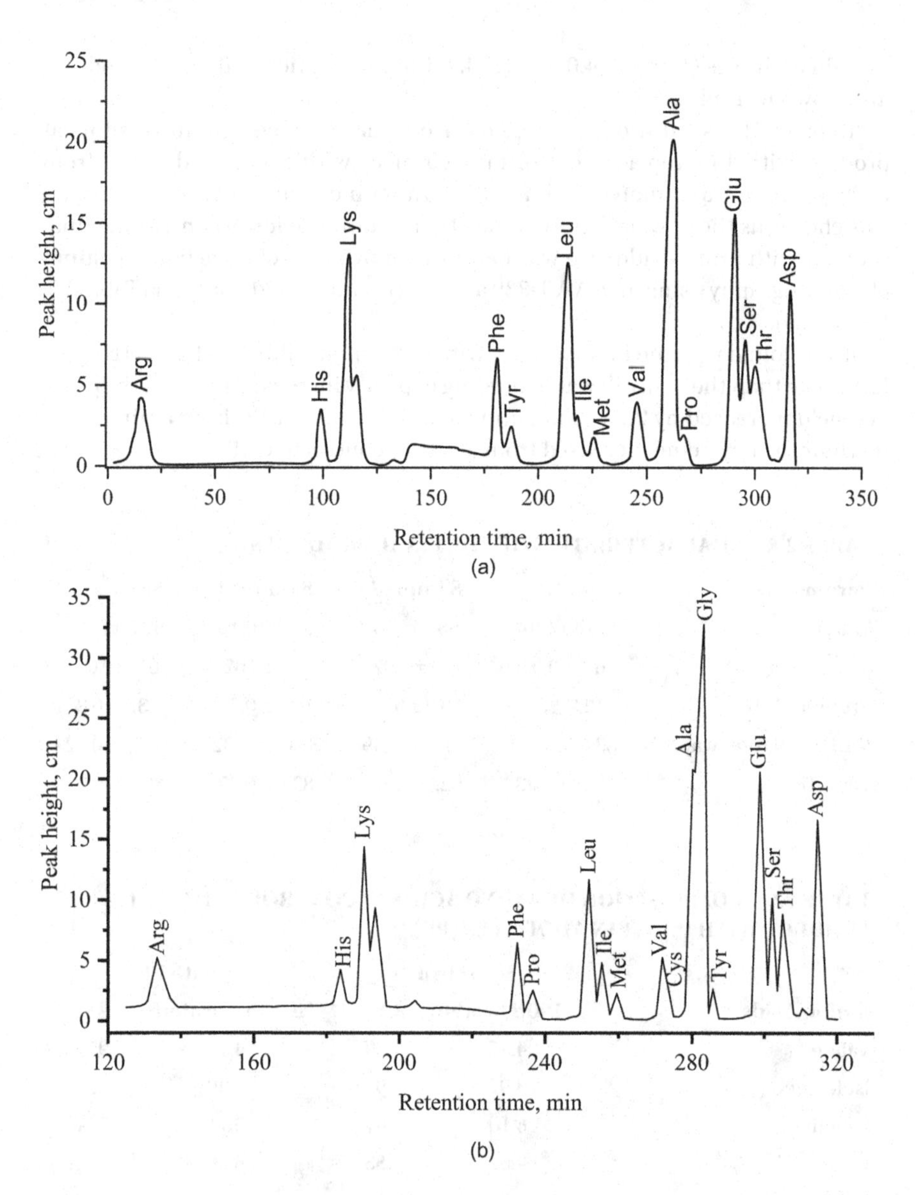

Figure 2.9 Amino acid composition of control (a) and final product with protein additives (b).

2.6 CONCLUSIONS

Collagen-containing meat materials could be bioconverted into valuable by-products. Enzymatic processing of collagen-containing meat materials made it possible to obtain a protein enricher, which can be used for partial (15%) replacement of minced meat in the preparation of chopped semifinished meat products; which, in terms of their physicochemical and sensory parameters were not worse than the control sample. However, a further increase in the percentage level of replacing minced meat with protein enricher is not convenient, since this leads to a deterioration in the consumer qualities of the finished product.

REFERENCES

Abdel-Naeem, H.H.S., & Mohamed, H.M.H. (2016). Improving the physicochemical and sensory characteristics of camel meat burger patties using ginger extract and papain. *Meat Science, 118*, 52–60. https://doi.org/10.1016/j.meatsci.2016.03.021

Adesogan, A.T., Havelaar, A.H., McKune, S.L., Eilitta, M.E., & Dahl, G.E. (2020). Animal source foods: Sustainability problem or malnutrition and sustainability solution? Perspective matters. *Global Food Security, 25,* 100325. https://doi.org/10.1016/j.gfs.2019.100325

Adhikari, B.B., Chae, M., & Bressler, D.C. (2018). Utilization of slaughterhouse waste in value-added applications: Recent advances in the development of wood adhesives. *Polymers, 10*(2), 176. https://doi.org/10.3390/polym10020176

Alexandretti, C., Verlindo, R., Hassemer, G.S., Manzoli, A., Roman, S.S., Fernandes, I.A., Backes, G.T., Cansian, R.L., Alvarado Soares, M.B., Schwert, R., & Valduga, E. (2019). Structural and techno-functional properties of bovine collagen and its application in hamburgers. *Food Technology and Biotechnology, 57*(3), 369–377. https://doi.org/10.17113/ftb.57.03.19.5896

Almeida, P.F., de Calarge, F.A., & Santana, J.C.C. (2013). Production of a product similar to gelatin from chicken feet collagen. *Engenharia Agricola, 33*(6), 1289. https://doi.org/10.1590/s0100-69162013000600021

Aráujo, Í.B.S., Lima, D.A.S., Pereira, S.F., Paseto, R.P., & Madruga, M.S. (2021). Effect of storage time on the quality of chicken sausages produced with fat replacement by collagen gel extracted from chicken feet. *Poultry Science, 100*(2), 1262–1272. https://doi.org/10.1016/j.psj.2020.10.029

Argel, N.S., Ranalli, N., Califano, A.N., & Andrés, S.C. (2020). Influence of partial pork meat replacement by pulse flour on physicochemical and sensory characteristics of low-fat burgers. *Journal of the Science of Food and Agriculture, 100*(10), 3932–3941. https://doi.org/10.1002/jsfa.10436

Arshad, M.S., Kwon, J.H., Imran, M., Sohaib, M., Aslam, A., Nawaz, I., Amjad, Z., Khan, U., & Javed, M. (2016). Plant and bacterial proteases: A key towards improving meat tenderization, a mini review. *Cogent Food & Agriculture, 2*(1), 1–10. https://doi.org/10.1080/23311932.2016.12617

Barzideh, Z., Latiff, A. A., Gan, C.Y., Abedin, M., & Alias, A.K. (2014). ACE inhibitory and antioxidant activities of collagen hydrolysates from the ribbon jellyfish (Chrysaora *sp.*). *Food Technology and Biotechnology, 52*(4), 495–504. https://doi.org/10.17113/b.52.04.14.3641

Bhat, Z.F., Morton, J.D., Mason, S.L., & Bekhit, A.E.D.A. (2018). Applied and emerging methods for meat tenderization: A comparative perspective. *Comprehensive Reviews in Food Science and Food Safety, 17*(4), 841–859. https://doi.org/10.1111/1541-4337.12356

Boye, J., Wijesinha-Bettoni, R., & Burlingame, B. (2012). Protein quality evaluation twenty years after the introduction of the protein digestibility corrected amino acid score method. *British Journal of Nutrition, 108*(Suppl 2), S183–S211. https://doi.org/10.1017/S0007114512002309

Cao, C., Xiao, Z., Ge, C., & Wu, Y. (2021). Animal by-products collagen and derived peptide, as important components of innovative sustainable food systems-a comprehensive review. *Critical Reviews in Food Science and Nutrition, 62*(31), 8703–8727. https://doi.org/10.1080/10408398.2021.1931807

Choi, K.H., & Laursen, R.A. (2000). Amino-acid sequence and glycan structures of cysteine proteases with proline specificity from ginger rhizome Zingiber officinale. *European Journal of Biochemistry, 267*(5), 1516–1526. https://doi.org/10.1046/j.1432-1327.2000.01152.x

Daigle, S.P., Schilling, M.W., Marriott, N.G., Wang, H., Barbeau, W.E., & Williams, R.C. (2005). PSE-like turkey breast enhancement through adjunct incorporation in a chunked and formed deli roll. *Meat Science, 69*(2), 319–324. https://doi.org/10.1016/j.meatsci.2004.08.001

Dong, K., Samarasinghe, Y.A.H., Hua, W., Kocherry, L., & Xu, J. (2017). Recombinant enzymes in the meat industry and the regulations of recombinant enzymes in food processing. In R.C. Ray & C.M. Rosell (Eds.), *Microbial Enzyme Technology in Food Applications*, 363–374. https://doi.org/10.1201/9781315368405

FAO. (2013). Dietary protein quality evaluation in human nutrition. Report of on FAO expert consultation, Rome. https://www.fao.org/ag/humannutrition/35978-02317b979a686a57aa4593304ffc17f06.pdf

Clifford, T., Ventress, M., Allerton, D., Stansfield, S., Tang, J., Fraser, W.D., Vanhoecke, B., Prawitt, J., & Stevenson, E. (2019). The effects of collagen peptides on muscle damage, inflammation and bone turnover following exercise: A randomized, controlled trial. *Amino Acids, 51*, 691–704. https://doi.org/10.1007/s00726-019-02706-5

González-Serrano, D.J., Hadidi, M., Varcheh, M., Jelyani, A.Z., Moreno, A., & Lorenzo, J.M. (2022). Bioactive peptide fractions from collagen hydrolysate of common carp fish byproduct: Antioxidant and functional properties. *Antioxidants (Basel, Switzerland), 11*(3), 509. https://doi.org/10.3390/antiox11030509

Ha, M., Bekhit, A.E.D.A., Carne, A., & Hopkins, D.L. (2012). Characterisation of commercial papain, bromelain, actinidin and zingibain protease preparations and their activities toward meat proteins. *Food Chemistry, 134*(1), 95–105. https://doi.org/10.1016/j.foodchem.2012.02.071

Hashim, P., Mohd Ridzwan, M.S., Bakar, J., & Mat Hashim, D. (2015). Collagen in food and beverage industries. *International Food Research Journal, 22*(1), 1–8. Collagen_in_food_and_beverage_industries.pdf

Hashimoto, A., Takeuti, Y., Kawahara, Y., & Yasumoto, K. (1991). Proteinase and collagenase activities in ginger rhizome. *Japan Society of Nutrition and Food Science, 44*(2), 127–132. https://doi.org/10.4327/jsnfs.44.127

Hayashi, Y., Ikeda, T., Yamada, S., Koyama, Z., & Yanagiguchi, K. (2014). The application of fish collagen to dental and hard tissue regenerative medicine. In S.K. Kim (Ed.), *Seafood Processing By-Products*. Springer, New York. https://doi.org/10.1002/9781118464717.ch8

Henchion, M., Hayes, M., Mullen, A. M., Fenelon, M., & Tiwari, B. (2017). Future protein supply and demand: Strategies and factors influencing a sustainable equilibrium. *Foods, 6*, 53. https://doi.org/10.3390/foods6070053

Hidaka, S., & Liu, S.Y. (2013). Effects of gelatins on calcium phosphate precipitation: A possible application for distinguishing bovine bone gelatin from porcine skin gelatin. *Journal of Food Composition and Analysis, 16*(3), 477–483. https://doi.org/10.1016/S0889-1575(02)00174-6

Holmbeck, K., & Birkedal-Hansen, H. (2013). Collagenases. *Encyclopedia of Biological Chemistry*, 542–544. https://doi.org/10.1016/B978-0-12-378630-2.00008-6

Husain, Q. (2018). Nanocarriers immobilized proteases and their industrial applications: An overview. *Journal of Nanoscience and Nanotechnology, 18*(1), 486–499. https://doi.org/10.1166/jnn.2018.15246

Ionescu, A., Aprodu, I., & Pascaru, G. (2008). Effect of papain and bromelin on muscle and collagen proteins in beef meat. *The Annals of the University Dunarea de Jos of Galati. Fascicle VI—Food Technology, New Series Year II* (31), 9–16. https://www.gup.ugal.ro/ugaljournals/index.php/food/article/view/3681/3268

Ivanov, V., Shevchenko, O., Marynin, A., Stabnikov, V., Gubenia, O., Stabnikova, O., Shevchenko, A., Gavva, O., & Saliuk, A. (2021). Trends and expected benefits of the breaking edge food technologies in 2021–2030. *Ukrainian Food Journal, 10*(1), 7–36. https://doi.org/10.24263/2304-974X-2021-10-1-3

Jayathilakan, K., Sultana, K., Radhakrishna, K., & Bawa, A.S. (2012). Utilization of byproducts and waste materials from meat, poultry and fish processing industries: A review. *Journal of Food Science and Technology, 49*(3), 278–293. https://doi.org/10.1007/s13197-011-0290-7

Jayawardena, S.R., Morton, J.D., Bekhit, A.E.D.A., Bhat, Z.F., & Brennan, C.S. (2022). Effect of drying temperature on nutritional, functional and pasting properties and storage stability of beef lung powder, a prospective protein ingredient for food supplements, *LWT, 161*, 113315. https://doi.org/10.1016/j.lwt.2022.113315

Jendricke, P., Kohl, J., Centner, C., Gollhofer, A., & König, D. (2020). Influence of specific collagen peptides and concurrent training on cardiometabolic parameters and performance indices in women: A randomized controlled trial. *Frontiers in Nutrition, 7*, 580918. https://doi.org/10.3389/fnut.2020.580918

Kaić, A., Janječić, Z., Žgur, S., Šikić, M., & Potočnik, K. (2021). Physicochemical and sensory attributes of intact and restructured chicken breast meat supplemented with transglutaminase. *Animals (Basel), 11*(9), 2641. https://doi.org/10.3390/ani11092641

Kim, M., Hamilton, S.E., Guddat, L.W., & Overall, C.M. (2007). Plant collagenase: Unique collagenolytic activity of cysteine proteases from ginger. *Biochimica et Biophysica Acta, 1770*(12), 1627–1635. https://doi.org/10.1016/j.bbagen.2007.08.003

Kirti, & Khora, S.S. (2020). Current and potential uses of marine collagen for regenerative medicines. In N.M. Nathani, C. Mootapally, I.R. Gadhvi, B. Maitreya, & C.G. Joshi (Eds.), *Marine Niche: Applications in Pharmaceutical Sciences*. Springer, Singapore, 437–458. https://doi.org/10.1007/978-981-15-5017-1_24

König, D., Kohl, J., Jerger, S., & Centner, C. (2021). Potential relevance of bioactive peptides in sports nutrition. *Nutrients, 13*(11), 3997. https://doi.org/10.3390/nu13113997

Lafarga, T., & Hayes, M. (2017a). Effect of pre-treatment on the generation of dipeptidyl peptidase-IV-and prolyl endopeptidase-inhibitory hydrolysates from bovine lung. *Irish Journal of Agricultural and Food Research, 56*(1), 12–24. https://doi.org/10.1515/ijafr-2017-0002

Lafarga, T., & Hayes, M. (2017b). Bioactive protein hydrolysates in the functional food ingredient industry: Overcoming current challenges. *Food Reviews International, 33*(3), 217–246. https://doi.org/10.1080/87559129.2016.1175013

Lee, C.H., Singla, A., & Lee, Y. (2001). Biomedical applications of collagen. *International Journal of Pharmaceutics, 221*(1–2), 1–22. https://doi.org/10.1016/s0378-5173(01)00691-3

Li, C., Fu, Y., Dai, H., Wang, Q., Gao, R., & Zhang, Y. (2022). Recent progress in preventive effect of collagen peptides on photoaging skin and action mechanism. *Food Science and Human Wellness, 11*(2), 218–229. https://doi.org/10.1016/j.fshw.2021.11.003

Lima, C.A., Filho, J.L.L., Neto, B.B., Converti, A., Carneiro da Cunha, M.G., & Porto, A.L.F. (2011). Production and characterization of a collagenolytic serine proteinase by Penicillium aurantiogriseum URM 4622: A factorial study. *Biotechnology and Bioprocess Engineering, 16*(3), 549–560. https://doi.org/10.1007/s12257-010-0247-0

Lima, J.L., Bezerra, T.K.A., Carvalho, L.M., Galvão, M.S., Lucena, L., Rocha, T.C., Estevez, M., &, Madruga, M.S. (2022). Improving the poor texture and technological properties of chicken wooden breast by enzymatic hydrolysis and low-frequency ultrasound. *Journal of Food Science, 87*(6), 2364–2376. https://doi.org/10.1111/1750-3841.16149

Limeneh, D.Y., Tesfaye, T., Ayele, M., Husien, N.M., Ferede, E., Haile, A., Mengie, W., Abuhay, A., Gelebo, G.G., Gibril, M., & Kong, F.A. (2022). Comprehensive review on utilization of slaughterhouse by-product: Current status and prospect. *Sustainability, 14*, 6469. https://doi.org/10.3390/su14116469

Lupu, M.A., Gradisteanu Pircalabioru, G., Chifiriuc, M.C., Albulescu, R., & Tanase, C. (2020). Beneficial effects of food supplements based on hydrolyzed collagen for skin care (review). *Experimental and Therapeutic Medicine, 20*(1), 12–17. https://doi.org/10.3892/etm.2019.8342

Lutfee, T., Alwan, N., Alsaffar, M.A., Ghany, M., Mageed, A., & Abdulrazak, A. (2021). An overview of the prospects of extracting collagens from waste sources and its applications. *Chemical Papers, 75*, 6025–6033. https://doi.org/10.1007/s11696-021-01768-8

Marques, A.Y., Maróstica, M.R., & Pastore, G.M. (2010). Some nutritional, technological and environmental advances in the use of enzymes in meat products. *Enzyme Research, 480923*. https://doi.org/10.4061/2010/480923

Matinong, A.M.E., Chisti, Y., Pickering, K.L., & Haverkamp, R.G. (2022). Collagen extraction from animal skin. *Biology, 11*, 905. https://doi.org/10.3390/biology11060905

McAlindon, T.E., Nuite, M., Krishnan, N., Ruthazer, R., Price, L.L., Burstein, D., Griffith, J., & Flechsenhar, K. (2011). Change in knee osteoarthritis cartilage detected by delayed gadolinium enhanced magnetic resonance imaging following treatment with collagen hydrolysate: A pilot randomized controlled trial. *Osteoarthritis and Cartilage, 19*, 399–405. https://doi.org/10.1016/j.joca.2011.01.001

Meyer, M. (2019). Processing of collagen based biomaterials and the resulting materials properties. *BioMedical Engineering OnLine, 18*(1), 24. https://doi.org/10.1186/s12938-019-0647-0

Miller, A.T. (1996). Current and future uses of limed hide collagen in the food industry. *Journal of American Leather and Chemists Association, 91*, 183–189.

Mokrejs, P., Langmaier, F., Mladek, M., Janacova, D., Kolomaznik, K., & Vasek, V. (2009). Extraction of collagen and gelatine from meat industry by-products for food and non food uses. *Waste Management Resources, 27*(1), 31–37. https://doi.org/10.1177/0734242X07081483

Mostafaie, A., Bidmeshkipour, A., Shirvani, Z., Mansouri, K., & Chalabi, M. (2008). Kiwifruit actinidin: A proper new collagenase for isolation of cells from different tissues. *Applied Biochemistry and Biotechnology, 144*, 123–131. https://doi.org/10.1007/s12010-007-8106-y

Nogueira, A.C., de Oliveira, R.A., & Steel, C.J. (2019). Protein enrichment of wheat flour doughs: Empirical rheology using protein hydrolysates. *Food Science and Technology, 40*(Suppl. 1), 97–105. https://doi.org/10.1590/fst.06219

Nollet, L.M., & Toldrá, F. (Eds.). (2011). *Handbook of Analysis of Edible Animal By-Products*, Boca Raton, London, New York, CRC Press.

Ntasi, G., Sbriglia, S., Pitocchi, R., Vinciguerra, R., Melchiorre, C., Ioio, L.D., Fatigati, G., Crisci, E., Bonaduce, I., Carpentieri, A., Marino, G., & Birolo, L. (2022). Proteomic characterization of collagen-based animal glues for restoration. *Journal of Proteome Research, 21*(9), 2173–2184. https://doi.org/10.1021/acs.jproteome.2c00232

Olsen, D., Yang, C., Bodo, M., Chang, R., Leigh, S., Baez, J., Carmichael, D., Perälä, M., Hämäläinen, E.R., Jarvinen, M., & Polarek, J. (2003). Recombinant collagen and gelatin for drug delivery. *Advanced Drug Delivery Reviews, 55*(12), 1547–1567. https://doi.org/10.1016/j.addr.2003.08.008

Oertzen-Hagemann, V., Kirmse, M., Eggers, B., Pfeiffer, K., Marcus, K., De Marées, M., & Platen, P. (2019). Effects of 12 weeks of hypertrophy resistance exercise training combined with collagen peptide supplementation on the skeletal muscle proteome in recreationally active men. *Nutrients, 11*(5), 1072. https://doi.org/10.3390/nu11051072

Owczarzy, A., Kurasiński, R., Kulig, K., Rogóż, W., Szkudlarek, A., & Maciążek-Jurczyk, M. (2020). Collagen – Structure, properties and application. *Engineering of Biomaterials, 156*, 17–23. https://doi.org/10.34821/eng.biomat.156.2020.17-23

Pal, G.K. & Suresh, P.V. (2016) Microbial collagenases: challenges and prospects in production and potential applications in food and nutrition. *RSC Advances*, 6(40), 33763–33780. https://doi.org/10.1039/c5ra23316

Palamutoglu, R.K.C. (2019). Antioxidant effect of fish collagen hydrolysate addition to meatballs. *Mugla Journal of Science and Technology, 2*, 56–61. https://doi.org/10.22531/muglajsci.576757

Park, S., Kim, Y.A., Lee, S., Park, Y., Kim, N., & Choi, J. (2021). Effects of pig skin collagen supplementation on broiler breast meat. *Food Science of Animal Resources, 41*(4), 674–686. https://doi.org/10.5851/kosfa.2021.e28

Peng, Y., Glattauer, V., Werkmeister, J.A., & Ramshaw, J.A. (2004). Evaluation for collagen products for cosmetic application. *International Journal of Cosmetic Science, 26*(6), 313. https://doi.org/10.1111/j.1467-2494.2004.00245_2.x

Prestes, R.C., Graboski, A., Roman, S.S., Kempka, A.P., Toniazzo, G., Demiate, I.M., & Di Luccio, M. (2013). Effects of the addition of collagen and degree of comminution in the quality of chicken ham. *Journal of Applied Poultry Research, 22*(4), 885–903. https://doi.org/10.3382/japr.2013-00809

Rachel, N.M. & Pelletier, J.N. (2013). Biotechnological applications of transglutaminases. *Biomolecules*, 3(4), 870–888. https://doi.org/10.3390/biom3040870

Rafikov, A.S., Khakimova, M.S., Fayzullayeva, D.A., & Reyimov, A.F. (2020). Microstructure, morphology and strength of cotton yarns sized by collagen solution. *Cellulose, 27*, 10369–10384. https://doi.org/10.1007/s10570-020-03450-w

Ran, X.G., & Wang, L.Y. (2014). Use of ultrasonic and pepsin treatment in tandem for collagen extraction from meat industry by-products. *Journal of the Science of Food and Agriculture, 94*, 585–590. https://doi.org/10.1002/jsfa.6299

Rastogi, H., & Bhatia, S. (2019). Future prospectives for enzyme technologies in the food industry. In P. Osborn (Ed.), *Enzymes in Food Biotechnology*. London, Academic Press, 845–860.

Pooja, K.M., Rani, S., Pal, P., & Pal, G.K. (2022). Application of microbial enzymes for the tenderization of meat. *Research and Technological Advances in Food Science*, 91–107. https://doi.org/10.1016/B978-0-12-824369-5.00001-4

Russ, W., & Pittroff, R.M. (2004). Utilizing waste products from the food production and processing industries. *Critical Reviews in Food Science and Nutrition, 44*(2), 57–62. https://doi.org/10.1080/10408690490263783

Ryder, K., Ha, M., Bekhit, A.E.D., & Carne, A. (2015). Characterisation of novel fungal and bacterial protease preparations and evaluation of their ability to hydrolyse meat myofibrillar and connective tissue proteins. *Food Chemistry, 172*, 197–206. https://doi.org/10.1016/j.foodchem.2014.09.061

Samara, C.S., Sinara, P.F., Cristiane, R.A.P., Narciza, M.O.A., Fábio, A.P.S., Valquíria, C.S.F., Barreto, M.D.S., & Íris, B.S.A. (2017). Quality parameters of frankfurter-type sausages with partial replacement of fat by hydrolyzed collagen. *LWT - Food Science and Technology, 76*, 320–325. https://doi.org/10.1016/j.lwt.2016.06.034

Scanlon, M.G., Henrich, A.W., & Whitaker, J.R. (2018). Factors affecting enzyme activity in food processing. In R.Y. Yada (Ed.), *Proteins in Food Processing*. Amsterdam, Woodhead Publishing, 337–365. https://doi.org/10.1016/B978-0-08-100722-8.00014-0

Schaafsma, G. (2012). Advantages and limitations of the protein digestibility-corrected amino acid score (PDCAAS) as a method for evaluating protein quality in human diets. *British Journal Nutrition, 108*(Suppl 2), S333–S336. https://doi.org/10.1017/S0007114512002541

Schilling, M.W., Mink, L.E., Gochenour, P.S., Marriott, N.G., & Alvarado, C.Z. (2003). Utilization of pork collagen for functionality improvement of boneless cured ham manufactured from pale, soft, and exudative pork. *Meat Science, 65*(1), 547–553. https://doi.org/10.1016/S0309-1740(02)00247-4

Schmidt, M.M., Dornelles, R.C.P., Mello, R.O., Kubota, E.H., Mazutti, M.A., Kempka, A.P., & Demiate, I.M. (2016). Collagen extraction process. *International Food Research Journal, 23*(3), 913–922. http://www.ifrj.upm.edu.my/23%20(03)%202016/(1).pdf

Schunck, M., Louton, H., & Oesser, S. (2017). The effectiveness of specific collagen peptides on osteoarthritis in dogs-impact on metabolic processes in canine chondrocytes. *Open Journal of Animal Sciences, 7*(3), 254–266. https://doi.org/10.4236/ojas.2017.73020

Seong, P.N., Kang, G.H., Park, K.M., Cho, S.H., Kang, S.M., Park, B.Y., Moon, S.S, & Ba, H.V. (2014). Characterization of hanwoo bovine by-products by means of yield, physicochemical and nutritional compositions. *Korean Journal for Food Science and Animal Resource, 34*(4), 434–447. https://doi.org/10.5851/kosfa.2014.34.4.434

Sharma, S., & Vaidya, D. (2018). Application of kiwifruit protease enzyme for tenderization of spent hen chicken. *Journal of Pharmacognosy and Phytochemistry, 7*(1), 581–584.

Shoulders, M.D., & Raines, R.T. (2009). Collagen structure and stability. *Annual Review of Biochemistry, 78*, 929–958. https://doi.org/10.1146/annurev.biochem.77.032207.120833

Shukurlu, Y., Salmanova, A., & Sharifova, M. (2022). Biotechnological aspects of the modification of secondary collagen-containing raw materials – Tripe for the production of cost-effective functional meat products. *Food Science and Technology, 42*. https://doi.org/10.1590/fst.85521

Silva, V.D.M., & Silvestre, M.P.C. (2003). Functional properties of bovine blood plasma intended for use as a functional ingredient in human food. *LWT-Food Science and Technology, 36*(5), 709–718. https://doi.org/10.1016/S0023-6438(03)00092-6

Singh, A., Negi, M.S., Dubey, A., Kumar, V., & Verma, A.K. (2018). Methods of enzyme immobilization and its applications in food industry. In R.J. Whitehurst & M. van Oort (Eds.), *Enzymes in Food Technology*. Singapore, Springer, 103–124.

Sorapukdee, S., Sumpavapol, P., Benjakul, S., & Tangwatcharin, P. (2020). Collagenolytic proteases from Bacillus subtilis B13 and B. siamensis S6 and their specificity toward collagen with low hydrolysis of myofibrils. *LWT, 126*, 109307. https://doi.org/10.1016/j.lwt.2020.109307

Stefan, D.S., Zainescu, G., Manea-Saghin, A.M., Triantaphyllidou, I.E., Tzoumani, I., Tatoulis, T.I., Syriopoulos, G.T., & Meghea, A. (2020). Collagen-based hydrogels composites from hide waste to produce smart fertilizers. *Materials, 13*(19), 4396. https://doi.org/10.3390/ma13194396

Sun, Q., Zhang, B., Yan, Q.J., & Jiang, Z.Q. (2016). Comparative analysis on the distribution of protease activities among fruits and vegetable resources. *Food Chemistry, 213*, 708–713. https://doi.org/10.1016/j.foodchem.2016.07.029

Tang, C., Zhou, K., Zhu, Y., Zhang, W., Xie, Y., Wang, Z., Zhou, H., Yang, T., Zhang, Q., & Xu, B. (2022). Collagen and its derivatives: From structure and properties to their applications in food industry. *Food Hydrocolloids, 11*, 504. https://doi.org/10.1016/j.foodhyd.2022.107748

Tantamacharik, T., Carne, A., Agyei, D., Birch, J., & Bekhit, A.E.D.A. (2018). Use of plant proteolytic enzymes for meat processing. In M.G. Guevara & G.R. Daleo (Eds.), *Biotechnological Applications of Plant Proteolytic Enzymes*. Springer, Cham, 43–67.

Tenorová, K., Masteiková, R., Pavloková, S., Kostelanská, K., Bernatonienė, J., & Vetchý, D. (2022). Formulation and evaluation of novel film wound dressing based on collagen/microfibrillated carboxymethylcellulose blend. *Pharmaceutics, 14*(4), 782. https://doi.org/10.3390/pharmaceutics14040782

Tiwari, U., & Cummins, E. (2021). Legume fiber characterization, functionality, and process effects. In *Pulse Foods (Second Edition), Processing, Quality and Nutraceutical Applications*, 147–175. https://doi.org/10.1016/B978-0-12-818184-3.00007-6

Troncoso, F.D., Sánchez, D.A., & Ferreira, M.L. (2022). Production of plant proteases and new biotechnological applications: An updated review. *ChemistryOpen, 11*(3), e202200017. https://doi.org/10.1002/open.202200017

Van Wart, H.E. (2013). Clostridium collagenases. In *Handbook of Proteolytic Enzymes (Third Edition)*, 1, 607–611. https://doi.org/10.1016/B978-0-12-382219-2.00126-5

Vidal, A.R., Cansian, R.L., Mello, R.D.O., Demiate, I.M., Kempka, A.P., Dornelles, R.C.P., Rodriguez, J.M.L., & Campagnol, P.C.B. (2022). Production of collagens and protein hydrolysates with antimicrobial and antioxidant activity from sheep slaughter by-products. *Antioxidants, 11*, 1173. https://doi.org/10.3390/antiox11061173

Wanderley, M.C., Neto, J.M., Filho, J.L., Lima, C.A., Teixeira, J.A., & Porto, A.L. (2017). Collagenolytic enzymes produced by fungi: A systematic review. *Brazilian Journal of Microbiology, 48*(1), 13–24. https://doi.org/10.1016/j.bjm.2016.08.001

Wang, H. (2021). A review of the effects of collagen treatment in clinical studies. *Polymers (Basel), 13*(22), 3868. https://doi.org/10.3390/polym13223868

Wang, W., Wang, X., Zhao, W., Gao, G., Zhang, X., Wang, Y., & Wang, Y. (2018). Impact of pork collagen superfine powder on rheological and texture properties of Harbin red sausage. *Journal of Texture Studies, 49*(3), 300–308. https://doi.org/10.1111/jtxs.12300

Westhoek, H., Rood, T., Van den Berg, M., Janse, J., Nijdam, D., Reudink, M., Stehfest, E., Lesschen, J.P., Oenema, O., & Woltjer, G.B. (2011). The protein puzzle: The consumption and production of meat, dairy and fish in the European Union (No. 500166001). Netherlands Environmental Assessment Agency.

Zhao, G.Y., Zhou, M.Y., Zhao, H.L., Chen, X.L., Xie, B. B., Zhang, X.Y., & Zhang, Y.Z. (2012). Tenderization effect of cold-adapted collagenolytic protease MCP-01 on beef meat at low temperature and its mechanism. *Food Chemistry, 134*(4), 1738–1744. https://doi.org/10.1016/j.foodchem.2012.03.118 MCP-01

Zhang, B., Sun, Q., Liu, H.J., Li, S.Z., & Jiang, Z.Q. (2017). Characterization of actinidin from Chinese kiwifruit cultivars and its applications in meat tenderization and production of angiotensin I-converting enzyme (ACE) inhibitory peptides. *LWT - Food Science and Technology, 78*, 1–7. https://doi.org/10.1016/j.lwt.2016.12.012

Zinina, O., Merenkova, S., Rebezov, M., Tazeddinova, D., Yessimbekov, Z., & Vietoris, V. (2019). Optimization of cattle by-products amino acid composition formula. *Agronomy Research, 17*(5), 2127–2138. https://doi.org/10.15159/AR.19.159

Chapter 3

The Use of Wine Waste as a Source of Biologically Active Substances in Confectionery Technologies

Nataliya Grevtseva
V.N. Karazin Kharkiv National University

Olena Gorodyska
T.H. Shevchenko National University «Chernihiv Colehium»

Tatiana Brykova
Institute of Trade and Economics of National University of Trade and Economics

Sergey Gubsky
State Biotechnology University

CONTENTS

DOI: 10.1201/9781003329671-3

3.1 INTRODUCTION

Grapes are one of the most cultivated fruit crops in the agricultural world market. According to the statistical database FAOSTAT (The Food and Agriculture Organization of the United Nations), the value of grapes produced in 2020 came to approximately 83.88 billion US dollars (*FAOSTAT*, 2020). Grapes are widely used as fresh fruit and for the production of wine, juice, vinegar, jam and jelly. The major quantity of grapes, nearly 71%, is used for making wine (Vorobiev & Lebovka, 2016). The industrial residues of wine production typically contain skins, pulp, stalks, and seeds called grape pomace or wine pomace. This by-product has attracted attention of food scientists and food industry because of its high content of nutrients and bioactive compounds (García-Lomillo & González-SanJosé, 2017) and there is a growing interest in grape pomace as a value-added product with a potential opportunity to be used in food technologies. Indeed, the search of literature references through Scopus with a combination of keywords "grape pomace" and "food technology" in title, abstract and keywords, generated 510 documents over the past ten years (2011–2021). An increasing interest related to the research on grape pomace is obvious within these years and it is followed by a similar trend in published works related to their valorization in food technologies (Figure 3.1a).

The issue of grape pomace valorization is paid more attention by researchers from countries with a traditionally developed wine industry (Figure 3.1b). This tendency is largely influenced by increasing requests for solution of environmental problems related to the management of winemaking industrial by-products. Environmental specialists are concerned that unprocessed pomace is a significant source of environmental pollution. A growth in the number

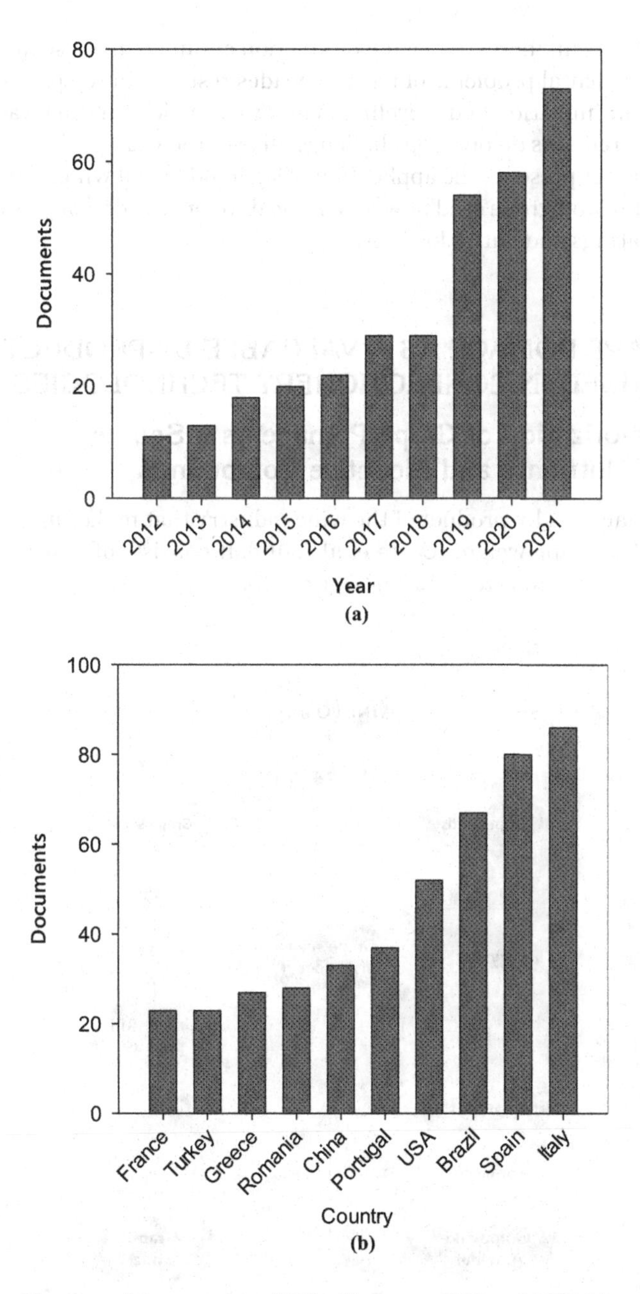

Figure 3.1 Number of documents published between 2012 and 2021 (a) and by country (b) in the databases Web of Science Core Collection with topics "Grape pomace" AND "Food" in categories "Food Science Technology."

of wineries and an increase in wine production entail an increase in waste too. The environmental problems of recent decades resulted in the passing of laws on waste minimization and recycling in many countries. Finding ways of recycling waste remains an ongoing challenge all over the world.

This chapter presents the application of such industrial winery by-products as grape skin and grape seed powders in the developed technologies of confectionery coatings and butter biscuits.

3.2 GRAPE POMACE AS A VALUABLE BY-PRODUCT TO BE USED IN CONFECTIONERY TECHNOLOGIES

3.2.1 Valorization of Grape Pomace as a Source of Nutrients and Bioactive Compounds

Grape pomace is a by-product of the wine industry that makes up about 20%–30% of initial grape weight (Costa et al., 2019). It consists of grape skin (50%), grape seeds (25%) and stems (25%) (Figure 3.2).

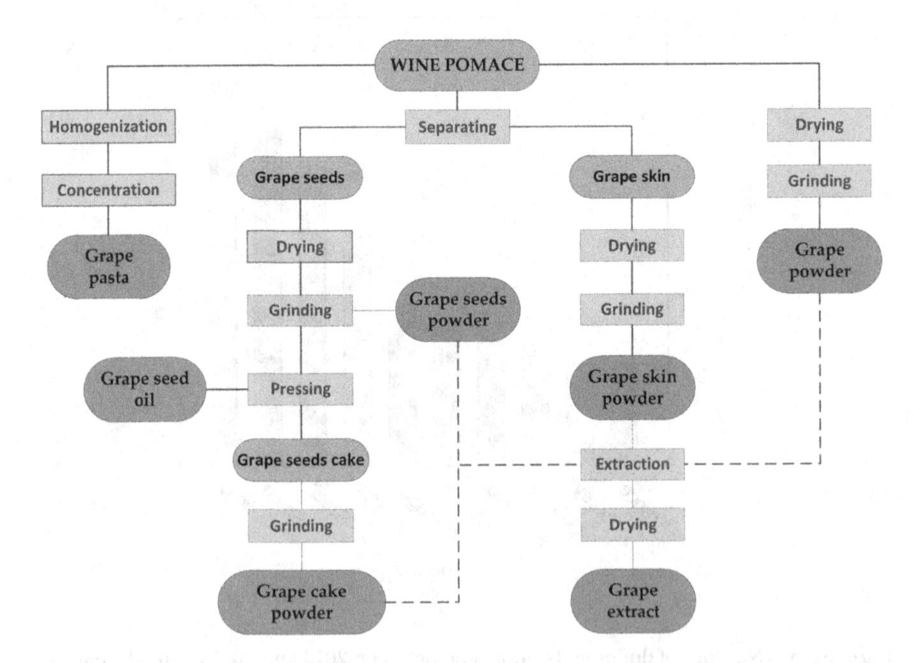

Figure 3.2 Chart of added-value by-products processing from grape pomace for further valorization.

The ideas of valorizing grape pomace are not new. At the end of the last century, various options for using valuable components of grape pomace were proposed. The most interesting proposal was the development of enocyanin. It is a red grade dye that has found use in food, pharmaceutical and cosmetic industries. At present, anthocyanins made from red grape pomace are known as food-grade dye (E163). The second most popular product is grape seed oil (Figure 3.2). When processing pomace, the extracts of bioactive compounds, tartaric acid and ethyl alcohol can be obtained. Over the past decades, alternative methods of pomace processing including those ones without steps of extraction have been developed. They allow to make better use of the potential of that product. The study of the composition of grape pomace showed its high value as a source of bioactive compounds (Baroi et al., 2022; Bharathiraja et al., 2020; Gupta et al., 2020). The most valuable of them are phenolic compounds and dietary fiber. This opened up new prospects for the rational use of grape pomace in medicine, pharmacology, cosmetology and the food industry, which are evidenced by numerous new technologies for processing by-product of the wine industry to create a wide range of various products (Chen & Yu, 2017; Dávila et al., 2017; Mandade & Gnansounou, 2022); numerous innovations for the production of new foods (Antonic et al., 2021; García-Lomillo & González-SanJosé, 2017; Kalli et al., 2018) and for the extraction of polyphenols (Chang et al., 2022; Lo et al., 2022; Pérez-Ramírez et al., 2018) have been generated as well.

The most useful winemaking waste for the extraction of bioactive compounds is grape skin and seeds obtained after processing pomace (Figure 3.2). These by-products are composed of proteins, lipids, carbohydrates, vitamins, minerals and compounds with important biological properties such as phenolic compounds and dietary fiber (Figure 3.3).

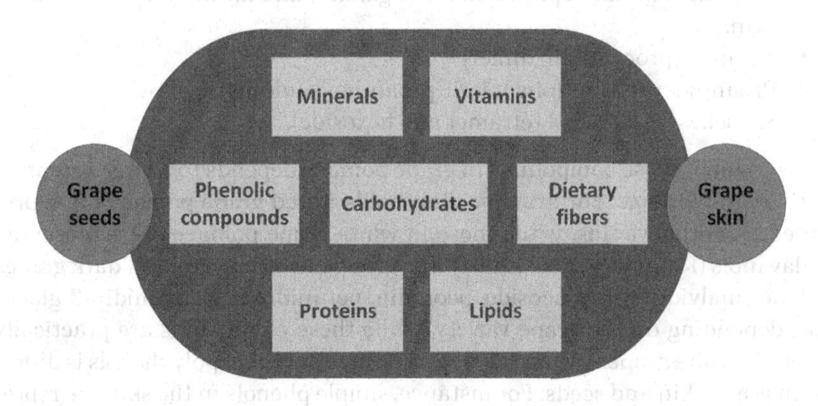

Figure 3.3 Chemical compounds of grape seeds and skin.

Grape is a polyphenols-rich fruit (Garrido & Borges, 2013; Tomaz et al., 2017). Polyphenolic compounds have the potential to exert antioxidant, anti-inflammatory, antimicrobial, anticancer and antithrombotic effects, and antidepressant activity, as a relevant therapy to diabetes and its complications (Kalli et al., 2018; Serina & Castilho, 2022). During the process of making wine, a part of the extractable polyphenolic compounds goes to the finished product. However, according to numerous reviews (Antonić et al., 2020; García-Lomillo & González-SanJosé, 2017; Kammerer et al., 2004; Peralbo-Molina & Luque de Castro, 2013; Teixeira et al., 2014), a significant amount of polyphenolic compounds remains in grape pomace. According to Georgiev et al., 2016, more than 70% of grape phenolics can be retained in skin and seeds. The extractable polyphenols in grapes are found in the seeds, 60%–70%, and in the skin, 28%–35% (Shi et al., 2003); it should be noted that total polyphenolic content (TPC) varies from 0.28 to 8.70 g/100 g of the dry sample (Antonić et al., 2020).

The phenolic composition of grape seeds has been extensively described. Numerous experimental data from the study of grape seed extracts in various solvents by chromatography, among other techniques (Beres et al., 2017; de Andrade et al., 2021; Di Stefano et al., 2022; Gil-Sánchez et al., 2018; Nassiri-Asl & Hosseinzadeh, 2016; Sagar et al., 2018; Sciortino et al., 2021), have shown the presence of different classes of polyphenolic compounds, which have been classified according to Liu and Felice (2007); the main of them are:

- Phenolic acids: hydroxybenzoic (gallic, ellagic, vanillic, syringic, p-hydroxybenzoic) and hydroxycinnamic acids (ferulic, caffeic, p-coumaric, syringic).
- Flavonoids, including flavonols (kaempferol, myricetin, quercetin and their derivatives); flavanols (catechin, epicatechin, gallocatechin, epigallocatechin and epicatechin 3-O-gallate) and anthocyanidins from skin.
- Tannins (procyanidin dimer).
- Proanthocyanidins (oligomeric proanthocyanidins).
- Stilbenes (resveratrol tetramer and hexoside).

The content of these compounds in grape pomace depends to a large extent on cultivar and vintage. For example, flavonoids in red grape pomace are represented by anthocyanins, while those in white grape pomace are represented by flavanols (Kammerer et al., 2004). The group of anthocyanins of dark grapes includes malvidin-3-O-glucoside, peonidin, petunidin or delphinidin-3-glucoside, depending on the grape variety, while these compounds are practically absent in white grapes (Amico et al., 2008). The content of polyphenols is different in grape skin and seeds. For instance, simple phenols in the skin are represented by hydroxycinnamic acids and esters of tartaric and hydroxycinnamic acids, mainly caftaric and coumaric acids. Gallic and protocatechuic acids predominate in seeds (Teixeira et al., 2014). Seeds are also rich in gallocatechins (Montealegre et al., 2006) and proanthocyanidins (Padilla-González et al., 2022). The proanthocyanidins of seeds have a lower degree of polymerization (10–20 units) than proanthocyanidins of skin (25–35 units) (Ky et al., 2014). A large amount of nonextractable phenolic compounds remains in grape pomace (Pérez-Jiménez et al., 2009). These are proanthocyanidins connected with other components, first of all, cellulose, as well as monomeric phenols connected with proteins, polysaccharides or cell walls by hydrophilic or hydrophobic interactions, and hydrogen or covalent bonds (Brenes et al., 2008).

The second important class of bioactive compounds of grape pomace is dietary fibers. Their total content can vary from 17.3% to 88.7% in dried grape pomace (Antonić et al., 2020). According to the same source, most of them exist in the form of insoluble fiber in the range of 16.4%–63.7%, while soluble fiber is available in the range of 0.7%–12.8%. The soluble dietary fiber mostly contains pectin, while the insoluble fiber contains cellulose, hemicelluloses and lignin (Beres et al., 2016, 2017, 2019; Deng et al., 2011; Spinei & Oroian, 2021). When including fiber-rich ingredients in the recipe, it should be taken into account the fact that they can modify the consistency, texture, rheological behavior and organoleptic characteristics of final products (Dhingra et al., 2012). The reason for this often lies in the interaction of fibers with water and fat. Therefore, an experimental study of functional properties of fibers such as water binding and holding capacity, and an oil retention capacity is a priority in the development of a food item. Health benefits of fiber are a fairly well-studied area of knowledge and need no special comments. It could be only mentioned that the consumption of dietary fiber reduces the risk associated with cardiovascular diseases, diabetes prevention, cancer protection, cholesterol reduction, constipation and obesity prevention (Benito-González et al., 2019; Dhingra et al., 2012).

3.2.2 Application of Grape Seed and Skin Powders in Confectionery Technologies

As indicated in the previous sections, the issue of valorization of two important by-products of grape pomace, namely, grape seeds and skin, remains relevant in terms of solving two important tasks: production waste management and utilization of a sufficiently valuable source of bioactive components and nutrients. It is the second factor that is the motive power behind the prospect of using these by-products in food technologies, and, in particular, in confectionery technology. Confectionery products cannot be classified as healthy food items, as they have a low nutritive value (Mazur et al., 2018).

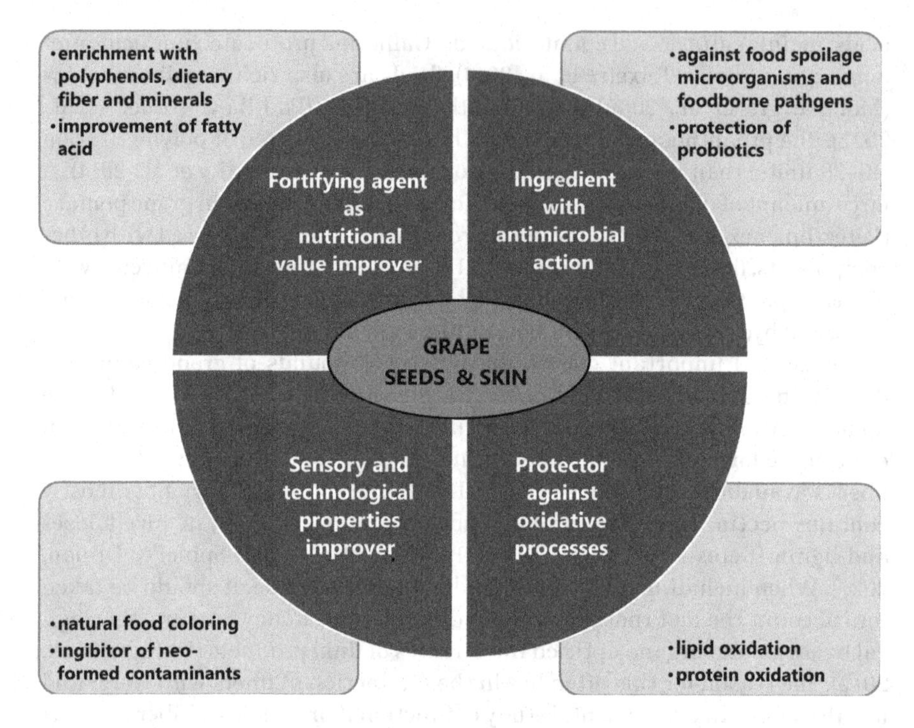

Figure 3.4 The main issues concerning grape seeds and skin application to foods. (Adopted from Antonić et al., 2020; García-Lomillo & González-SanJosé, 2017.)

Modern trends in the development of food technologies require more attention to functional foods and nutraceuticals when developing confectionery products. With that point of view, the presence of bioactive compounds and nutrients in grape seeds and skin allows them to be considered as a fortifying agent for improving nutritive value (Figure 3.4). The addition of processing products of grape pomace in traditional confectionery allows solving some dietary issues associated with insufficient intake of antioxidants, dietary fiber and mineral compounds. The review (Antonic et al., 2021) highlighted numerous examples of fortification of plant foods, meat, fish and dairy products, which were deemed successful. Basically, there is an improvement in sensory and technological properties of foods. However, the authors also specified the presence of side effects of enrichment, such as discoloration caused by polyphenolic compounds, and undesirable changes in texture, among others (Figure 3.4).

The main drivers of food fortification, as shown above, were polyphenols and dietary fiber. Their higher content increases the oxidative stability of fortified foods, especially fat-containing foods. To a greater extent, this is due to the display of powerful antioxidant properties of polyphenolic compounds. There are two stages of lipid oxidation: primary oxidation, inducing the formation of lipid hydroperoxides, diene and triene conjugates; and secondary oxidation, leading to the formation of volatile compounds (Porter, 1986). The result of these processes could be a deterioration in sensory characteristics of products and a decrease in their nutritional value due to the destruction of polyunsaturated fatty acids, vitamins and others. In the food industry, synthetic antioxidants are traditionally used to slow down the process of fat oxidation in products, which can display toxic and carcinogenic properties (Liu & Mabury, 2020). The best solution to reduce these risks is replacing them with alternative natural antioxidants having similar efficiency (Taghvaei & Jafari, 2015). In this regard, by-products of winemaking having a high content of compounds with proven antioxidant properties should be considered as one of the actual choices for this substitution. Two nuances regarding the antioxidant properties of grape pomace by-products should be noted. According to this study (Sánchez-Alonso et al., 2007), they slow down the second stage of lipid oxidation more than the first one. In addition, due to the display of a synergistic effect between different polyphenols, grape pomace products usually display higher activity than isolated compounds (Maestre et al., 2010; Shaker, 2006).

Another important function of wine pomace is their antimicrobial activity against food spoilage microorganisms and foodborne pathogens as well as protection of probiotic microorganisms presented in food products (García-Lomillo & González-SanJosé, 2017; Stabnikova et al., 2021). It is proved that grape pomace extracts have a bactericidal effect on yeasts and molds (Corrales et al., 2010; Sagdic et al., 2011). The incorporation of grape seed extracts in food products slowed down the growth of mesophilic aerobic bacteria, lactic acid bacteria, bacteria of the genus *Pseudomonas* and psychotrophic bacteria (Boban et al., 2021; Lorenzo et al., 2014; Santoro et al., 2020). The antimicrobial effect of grape pomace processing products is caused by the presence of phenolic acids in them, mainly gallic acid, as well as p-hydroxybenzoic and vanillic acids (Baroi et al., 2022).

The utilization of grape skin and seeds suggests two approaches for enrichment of food products with bioactive compounds. One approach includes the utilization of extracts obtained by various types of extraction, as a rule using organic solvents of different nature and water (Figure 3.2). In this case, a large portion of polyphenols cannot be extracted and, therefore, can be lost. Another approach involves the use of grape skin and seeds without prior extraction. According to this, nonextractable polyphenols make up a basic part of all polyphenols with significant bioactivities (Saura-Calixto, 2012). This method should be considered as a more efficient way for utilization of by-products for enrichment of food with dietary fiber, mineral, protein, fat and polyphenols, including nonextractable polyphenolic compounds in form of powder or flour. Powder does not require any special shelf-life conditions; they are highly adaptable, and well mixed with other types of food raw materials, especially with loose materials. They are widely used in such flour-containing foods as bread, biscuits, pasta and muffins (Aksoylu et al., 2015; Antonic et al., 2021; Bender et al., 2017; Koca et al., 2018; Kuchtová et al., 2018; Meral & Doğan, 2013; Mironeasa et al., 2019; Ortega-Heras et al., 2019; Pečivová et al., 2014; Soto et al., 2012; Theagarajan et al., 2019; Walker et al., 2014). Addition of grape powder improves dough structure, slows down the oxidation of fats, for instance, in biscuits, and increases the nutritive value of flour-based products (Iuga & Mironeasa, 2020).

The search for information on the utilization of grape powder in the recipe for confectionery coating did not give any positive results. However, there are a number of publications covering fortification of chocolate, one of the basic components of chocolate-enrobed confectionery products. According to Barišić et al. (2021), the nutritive value of chocolate can be improved by addition of fiber and polyphenols. As a rule, the implementation of this approach lies in the partial substitute of one or several main components of the chocolate recipe, such as sugar, milk powder and whey powder with grape pomace powder (Altınok et al., 2022; Bursa et al., 2022); it allows to improve taste and increase the amount of polyphenols, the content of which is low or they are totally absent, such as in white chocolate (Rimbach et al., 2011). To modify a traditional recipe for confectionery coatings and butter biscuits, the partial substitution of one or several main components with grape powder is usually used (Figure 3.5).

When choosing a food item for fortification, along with the factors and requirements, which were discussed above, it is necessary to take into account the following processing factors:

- Physicochemical and sensory compatibility of the enriching additive with the enriched product.
- Possible changes in physicochemical and sensory properties of the product in case of the partial substitute of one of the main ingredients with the enriching additive.
- Possibility to create a sufficiently simple and reliable technology for product enrichment, which ensures a uniform distribution of added ingredients throughout the product.
- Safety of the finished product.

The consideration of functional and processing properties of the enriching agent is a main factor in the creation of a functional food item (Evlash et al., 2022).

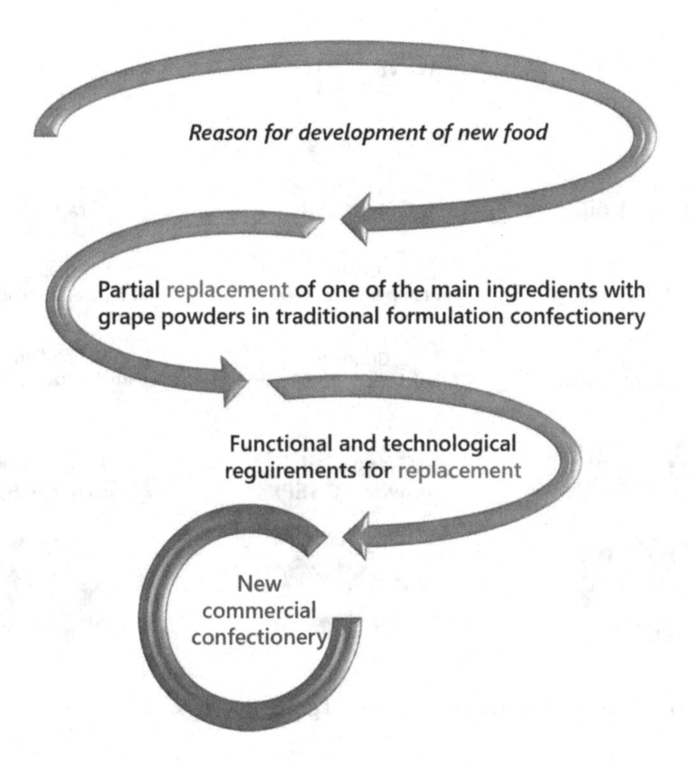

Figure 3.5 Model for developing new commercial confectionery with use of grape powders.

3.3 CHARACTERISTICS OF GRAPE SEED AND SKIN POWDERS AS FORMULATION INGREDIENTS

Commercially available OleoVita™ (Orion, Odessa, Ukraine) powders were used in the developed confectionery technologies: grape seed powder (GSEP); defatted grape seed cake powder (GSCP); grape skin powder (GSKP). These grape powders are made from grape pomace of grape varieties (*Vitis vinifera* L.) of *Cabernet Sauvignon, Cabernet, Muscat Blanc à Petits Grains, Muscat* varieties harvested in 2015–2017. These grape varieties are widely cultivated in the Odessa region, Ukraine, to be used in winemaking enterprises. The technology of powder production includes the drying of grape pomace at a temperature not exceeding 60°C, their purification and separation into seeds and skin with grape stalk fractions (Figure 3.6).

Grape oil is obtained from the seeds by cold pressing; the grape cake remaining after pressing is carefully crushed and a fine powder of defatted grape seeds cake with a fat content of 8%–9% or less is obtained. The fraction, consisting

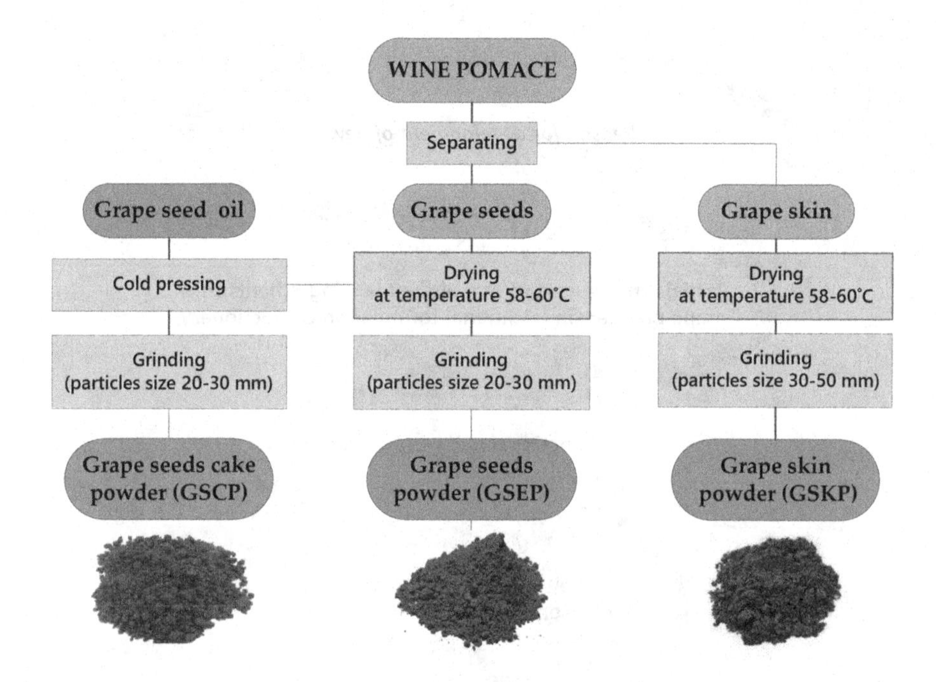

Figure 3.6 Chart of industrial production of grape powders.

of skin and stalks, is also dried and finely ground, obtaining a powder of grape skins. Grape powders are fine-grained and homogeneous with the main fraction (about 80%–82%) of the granulometric composition of particles in size from 20 to 30 μm for grape seed powder and grape seed cake powder, and from 30 to 50 μm for grape skin powder. The sensory characteristics of grape seed and skin powders are presented in Table 3.1.

The brown color, fine dispersed homogeneous structure and no taste make it possible to consider grape powders as an ingredient for the partial substitute of cocoa powder in confectionery coating or wheat flour of butter biscuits with an additional colorant function.

In the previous section, it was indicated that a more efficient option for using all functional components for food fortification is the utilization of by-products of grape pomace without prior extraction. The results of the chromatographic studies of extracts of grape seed powder and grape seed cake powder in water–alcohol systems show the presence of various types of phenolic compounds, where the main ones are shown in Figure 3.7.

The content of phenolic compounds in grape seed cake powder was higher per unit mass of dry matter than in grape seed powder. The identified types

TABLE 3.1 SENSORY CHARACTERISTICS OF GRAPE SEED AND SKIN POWDERS

Characteristic	Grape Seed Powder	Grape Seed Cake Powder	Grape Skin Powder
Appearance	Finely ground, homogeneous, crumbly powder without lumps or granules dry to the touch		
Color	Brown	Brown slightly lighter than grape seed powder	Brown with a purple tint
Taste	Neutral with slightly acidic tart	Neutral with slightly acidic tart, woody taste	Sweet and sour
Flavor	Inherent in this product without odors		Pleasant with the aroma of prunes

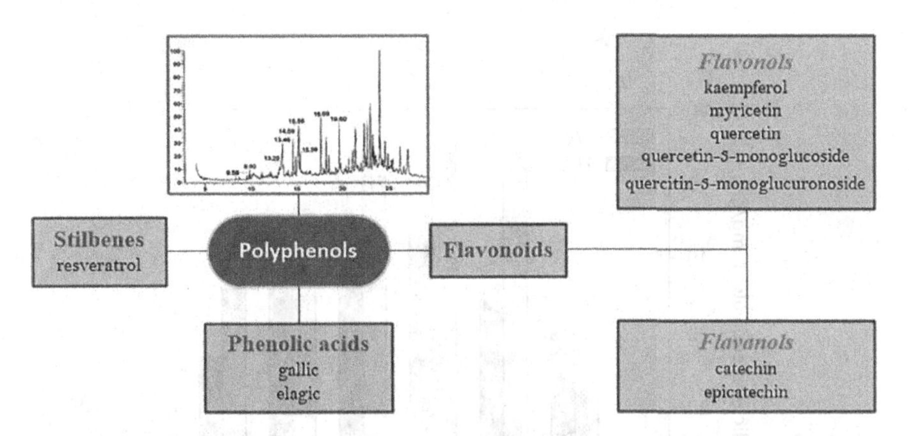

Figure 3.7 Main polyphenols in grape seed powder and grape seed cake powder determined by high-performance liquid chromatography–mass spectrometry (HPLC–MS).

of polyphenols correspond to numerous literature references given in the previous section. All identified phenolic compounds have significant antioxidant properties, especially resveratrol, quercetin and catechin and the results of chromatographic analysis showed the presence of a significant amount of polyphenolic antioxidants in the studied powders.

However, information on the content of specific compounds of antioxidants is not always an informative indicator. The determination of a general or integral indicator corresponding to a total antioxidant potential of all components in their interaction (synergistic or antagonistic) is often more informative for practical purposes. In the assessment, it is important that an analytical signal was due to the availability of the same type of compounds related to the structure–function relationship. In our case, this condition is real, taking into account the advantage in the content of flavonoids. The total antioxidant capacity of samples was determined by coulometric titration with electrogenerated bromine (Gubsky et al., 2016, 2018; Mazur et al., 2018), which included the concentration of polyphenolic compounds by extraction. The extraction efficiency depends on the selection of solvent. But for food purposes, this choice is limited, as a rule, to water, ethanol or a mixture of both. The study was carried out with grape seed cake powder; the extraction of polyphenolic compounds from this sample should be better in these polar solvents on account of the low fat content. The dependence of total antioxidant capacity (TAC) and total content of polyphenols (TPC) in extracts on the composition of the aqueous–alcoholic solvent at a temperature of 60°C (maximum achievable for alcoholic extracts) is given in Figure 3.8.

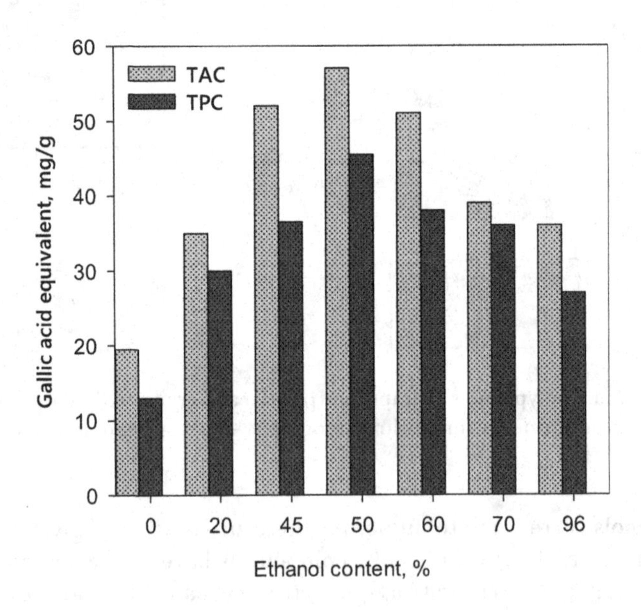

Figure 3.8 Effect of ethanol content on total antioxidant capacity (TAC) and total polyphenol content (TPC) of the water–ethanol extract of grape seed cake powder (GSCP) at a temperature of 60°C (extraction time 100 min and ratio liquid to solid 1:30 for all extracts).

The maximum values of TAC and TPC are observed in the aqueous–alcoholic solvent with an ethanol content of 50% wt. TAC and TPC values for this extract are equal to 57.2 and 45.5 GAE mg/g, respectively, where GAE is a gallic acid equivalent. Taking into account the absolute values, it should be noted that the content of polyphenols is quite high, amounting to almost 4.5% of the mass of dry matter of the powder in a gallic acid equivalent. This value corresponds to the range of variation of the total polyphenolic content in by-products of grape pomace (Antonić et al., 2020). The results obtained support the conclusion that a grape seed extract is one of the strongest antioxidants as free-radical scavenger in nature.

3.4 APPLICATION OF GRAPE SEED POWDER IN CONFECTIONERY COATING TECHNOLOGIES

Chocolate and compound coatings are the basis of all chocolate-enrobed confectionery products (Taghvaei & Jafari, 2015; Talbot, 2009a) or confectionery coatings (Urbanski, 2009). The latter is a product made from sugar, fats, cocoa products, stabilizers, emulsifiers with or without the addition of powdered milk, flavoring and aromatic substances. Compound coatings or confectionery coatings contain an ingredient (or ingredients) of the fat phase which may be a different fat than cocoa butter (Hartel et al., 2018). Coatings play a leading role in improving the appearance of confectionery. Most often, confectionery coatings are used in the technology of candies, products with a foam-like structure (marshmallows, whipped candies), as well as in the production of flour confectionery (cakes, pastries, cookies, waffles). Confectionery coating has different functions in enrobed confectionery: (a) it slows down the oxidation processes, in particular, fats; (b) regulates the moisture content, preventing staleness or conversely, moisture barriers for products; (c) provides an attractive appearance and compositional completeness; (d) improves the product flavor palette, and participates in the formation of its nutritive value.

The demand for confectionery coatings is constantly growing. One of the reasons for this has been the continual consumer demand for new taste and textural sensations. So the requirements for its range quality and functionality are changing. The assortment of confectionery coatings is very wide and depends on its purpose. Recently, plant cocoa butter alternative (CBA) has become increasingly popular (Naik & Kumar, 2014). The abbreviation CBA refers to the general name for fats fulfilling the function of cocoa butter completely or in parts (Lipp & Anklam, 1998). The main goal of the presented technology for confectionery coating was the utilization of CBA instead of cocoa butter and grape seed powder and grape seed cake powder with a partial substitute of cocoa powder in the traditional recipe. This technology with grape powder as a fortifying agent made it possible to improve the nutritional value of the finished product. In addition, the incorporated ingredients served as both a protector against fat oxidation and an antimicrobial agent to improve product storage stability.

3.4.1 Choice of Fats for Confectionery Coatings

The quality and functional properties of confectionery coating depend on the nature and characteristics of the fatty phase as a dispersion medium. The type of fat determines structural–mechanical and sensory properties of the ready-made coating and affects the coating process with some possible issues such as fat bloom, fat migration between component parts of the product, coating thickness, bond strength between a coating and a body, and the appearance of the product (Beckett et al., 2017; Talbot, 2009b). Selecting a coating with a specific fat composition can increase the setting rate and control the hardness of the coating. Two types of cocoa butter were used for the development of the coating technology for confectionery:

- Cocoa butter substitute CEBESTM MC 80 (AAK, Sweden) is a lauric type (LF), having the same physical properties as cocoa butter, but differing from it in fatty acid composition. Due to a different structure of triacylglycerol, lauric fats are used to substitute almost all or most of cocoa butter. Coatings on such fats do not require tempering, because they crystallize immediately into a stable form. Cocoa powder with a fat content of not more than 12% is used in the recipe. Coatings have a pleasant taste and aroma, a shiny surface, and desired product consistency; they are resistant to oxidation and fat bloom and have high rigidity and brittleness. But their main disadvantage is a lauric high content. That can cause a soapy taste.
- Cocoa butter replacer Olivia Glaze Lux (Torhovyy Dim Shchedro Ltd., Ukraine) is of a nonlauric type (NLF). Nonlauric fats can be used to partially substitute cocoa butter in combination with vegetable oils, cocoa and milk powder. Coatings on nonlauric fats do not need to be tempered; they are resistant to oxidation and have a long shelf life without any losses of quality; and they have no soapy taste. These coatings can be used for any products, including those with a high nut butter content and high moisture content.

3.4.2 Comparison of Grape Seed and Cocoa Powders

An important raw material component of confectionery coatings is cocoa powder. Grape powder is identical in structure and color of cocoa powder, but they differ in chemical composition (Table 3.2).

TABLE 3.2 COMPARISON OF SOME CHARACTERISTICS OF GRAPE SEED AND COCOA POWDERS

Characteristics	Grape Seed Powder	Grape Seed Cake Powder	Cocoa Powder Alkalized[a]	Cocoa Powder Natural[b]
The proximate composition, g/100 g				
Fat	18.1±0.9	8.26±0.34	11.0±0.5	11.2±0.5
Protein	9.49±0.45	11.80±0.48	15.1±0.6	15.6±0.7
Mono-and disaccharides	1.00±0.04	1.14±0.03	1.15±0.04	1.00±0.04
Dietary fiber	47.1±2.2	51.3±2.5	43.9±2.2	44.0±2.2
Total polyphenol content	3.00	3.30	1.50	1.65
Ash content	2.90±0.12	3.10±0.14	13.61±0.56	11.71±0.48
Vitamins content, mg/100 g				
B_1	0.18±0.01	0.20±0.01	0.11±0.01	0.10±0.01
B_2	0.49±0.01	0.54±0.01	0.46±0.01	0.24±0.01
B_6	0.19±0.01	0.21±0.01	0.12±0.01	0.12±0.01
B_{12}	5.10±0.21	5.67±0.18	-	-
PP	3.30±0.14	3.33±0.15	2.40±0.10	2.20±0.09
Minerals content, mg/100 g				
Iron	18.08±0.88	20.09±0.78	15.52±0.68	13.86±0.57
Calcium	350±16	389±19	111±16	128±16
Magnum	349±17	388±19	276±13	299±14
Sodium	8.53±0.40	9.51±0.41	19.02±0.85	21.07±1.03
Potassium	1390±63	1544±76	1803±70	1524±76
Phosphorus	293±14	325±15	238±11	234±11
Zinc	7.50±0.35	8.21±0.38	1.58±0.06	6.80±0.24
Magnesium	1.35±0.04	1.50±0.05	0.63±0.02	0.84±0.02
Physicochemical properties				
Moisture content, %	6.0	6.0	5.0	5.0
pH of aqueous extract	4.36	4.02	6.94	5.12
Water-holding capacity, g/g	2.23±0.10	2.41±0.11	1.98±0.09	1.86±0.09
Fat-retaining capacity, g/g	1.31±0.06	1.46±0.07	1.19±0.05	1.10±0.06

[a] Cocoa powder M75 (Olam Food Ingredients Spain S.L., Spain).
[b] Cocoa powder of ADM Polska Sp. z o.o. Poland.

For instance, the content of polyphenols and dietary fiber in grape powder is higher by 6.7%–14.4% than in cocoa powder. Grape powder also contains more micro and macro elements: iron—by more than 14%–31%, calcium—by more than 63%–71%, magnesium—by more than 14%–29%, phosphorus—by more than 19%–28%, zinc—by more than 9%–81% and manganese—by more than 38%–58%. Grape powder has a richer vitamin composition; it contains vitamin B12, which is not found in cocoa powder. The amount of vitamins B1 and PP in grape powder is higher by 39%–50% and 27%–34% than in cocoa powder. Cocoa powder contains a large mass percent of proteins (by more than 22%–39%), but it is not advisable to consider it as a protein fortifier, because vegetable proteins are not complete. The amount of ash in cocoa powder exceeds its content in grape seed powder and grape seed cake powder by about four times, which may indicate the presence of shell particles from cocoa beans in it. Thus, grape powder has a richer chemical composition than that from cocoa, which makes it an effective substitute for cocoa powder in confectionery coating technology.

Grape powder can absorb and retain fat; therefore, their incorporation can lead to changes in the properties of the mass. The research of Grevtseva et al., 2018, showed that seeds powder and grape seed cake powder have higher water and fat-holding capacity than cocoa powder (Table 3.2). The difference in the above-specified capacity will affect the viscosity of coating, which is one of the main indicators of the quality of the confectionery coating. This indicator determines the appearance of coated items, their thickness and condition of the coating, and the behavior of the coating during cooling, and it can complicate the manufacturing process and application of the product.

The particle size is important for flow properties of chocolate. In order for the coating to have high sensory indicators of quality, namely, a homogeneous structure, a delicate flavor and a high melting quality, and no gritty feel all over the mouth, the particle size should be in the range of 25–35 µm. The study of particle size distributions shows that the proportion of particles in grape seed powder and grape seed cake powder with a size of 20–30 µm is 80%–82%. Cocoa powder is characterized by a larger grinding size: a fractional ratio of particles with a size of 10–22 µm is from 8% to 12%, with a size of 22–45 µm from 45% to 44%, and particles with a size of 45–70 µm from 4% to 22%. Cocoa powder is characterized by greater polydispersity. The particle size values of the powder correlate with each other, and they are in the range of 20–30 µm in grape powder and in the range of 30–42 µm in cocoa powder. The particles of natural cocoa powder are characterized by a large size. *Thus, grape powder can be used for a partial replacement of cocoa powder, as they have a similar fine structure and similar sensory qualities.*

3.4.3 Influence of Grape Seed Powders on the Oxidation and Hydrolysis of Fats

An important process that affects the quality of fat-containing confectionery products is hydrolytic rancidity. This process runs under the action of enzyme lipase, which may be present in raw material or synthesized by microorganisms. The lipase activity of grape seed powder and some samples of cocoa powder used in confectionery glaze formulations were determined with the titrimetric method (Pinsirodom & Parkin, 2005). The lipase activity of cocoa powders was in the range from 0.84 to 1.87 mL of 0.01 KOH per g; meanwhile, lipase activities of grape powders were 1.03 and 1.12 mL of 0.01 KOH for grape seed powder and grape seed cake powder, respectively. Thus, the activity of lipase in the grape powder was almost the same as in the cocoa powder. As a result of fat hydrolysis, fatty acids are released, which leads to rancidity, and an increased acid value of fats is observed. The acid value of fats in the model systems and its change during the shelf life were determined to explore the influence of fats hydrolysis on product quality. The samples with incorporated powder had significantly higher acid values of 1.6–1.9 mg of KOH compared to test samples without powder, which were 0.2–0.4 mg of KOH (Gorodyska et al., 2018). The higher initial acid values in the samples with grape seed powder and grape seed cake powder are related to the fact that the grape seed powder contains organic acids in its composition. This should be judged on pH value of the aqueous extract in the range of 4.0–4.5 (Table 3.3). The analysis of the influence of the process of hydrolysis on oxidation of fats is more expedient to consider the dynamics of acid value instead of absolute values. Thus, for all samples, the relative change of acid value over time during the shelf life was used (Gorodyska et al., 2018). The acid value for samples with the addition of grape seed powder with both lauric and nonlauric fats increased after 28 days of storage time (Figure 3.9).

The reaction of fat autoxidation involving oxygen and unsaturated fatty acids is the most vulnerable to product quality. This autocatalytic process on a free-radical chain mechanism leads to the formation of primary products of fat oxidation—peroxides (Taghvaei & Jafari, 2015). However, the incorporation of synthetic or natural antioxidants into the composition of food items inhibits this process. When considering grape powder as natural antioxidants, one should expect its inhibitory effect on the processes of fat oxidation in the glaze. This will have a positive effect on the shelf life of coated products. As a marker of the influence of grape powder on the process of fat oxidation, a peroxide value was used. This indicator characterizes the specific content of peroxide compounds in fat and allows modeling the effect of grape seed powder on the oxidation of lauric and nonlauric fats. Determination of peroxide values by iodometric titration confirmed the hypothesis of the effect of antioxidants in grape seed powder on the process of fat autoxidation (Figure 3.10).

TABLE 3.3 FORMULATION OF CONFECTIONERY COATINGS

	Confectionery Coatings[b]					
Ingredient, g/100 g of Coating	**CC1**	**CC2**	**CC3**	**CC4**	**CC5**	**CC6**
Lauric fat[a]	33.0		33.00		33.00	
Nonlauric fat[a]		33.0		33.0		33.0
Sugar	53.0	53.0	53.0	53.0	53.0	53.0
Cocoa-powder alkalized	12.0	12.0	10.0	12.0	12.0	12.0
Cocoa-powder natural	3.0	3.0				
Lecithin	0.2	0.3	0.3	0.4	0.3	0.4
Grape seed powder			5.0		3.0	
Grape seed cake powder				3.0		3.0

[a] Lauric fat is cocoa butter substitute CEBESTM MC 80 (AAK, Sweden); nonlauric fat is cocoa butter replacers Olivia Glaze Lux (Torhovyy Dim Shchedro Ltd., Ukraine).

[b] CC1 and CC2 are acronyms of control confectionery coatings; CC3 and CC5 are acronyms of confectionery coatings with grape seed powder; CC4 and CC6 are acronyms of confectionery coatings with grape seed cake powder.

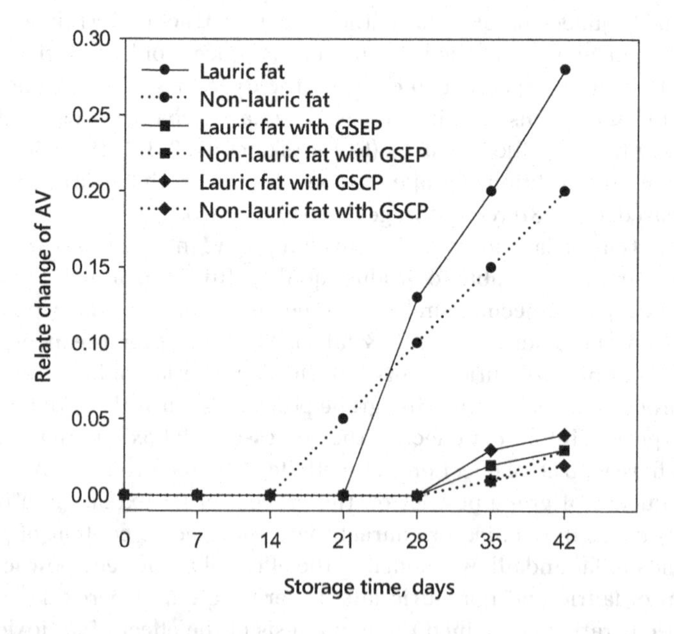

Figure 3.9 Effect of antioxidants of grape seed powder (GSEP) and grape seeds cake powder (GSCP) on acid value of lauric and nonlauric fats during storage time.

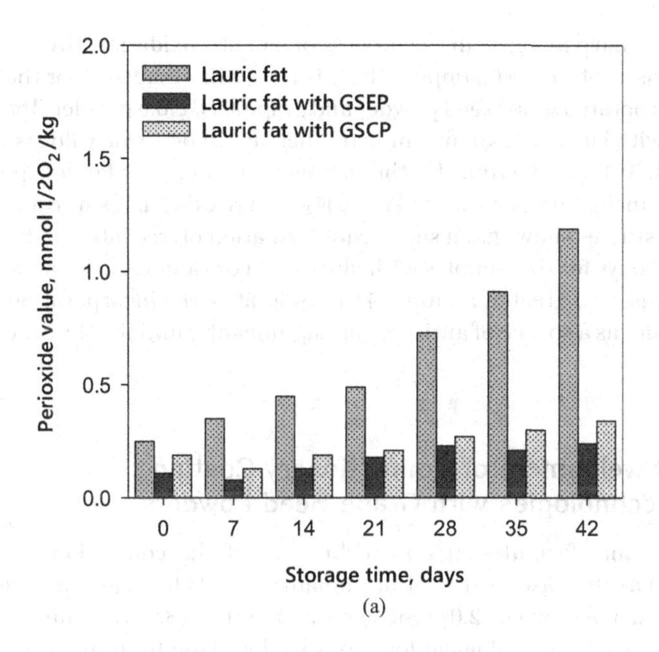

(a)

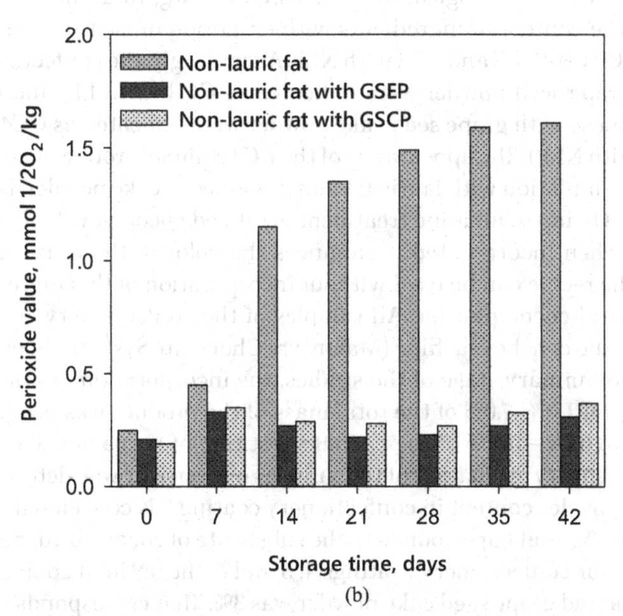

(b)

Figure 3.10 Effect of antioxidants of grape seed powder (GSEP) and grape seed cake powder (GSCP) on peroxide value of lauric fat (a) and nonlauric fat (b) during storage time.

While a sharp increase in the process of fat auto-oxidation after 14 days is a characteristic of the test samples, there is no similar tendency for the samples with incorporated grape seed powder and grape seed cake powder. Thus, for the samples with lauric fat, an insignificant increase in peroxide value is observed in 21 days. At the same time, for the sample with nonlauric fat, it is possible to state the unchanging character ($p<0.05$) of peroxide values during shelf life. Practical studies show that a significant formation of free fatty acids is slowed down by 7 days for the samples with lauric fat. For samples from nonlauric fat, this delay is even greater, mainly, 14 weeks. That is, the incorporation of grape seed powder as a source of antioxidants significantly inhibits the process of fat hydrolysis.

3.4.4 Development of Confectionery Coating Technologies with Grape Seed Powders

The traditional formulation and technology of the confectionery coatings were used as the base: lauric fat or nonlauric fat, 33.0 g; sugar powder, 53.0 g; cocoa powder alkalized, 12.0 g; cocoa powder natural, 3.0 g; lecithin, 0.2 for CC1 and 0.3 for CC2, technological loss consisted of 2.0 g; total mass was 100.0 g. Therefore, the content of ingredients with exception of lecithin was the same for control CC1 with LF and CC2 with NLF. Accordingly, the confectionery coatings with grape seed powder were indicated as CC3 (with LF) and CC5 (with NLF), and those with grape seed cake powder were indicated as CC4 (with LF) and CC6 (with NLF). The appearance of the CC4 sample, produced according to the above formulation with lauric fat and grape seeds cake powder, is shown in Figure 3.11. Owing to alkaline treatment alkalized cocoa powder has a darker color, and when incorporated it enhances the color of the finished product. However, the recipe can be used without incorporation of this ingredient, limited to natural cocoa powder. All samples of the confectionery coating were prepared in the conche machine (MacIntyre Chocolate Systems, Scotland, UK).

At the preliminary stage of the studies, the incorporation of grape powder in the range of 1.0%–7.0% of the total mass of the product was used. That corresponded to 6.7%–46.7% of the partial substitute of cocoa powder in the recipe. After a sensory evaluation of the samples obtained, it was determined that grape seed powder content in confectionery coating 3 is considered optimal in the range of 5%. That corresponds to the substitute of cocoa powder at the level of 33.3%. As for confectionery coatings 4, 5 and 6, the optimal content of grape seed powder and grape seed cake powder was 3%. That corresponds to the substitute of cocoa powder at the level of 20%. Coating samples with a partial substitute of cocoa powder with grape powder had a dark brown color, a pleasant taste with a slight fruity smell and a cooling aftertaste. A light fruity aftertaste

Figure 3.11 The appearance of the confectionery coatings sample (CC4).

and a cooling effect of the coating are given by tannins and terpenes, which are presented in the grape powder composition. When a big amount of powder is incorporated, the color becomes lighter and a "woody" aftertaste appears, which should be considered as a deterioration in sensory characteristics.

Chocolate is a food item with sensitive rheological properties (Barišić et al., 2021). Therefore, any addition of components that are not included in the standard recipe can have a significant impact on its physicochemical and sensorial properties. The incorporation into standard confectionery coatings recipe of grape powder increased the viscosity of the samples by almost two-fold. Thus, the viscosity of confectionery coatings 1 and 2 at an optimum temperature of 40°C was 10.2 and 11.2 Pa·s, respectively, and for confectionery coatings with grape powder, it was in the range of 16.6–23.3 Pa·s. Viscosity of confectionery coatings affects the appearance of chocolate-enrobed confectionery products by determining a coating thickness. Therefore, such an increase in the viscosity of confectionery coatings with grape powder should be considered an undesirable fact, and it requires adjustment to a desired value. The values of viscosity of confectionery coatings 1 and 2 were chosen. To effectively reduce their viscosity, an approach with an increase in the emulsifier content was used.

The emulsifier acts as a surfactant and changes the behavior of the continuous phase in the confectionery in the desired direction. For instance, the use of an emulsifier such as lecithin in chocolate and compound coatings results in a reduction in product viscosity and ease of processing in manufacturing of all chocolate-enrobed confectionery products (Hartel & Firoozmand, 2019). Initially, according to the standard recipe for confectionery coating, both in control and in samples with grape powder, the content of lecithin was 0.2% for samples with LF, and it was 0.3% for samples with NLF. The study of the dependence of the viscosity of confectionery coatings on the content of lecithin, followed by optimization of the experimental results to the optimal viscosity value in the range of 9.0–11.0 Pa·s, made it possible to make an appropriate correction in the recipe (Table 3.3).

An important indicator of the quality of confectionery coating, which makes its consumer properties, is the grinding size. This indicator was determined at the end as a marker of the end of the grinding process and completion of coating. The grinding size of both test samples and samples with the addition of grape seed powder was the same, and it was within 20–25 μm.

3.4.5 Quality Characteristics of Confectionery Coatings

The content of vitamins, minerals and polyphenols in grape powder was higher than in cocoa powder (Table 3.4), which makes it possible to predict an increase in the nutritional and biological value of the developed confectionery coatings. The calculations performed show a slight change of up to 1% in the energy value of confectionery coatings 4, 5 and 6 compared to confectionery coatings 1 and 2

TABLE 3.4 SENSORY CHARACTERISTICS OF TRADITIONAL AND ENRICHED WITH GRAPE POWDER BUTTER BISCUITS

Characteristics	Control (Without Grape Powder)	Butter Biscuits With	
		Grape Seeds Cake Powder	Grape Skin Powder
Appearance	The set form is preserved without fractures bubbles or cracks		
Color	Golden	Chocolate	Chocolate with a violet tone
Texture	Crumbly loose free from undispersed traces with an even fracture porosity		
Taste, smell	Pleasant, typical of freshly baked butter biscuits, free from foreign tastes and odors		Pleasant, typical of freshly baked butter biscuits with a light plum flavor

due to the incorporation of an additional amount of lecithin. At the same time, the integrated score of minerals and vitamins for confectionery coatings 3 and 5, according to calculations, increased by 7% for phosphorus, 18% for manganese, 62% for calcium, 7% for magnesium and up to 25% for vitamins B1, B2, B6 and B12 compared with control, and for confectionery coatings 4 and 6, according to similar indicators, the integral score increased to 43% for minerals and up to 25% for vitamins compared to control.

A significant increase in the content of polyphenolic compounds in the recipe of confectionery coating because of the incorporation of grape powder should have a positive effect on the physicochemical and microbiological stability of finished products during the shelf life caused by the ability of polyphenols to suppress oxidative processes and inhibit microbial activity (Figure 3.4). The modeling of the effect of the antioxidant properties of grape powder polyphenols on the oxidation and hydrolysis of confectionery coating fats confirmed the effectiveness of inhibition of fat oxidation and hydrolysis processes (Figures 3.9 and 3.10). Further studies were carried out on the developed samples of confectionery coating when stored for 12 months in a dry, clean and ventilated room at a temperature of 18±5°C and a relative humidity of not higher than 75%. The guaranteed shelf life of confectionery coating is 8 months, while the first signs of deterioration began to appear in the test samples after 9 months of storage. In confectionery coating 1, there was a loss of gloss, a decrease in the intensity of colors, and white spots appeared on the surface of the "monolith." When chewed, the coating became softer, lost its crunchiness, and a greasy taste and an unpleasant odor appeared. In confectionery coating 2, similarly, there was a decrease in color intensity, and the appearance of white spots; the surface of the monolith became crumbly with a large number of small crumbs. When chewed, the coating became very soft with a soapy aftertaste. The deterioration of the sensorial characteristics of the test samples of coating during storage is obviously associated with the accumulation of low molecular weight protein compounds and hydrolysis of the fat fraction. Confectionery coatings 4, 5 and 6 had high sensorial characteristics during the entire guaranteed shelf life of 9 months. Only after 12 months of storage, the decrease in the intensity of colors, taste and aroma, as well as slight softening and crumbling, began to occur in these coatings. Unlike control, no white spots and greasy or soapy aftertaste were seen during the entire shelf life of 12 months. Furthermore, various sweets (roasted, jelly and milky jelly), marshmallow, marmalade and pastille were coated with confectionery coatings 3 and 5, and shortbread cookies, gingerbread and waffles were coated with confectionery coatings 4 and 6. Coatings with grape powder incorporated differed from control with indicators of quality which are higher than those of control due to less crumbling, no traces of graying, and preservation of taste and aroma. The products had a brown color, a smooth matte surface without flashes, cracks, breaks, as well as

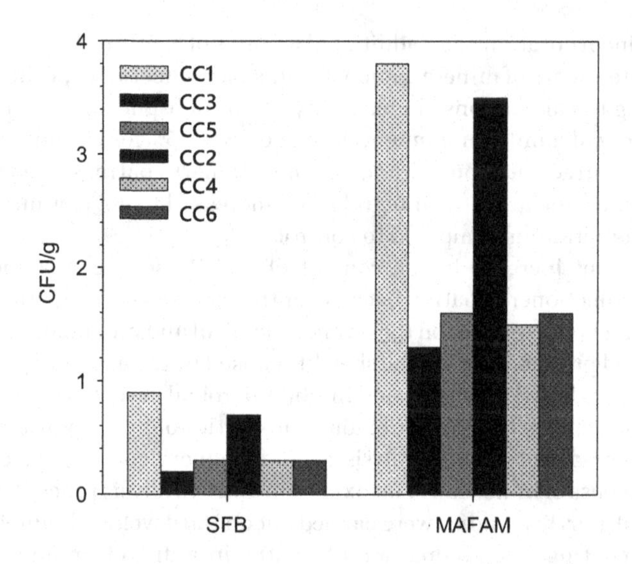

Figure 3.12 Number of spore-forming bacteria (SBF, CFU·10^2) and mesophilic aerobic and facultative anaerobic microorganisms (MAFAM, CFU·10^4) in the samples of confectioner's coatings.

a taste and smell of the prescription composition. When biting, the coating was soft and brittle. During shelf life, the physicochemical properties of the coating and coated products changed within the framework of the requirements of regulatory documentation.

Coating and coated products are not heat treated before packaging, so they cannot be made completely free from the presence of microorganisms. However, grape powder polyphenols have antimicrobial activity (Figure 3.4). It was shown that the amount of mesophilic aerobic and facultative anaerobic microorganisms (MAFAM), and spore-forming bacteria (SFB) in coatings with grape powder was reduced by more than three and three times, respectively, compared with control (Figure 3.12). Yeasts, mold (fungi) and *Escherichia coli* were not detected in any of the coatings.

3.5 APPLICATION OF GRAPE SEED AND GRAPE SKIN POWDERS IN BUTTER BISCUIT TECHNOLOGIES

Butter biscuits are one of the most popular confectionery products. They are prepared according to various recipes, but usually they consist of premium wheat flour, margarine or another fat component, sugar, sometimes eggs and baking powder. Therefore, traditional cookies contain a lot of fat and carbohydrates and few bioactive compounds. On the one hand, incorporation of grape powder in the recipe will enrich it with dietary fiber, polyphenolic compounds, minerals and vitamins. It will also have a stabilizing effect on the quality of cookies during shelf life due to antioxidant properties of grape polyphenols. On the other hand, a significant increase in additional chemical compounds in the food matrix of the components can have both positive and negative effects on dough properties and indicators of the finished product.

3.5.1 Effect of Addition of Grape Seed and Skin Powders on Dough Properties

The necessary condition for a delicate and crumbly structure of the finished product is the preparation of loose and viscous-plastic dough for butter cookies. The dough for butter biscuits is a complex system and its preparation is accompanied by the interaction between all types of raw materials, including added grape powder instead of wheat flour. The consequences of such an interaction can be various physical and chemical processes and, as a result, textural changes, at the stage of dough kneading or in the process of forming its structural and mechanical properties. Thus, it was determined that the interaction of grape seed lipids with flour gluten proteins, starch and hydrophobic components leads to a weaker dough consistency, increased viscosity and slowing down the process of starch gelatinization (Mironeasa et al., 2012, 2016). The opposite result was obtained in the study by Meral and Doğan (2013). The authors claim that the dough structure is strengthened, and its stability is increased when grape seed powder is added, and this fact is explained by the possible formation of covalent or noncovalent bonds between gluten proteins and phenols. The strengthening of gluten when adding powder from grape seeds and skins to wheat flour by a partial substitute (1%–5% by weight) is also reported by the authors (Lisjuk et al., 2011).

The effect of addition of grape powder on dough properties was studied by Goralchuk et al., 2020. As it was shown above, the impact of incorporation of wine pomace for partial substitution of wheat flour to the flour-containing foods can lead to a change in the structural and mechanical properties of the dough. In turn, that changes the flow of many processes, such as processing modes for dough mass, shaping of products and their changes during baking. One of the most important chemical components of any foodstuffs is water. Formulation of butter biscuits does not contain water but water is present in some of their components such as eggs and butter. Meanwhile, grape powder contains dietary fiber (Table 3.3), which has a good water-holding ability and can affect the rheological properties of dough. Compared to wheat flour,

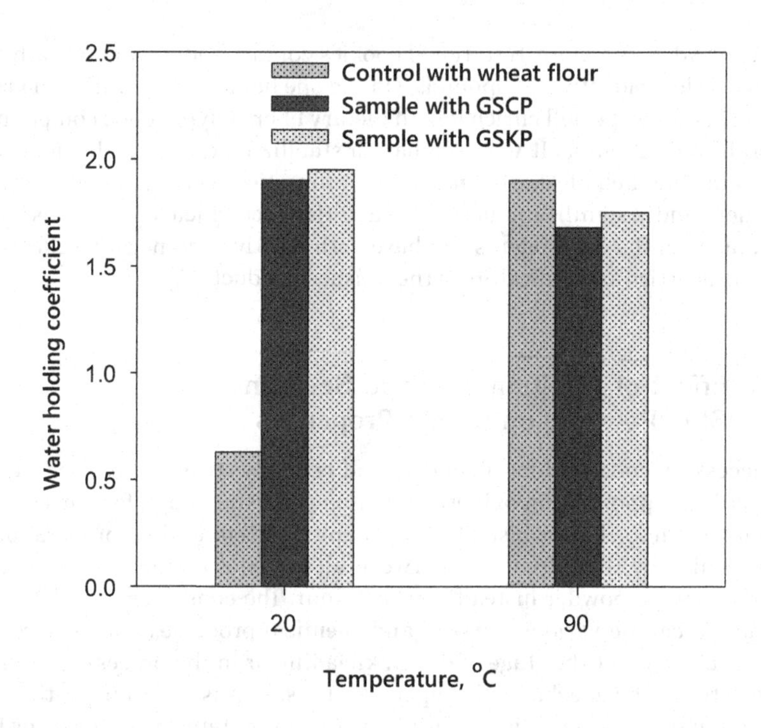

Figure 3.13 Dependence of the water-holding coefficient on temperature for wheat flour, grape seed cake powder (GSCP) and grape skin powder (GSKP).

modeling of the water absorption capacity of grape seed and grape skin powder was carried out at 20°C, which is a kneading temperature, and also at 90°C, as an approximate temperature at the center of butter biscuits during baking. The quantitative comparison parameter was a water absorption coefficient (the ratio of the sample mass after and before water absorption) during swelling for 10 min. The experimental data obtained indicate that at a temperature of 20°C, a water absorption coefficient of wheat flour 0.63 is three-fold lower than that of grape seed and skin powder 1.90 and 1.95, respectively (Figure 3.13).

Thus, adding grape seed powder or grape skin powder to the cookie dough will cause its compaction during the kneading step. This fact should be taken into account in its mechanical processing. At a temperature of 90°C, due to starch gelatinization, a water absorption coefficient of wheat flour increases by almost three times, and it is comparable to the water absorption coefficients of powder. That is, the addition of grape seed cake powder and grape skin powder will not affect the structure of baked products.

The butter cookie dough contains a large amount of fat. Therefore, an important property of new dough components, which can also affect the process of dough formation, is their fat-binding capacity. The grape powder demonstrated a better ability to bind fat than wheat flour, and the fat-retaining capacity of grape seed cake powder and grape skin powder is 97.3% and 90.3%, respectively, versus 60.6 for wheat flour. This is a prerequisite for better binding of fat during the preparation of the dough and its retention during shelf life of cookies.

An important role in the formation of the structure of dough and finished flour products belongs to the gluten of wheat flour. Butter biscuits require flour with weak gluten, which ensures its delicate and crumbly structure. It was shown that the addition of grape powder in the amount of more than 5% to the flour leads to strengthening of gluten, and, accordingly, to the compaction of biscuits (Lisjuk et al., 2011). Subsequent studies with a wider range of adding grape powder of 0%–20% by weight of flour found that the addition of grape powder in the amount of more than 13% to the flour strengthened gluten so much that it cannot be washed off (Samohvalova et al., 2016). In this case, gluten crumbles and does not form a lump. Therefore, it can be concluded that a partial substitution of flour in the recipe with grape powder in the range of 0%–13% leads to the compaction of the cookie structure, and when substituting 13%–20%, more crumbly cookies are obtained.

The addition of grape pomace powder affects wheat flour dough rheology. According to Iuga & Mironeasa, 2020, the viscoelastic behavior of dough and the textural and sensory characteristics of baked goods and pasta containing grape by-products depend on the additional amount and particle size. Indeed, smaller particle size of grape pomace powder can lead to better bread quality (Iuga & Mironeasa, 2019). The authors recommend using grape powder with a particle size of less than 200 μm. In the previous section (3.4.2), it was specified that, according to the data of studies of particle size distributions, grape powder used in the studies had a main fraction size (80%–82%) of 20–30 μm. The study of rheological properties of dough showed that the substitution of the part of the flour with grape skin and seed powder led to a decrease in the springiness and elasticity of dough (Figure 3.14a and b), but an increase in the apparent viscosity of the dough, its plasticity increased (Figure 3.14c).

The obtained results indicate that the strengthening of the dough texture ensures its stability during the formation of cookies and better preservation of the shape of cookies and the pattern on its surface. Moreover, the adhesive strength of dough decreases. This is a factor that reduces the stickiness of dough to working bodies of shapers. Thus, fine grinding of grape skin and seed powder improves rheological properties of cookie dough. In general, the experimental studies confirmed the feasibility of the incorporation of grape seed powder and grape skin powder into the recipe of butter biscuits for partial substitution of wheat flour.

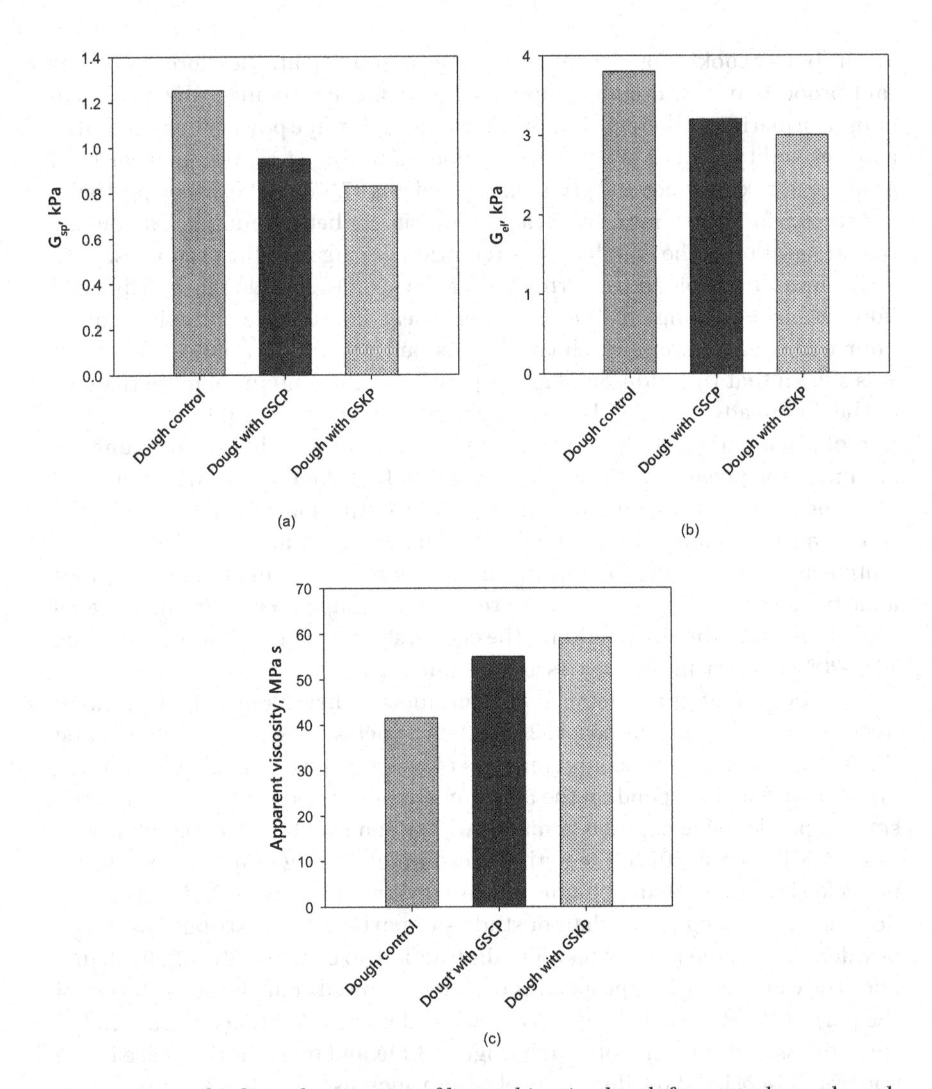

Figure 3.14 Rheological properties of butter biscuits dough for control sample and samples with grape seeds cake powder (GSCP) and grape skin powder (GSKP): springiness module G_{sp} (a), elasticity module G_{el} (b) and appearance viscosity η (c).

3.5.2 Formulation of Butter Biscuits

The traditional formulation and technology of the butter biscuits were used as control (BB1): composite flour or wheat flour, 61.4 g; sugar, 12.3; butter, 39.9 g; mélange, 3.1 g; vanilla powder, 0.3 g; technological loss consisted of 17.0 g, so,

Figure 3.15 The appearance of butter biscuits: with grape seeds cake powder (a) and with grape skin powders (b).

total mass was 100.0 g. Accordingly, the butter biscuits with grape seed cake powder and grape skin powder were referred to as BB2 and BB3, respectively. The process of making cookies included the following stages: preparation of raw materials, preparation of emulsion from butter, sugar and eggs, addition of flour and quick kneading of dough. The grape seed powder and grape skin powder were added to the oil as antioxidants during the emulsification step. Considering the effect of the amount of grape powder on gluten strengthening, the partial replacement of wheat flour in the butter biscuits with grape seed cake powder (BB2) and with grape skin powder (BB3) was 20% and 16%, respectively (Figure 3.15).

The addition of grape seed cake powder gave the cookies a chocolate color, and the addition of the grape skin powder added chocolate with a purple tint. According to sensorial assessment, the biscuits with these amounts of grape seed cake powder or grape skin powder were identical to the control biscuit except for the color (Table 3.4).

According to the physicochemical properties, biscuit samples with the addition of such an amount of grape seed powder or grape skin powder were close to a test sample (Table 3.5).

TABLE 3.5 PHYSICOCHEMICAL PROPERTIES OF TRADITIONAL AND ENRICHED WITH GRAPE POWDER BUTTER BISCUITS

		Butter Biscuits With	
Property	Control (Without Grape Powder)	Grape Seeds Cake Powder	Grape Skin Powder
Moisture, %	5.5	5.6	5.6
Specific volume, cm^3/g	1.78	1.72	1.85
Wetting ability, %	150	146	160

3.5.3 Impact of Grape Seed and Skin Powders on Oil Oxidation in Butter Biscuits

When storing biscuits, physicochemical, biochemical and microbiological processes occur that affect their quality and consumer value. Changes in the lipid complex connected mainly with the processes of oxidation and hydrolysis of fat are of the greatest importance. Peroxide and acid values were used to monitor these processes. The butter biscuits were stored in plastic bags at a temperature of 18±3°C and relative air humidity not higher than 75% for 60 days. The initial acid values of BB2 and BB3 were slightly higher than those of control (BB1) (Figure 3.16a). This fact characterizes an increase in the content of organic acids upon grape seed cake powder and grape skin powder incorporation into butter biscuits 2 and 3. During shelf lifetime of cookies, the accumulation of free fatty acids in control sample proceeds faster than in the test samples, and after 10 days the acid value of the test sample exceeds that in BB2 samples with grape seed cake powder, and after about 20 days—in BB3 samples with grape skin powder.

The degree of accumulation of primary oxidation products (peroxides and hydroperoxides), which was assessed by peroxide values, increased in BB1 test sample approximately 2.5 times faster than in the samples with BB2 and BB3 grape powder (Figure 3.16b). Thus, the obtained results prove a significant antioxidant potential of grape powder polyphenols in the processes of inhibition of fat oxidation and accumulation of primary oxidation products that affect sensory characteristics.

3.5.4 Quality Indicators of Butter Biscuits

The nature of the interaction of water with the components of the foods and the surrounding atmosphere affects the physical or textural characteristics of foods as well as their stability and shelf life (Ergun et al., 2010). In general, for

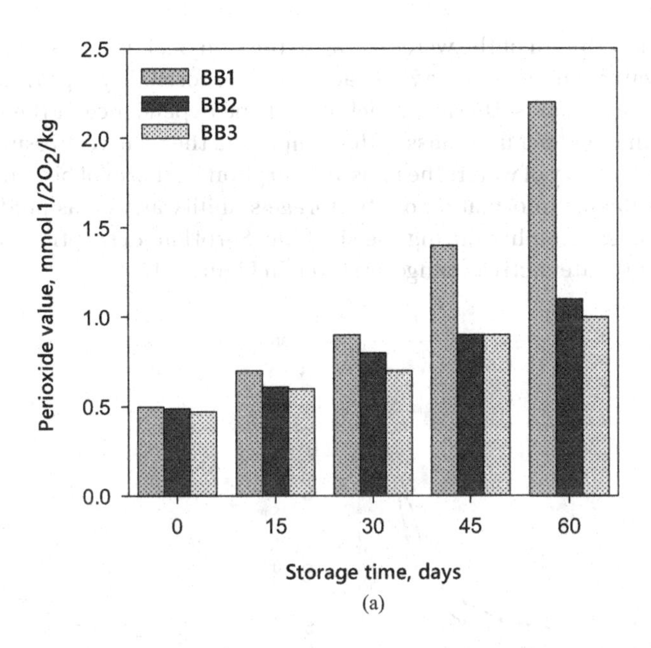

(a)

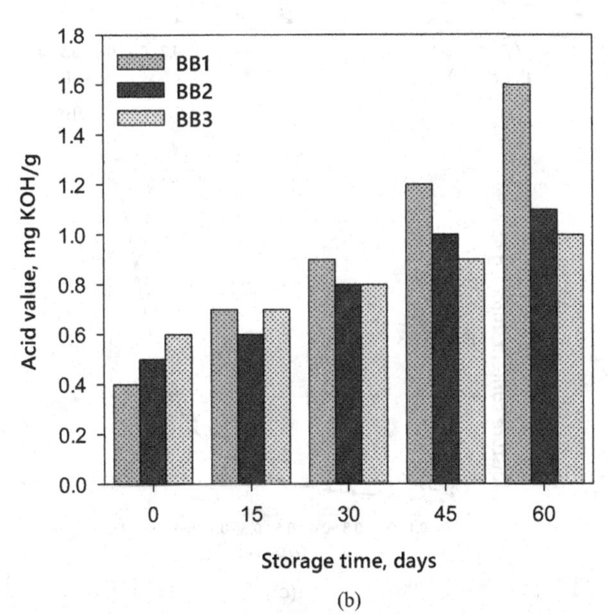

(b)

Figure 3.16 Changes of fat peroxide (a) and acid values (b) of butter biscuits: control only with wheat flour (BB1); with grape seed cake powder (BB2); with grape skin powder (BB3) during storage time.

a correct description of the water status in the sample, it is necessary to know both the water content and the water activity. This possibility is provided by the analysis of sorption isotherms, which reflect the dependence of the moisture content n in mmol per unit mass of the sample and the relative pressure p/p_o, in fact, on the activity of water. The moisture sorption isotherm of biscuit samples could be valuable information on its storage stability as well as prediction of microbiological stability during the shelf life. Sorption–desorption isotherms over 0.05–1.0 water activity range are given in Figure 3.17.

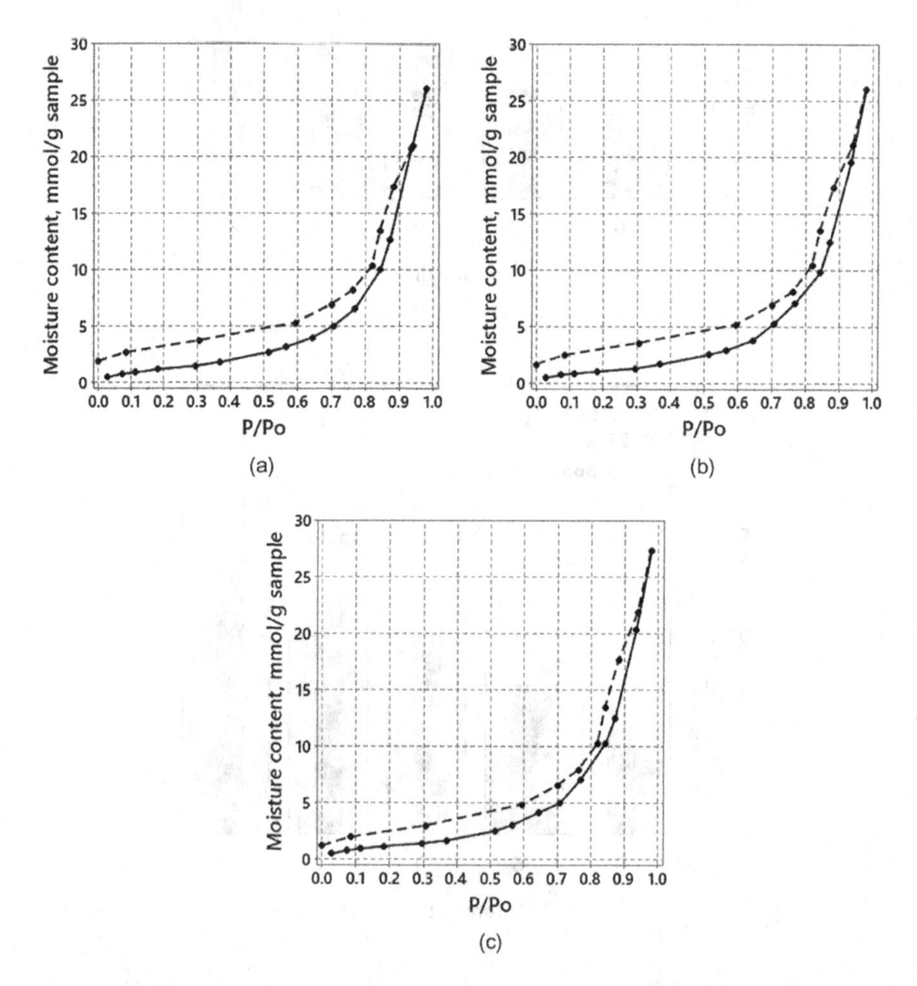

Figure 3.17 Moisture sorption isotherms (adsorption—solid line, desorption—dashed line) and pores distribution by radius for butter biscuits: control (a); BB2 with grape seed cake powder (b); BB3 with grape skin powder (c).

The moisture sorption isotherms of various samples presented a sigmoid shape with hysteresis. According to the classification of physisorption isotherms of IUPAC Technical Report (Thommes et al., 2015), these curves refer to Type II isotherms. Such isotherms correspond to materials that are nonporous or macroporous and unrestricted monolayer–multilayer adsorption up to high p/p_o, where p is the vapor pressure of the sample, and p_o is the vapor pressure of pure water at the same temperature and external pressure (Barbosa-Cánovas et al., 2020). The curves are characterized by the presence of a less distinct point B (as a point at the beginning of the middle almost linear section and usually corresponds to the completion of monolayer). This fact indicates a significant amount of overlap of monolayer coverage, and the onset of multilayer sorption and the thickness of the adsorbed multilayer generally appear to increase value p/p_o (Thommes et al., 2015). All the curves are characterized by the presence of hysteresis loops, complying with Type H3 classification according to IUPAC hysteresis loops. The sorption curves of the samples match in shape which is evidence of similar adsorption structures.

The Brunauer, Emmett and Teller (BET) isotherm equation (Brunauer et al., 1938) has been applied successfully over the past decade to a wide variety of foods. One of the main applications of the BET isotherm is the estimation of specific surface area a_s and average pore size (Majd et al., 2022). The sorption curves are used to estimate the coefficients of the BET sorption model. The regression analysis results for sorption isotherm are presented in Table 3.6 along with statistical parameters and estimated model coefficients.

According to the BET model, the monolayer moisture capacity of all samples is almost the same and is within 75–88 m^2/g (Table 3.6). Average pore sizes also vary slightly in the range of 21–25 nm. These calculations confirm the conclusion that the incorporation of grape powder does not change the structure of wine pomace pores and a similar adsorption behavior of samples is observed

TABLE 3.6 CALCULATED BET MODEL PARAMETERS FOR BUTTER BISCUITS

Butter Biscuits	Monolayer Moisture Capacity, mmol/g	Adjusted Constant C	Specific Surface Area, m^2/g	R^2
Control (without grape powder)	1.17±0.03	18.8±0.5	88±2	0.9969
With grape seeds cake powder	0.99±0.01	26.2±0.3	75±1	0.9996
With grape skin powder	1.08±0.02	22.0±0.4	81±2	0.9987

regardless of the ingredients. Hence, it can be assumed that sorption–desorption processes occurring during the storage of biscuits will proceed almost identically.

The incorporation of grape powder made it possible to significantly increase the nutritional value of biscuits, as evidenced by the increase in the integral score of such substances as dietary fiber, polyphenolic compounds, PP vitamin, magnesium and iron. Thus, the percentage of compliance of the dietary fiber content with a balanced nutrition recipe in cookies with the addition of grape seed powder increases from 0.3% to 31.2%, PP vitamin—from 5.25% to 10.88%, magnesium—from 2.2% to 9.03% and iron—from 5.06% to 22.71%. When adding grape skin powder, satisfaction with the need for dietary fiber increases from 0.3% to 24.5%, PP vitamin—from 5.25% to 12.75%, magnesium—from 2.2% to 7.43% and iron—from 5.06% to 21.18%. All types of butter biscuits are significantly enriched with polyphenolic compounds.

3.6 CONCLUSION

This chapter presents the use of grape seeds and skins in confectionery technologies. All by-products of winemaking have great potential for innovation in value-added food technology due to the presence of numerous biologically active compounds in their composition, including fiber, minerals and phenolic compounds. Thanks to the latter, the food industry receives natural by-products that can inhibit various microbiological and chemical reactions as food preservatives and antioxidants without compromising the stability of the final product. The developed successful new applications of winemaking by-products in food technology can be seen as a model of their use for fortification of other food products. This is a good reason for further research on this topic.

REFERENCES

Aksoylu, Z., Çağindi, Ö., & Köse, E. (2015). Effects of blueberry, grape seed powder and poppy seed incorporation on physicochemical and sensory properties of biscuit. *Journal of Food Quality*, *38*(3), 164–174. https://doi.org/10.1111/jfq.12133

Altınok, E., Kurultay, S., Konar, N., Toker, O. S., Kopuk, B., Gunes, R., & Palabiyik, I. (2022). Utilising grape juice processing by-products as bulking and colouring agent in white chocolate. *International Journal of Food Science & Technology*, *57*(7), 4119–4128. https://doi.org/10.1111/ijfs.15728

Amico, V., Chillemi, R., Mangiafico, S., Spatafora, C., & Tringali, C. (2008). Polyphenol-enriched fractions from Sicilian grape pomace: HPLC–DAD analysis and antioxidant activity. *Bioresource Technology*, *99*(13), 5960–5966. https://doi.org/10.1016/j.biortech.2007.10.037

Antonic, B., Dordevic, D., Jancikova, S., Holeckova, D., Tremlova, B., & Kulawik, P. (2021). Effect of grape seed flour on the antioxidant profile, textural and sensory properties ofwaffles. *Processes*, *9*(1), 1–9. https://doi.org/10.3390/pr9010131

Antonić, B., Jančíková, S., Dordević, D., & Tremlová, B. (2020). Grape pomace valorization: A systematic review and meta-analysis. *Foods*, *9*(11), 1627. https://doi.org/10.3390/foods9111627

Barbosa-Cánovas, G. V., Fontana, A. J., Schmidt, S. J., & Labuza, T. P. (2020). Water activity in foods. In G. V. Barbosa-Cánovas, A. J. Fontana, S. J. Schmidt, & T. P. Labuza (Eds.), *Water Activity in Foods*. Wiley. https://doi.org/10.1002/9781118765982

Barišić, V., Jozinović, A., Flanjak, I., Šubarić, D., Babić, J., Miličević, B., Jokić, S., Grgić, I., & Ačkar, Đ. (2021). Effect of addition of fibres and polyphenols on properties of chocolate – A review. *Food Reviews International*, *37*(3), 225–243. https://doi.org/10.1080/87559129.2019.1701008

Baroi, A. M., Popitiu, M., Fierascu, I., Sărdărescu, I. D., & Fierascu, R. C. (2022). Grapevine wastes: A rich source of antioxidants and other biologically active compounds. *Antioxidants 11*(2), 393. https://doi.org/10.3390/antiox11020393

Beckett, S. T., Fowler, M., & Ziegler, G. R. (2017). Traditional chocolate making. In S. T. Beckett, M. S. Fowler & G. R. Ziegler (Eds.), *Beckett's Industrial Chocolate Manufacture and Use* (pp. 1–8). https://doi.org/10.1002/9781118923597

Bender, A. B. B., Speroni, C. S., Salvador, P. R., Loureiro, B. B., Lovatto, N. M., Goulart, F. R., Lovatto, M. T., Miranda, M. Z., Silva, L. P., & Penna, N. G. (2017). Grape pomace skins and the effects of its inclusion in the technological properties of muffins. *Journal of Culinary Science and Technology*, *15*(2), 143–157. https://doi.org/10.1080/15428052.2016.1225535

Benito-González, I., Martínez-Sanz, M., Fabra, M. J., & López-Rubio, A. (2019). Health effect of dietary fibers. In C. M. Galanakis (Ed.), *Dietary Fiber: Properties, Recovery, and Applications* (pp. 125–163). Academic Press. https://doi.org/10.1016/B978-0-12-816495-2.00005-8

Beres, C., Costa, G. N. S., Cabezudo, I., da Silva-James, N. K., Teles, A. S. C., Cruz, A. P. G., Mellinger-Silva, C., Tonon, R. V., Cabral, L. M. C., & Freitas, S. P. (2017). Towards integral utilization of grape pomace from winemaking process: A review. *Waste Management*, *68*, 581–594. https://doi.org/10.1016/j.wasman.2017.07.017

Beres, C., Freitas, S. P., de O Godoy, R. L., de Oliveira, D. C. R., Deliza, R., Iacomini, M., Mellinger-Silva, C., & Cabral, L. M. C. (2019). Antioxidant dietary fibre from grape pomace flour or extract: Does it make any difference on the nutritional and functional value? *Journal of Functional Foods*, *56*, 276–285. https://doi.org/10.1016/j.jff.2019.03.014

Beres, C., Simas-Tosin, F. F., Cabezudo, I., Freitas, S. P., Iacomini, M., Mellinger-Silva, C., & Cabral, L. M. C. (2016). Antioxidant dietary fibre recovery from Brazilian Pinot noir grape pomace. *Food Chemistry*, *201*, 145–152. https://doi.org/10.1016/j.foodchem.2016.01.039

Bharathiraja, B., Iyyappan, J., Jayamuthunagai, J., Kumar, R. P., Sirohi, R., Gnansounou, E., & Pandey, A. (2020). Critical review on bioconversion of winery wastes into value-added products. *Industrial Crops and Products*, *158*(January), 112954. https://doi.org/10.1016/j.indcrop.2020.112954

Boban, M., Boban, N., Tonkić, M., Grga, M., Milat, A. M., & Mudnić, I. (2021). Antimicrobial activity of wine in relation to bacterial resistance to medicinal antibiotics. *OENO One*, *55*(1), 45–48. https://doi.org/10.20870/oeno-one.2021.55.1.2349

Brenes, A., Viveros, A., Goñi, I., Centeno, C., Sáyago-Ayerdy, S. G., Arija, I., & Saura-Calixto, F. (2008). Effect of grape pomace concentrate and vitamin E on digestibility of polyphenols and antioxidant activity in chickens. *Poultry Science, 87*(2), 307–316. https://doi.org/10.3382/ps.2007-00297

Brunauer, S., Emmett, P. H., & Teller, E. (1938). Adsorption of gases in multimolecular layers. *Journal of the American Chemical Society, 60*(2), 309–319. https://doi.org/10.1021/ja01269a023

Bursa, K., Kilicli, M., Toker, O. S., Palabiyik, I., Gulcu, M., Yaman, M., Kian-Pour, N., & Konar, N. (2022). Formulating and studying compound chocolate with adding dried grape pomace as a bulking agent. *Journal of Food Science and Technology, 59*(5), 1704–1714. https://doi.org/10.1007/s13197-021-05180-8

Chang, Y., Shi, X., He, F., Wu, T., Jiang, L., Normakhamatov, N., Sharipov, A., Wang, T., Wen, M., & Aisa, H. A. (2022). Valorization of food processing waste to produce valuable polyphenolics. *Journal of Agricultural and Food Chemistry, 70*(29), 8855–8870. https://doi.org/10.1021/acs.jafc.2c02655

Chen, M., & Yu, S. (2017). Lipophilized grape seed proanthocyanidin derivatives as novel antioxidants. *Journal of Agricultural and Food Chemistry, 65*(8), 1598–1605. https://doi.org/10.1021/acs.jafc.6b05609

Corrales, M., Fernandez, A., Vizoso Pinto, M. G., Butz, P., Franz, C. M. A. P., Schuele, E., & Tauscher, B. (2010). Characterization of phenolic content, in vitro biological activity, and pesticide loads of extracts from white grape skins from organic and conventional cultivars. *Food and Chemical Toxicology, 48*(12), 3471–3476. https://doi.org/10.1016/j.fct.2010.09.025

Costa, G. N., Tonon, R. V., Mellinger-Silva, C., Galdeano, M. C., Iacomini, M., Santiago, M. C., Almeida, E. L., & Freitas, S. P. (2019). Grape seed pomace as a valuable source of antioxidant fibers. *Journal of the Science of Food and Agriculture, 99*(10), 4593–4601. https://doi.org/10.1002/jsfa.9698

Dávila, I., Robles, E., Egüés, I., Labidi, J., & Gullón, P. (2017). The biorefinery concept for the industrial valorization of grape processing by-Products. In C. M. Galanakis (Ed.), *Handbook of Grape Processing By-Products: Sustainables Solutions* (pp. 29–53). Elsevier. https://doi.org/10.1016/B978-0-12-809870-7.00002-8

de Andrade, R. B., Machado, B. A. S., Barreto, G. D. A., Nascimento, R. Q., Corrêa, L. C., Leal, I. L., Tavares, P. P. L. G., Ferreira, E. D. S., & Umsza-Guez, M. A. (2021). Syrah grape skin residues has potential as source of antioxidant and anti-microbial bioactive compounds. *Biology, 10*(12). https://doi.org/10.3390/biology10121262

Deng, Q., Penner, M. H., & Zhao, Y. (2011). Chemical composition of dietary fiber and polyphenols of five different varieties of wine grape pomace skins. *Food Research International, 44*(9), 2712–2720. https://doi.org/10.1016/j.foodres.2011.05.026

Dhingra, D., Michael, M., Rajput, H., & Patil, R. T. (2012). Dietary fibre in foods: A review. *Journal of Food Science and Technology, 49*(3), 255–266. https://doi.org/10.1007/s13197-011-0365-5

Di Stefano, V., Buzzanca, C., Melilli, M. G., Indelicato, S., Mauro, M., Vazzana, M., Arizza, V., Lucarini, M., Durazzo, A., & Bongiorno, D. (2022). Polyphenol characterization and antioxidant activity of grape seeds and skins from Sicily: A preliminary study. *Sustainability, 14*(11), 6702. https://doi.org/10.3390/su14116702

Ergun, R., Lietha, R., & Hartel, R. W. (2010). Moisture and shelf life in sugar confections. *Critical Reviews in Food Science and Nutrition, 50*(2), 162–192. https://doi.org/10.1080/10408390802248833

Evlash, V., Pogozhikh, M., Aksonova, O., & Gubsky, S. (2022). Heme iron–containing dietary supplements and their application in fortified foods. In O. Paredes-López, O. Shevchenko, V. Stabnikov & V. Ivanov (Eds.), *Bioenhancement and Fortification of Foods for a Healthy Diet* (pp. 237–268). https://doi.org/10.1201/9781003225287-16

FAOSTAT. (2020). Value of Agricultural Production. https://www.fao.org/faostat/en/#data/QV/visualize

García-Lomillo, J., & González-SanJosé, M. L. (2017). Applications of wine pomace in the food industry: Approaches and functions. *Comprehensive Reviews in Food Science and Food Safety, 16*(1), 3–22. https://doi.org/10.1111/1541-4337.12238

Garrido, J., & Borges, F. (2013). Wine and grape polyphenols – A chemical perspective. *Food Research International, 54*(2), 1844–1858. https://doi.org/10.1016/j.foodres.2013.08.002

Georgiev, V., Ananga, A., & Tsolova, V. (2016). Dietary supplements/nutraceuticals made from grapes and wines. In M. V. Moreno-Arribas & B. B. Suáldea (Eds.), *Wine Safety, Consumer Preference, and Human Health* (pp. 201–227). https://doi.org/10.1007/978-3-319-24514-0_10

Gil-Sánchez, I., Cueva, C., Sanz-Buenhombre, M., Guadarrama, A., Moreno-Arribas, M. V., & Bartolomé, B. (2018). Dynamic gastrointestinal digestion of grape pomace extracts: Bioaccessible phenolic metabolites and impact on human gut microbiota. *Journal of Food Composition and Analysis, 68*, 41–52. https://doi.org/10.1016/j.jfca.2017.05.005

Goralchuk, A., Gubsky, S., Omel'chenko, S., Riabets, O., Grinchenko, O., Fedak, N., Kotlyar, O., Cheremska, T., & Skrynnik, V. (2020). Impact of added food ingredients on foaming and texture of the whipped toppings: A chemometric analysis. *European Food Research and Technology, 246*(10), 1955–1970. https://doi.org/10.1007/s00217-020-03547-3

Gorodyska, O., Grevtseva, N., Samokhvalova, O., Gubsky, S., Gavrish, T., Denisenko, S., & Grigorenko, A. (2018). Influence of grape seeds powder on preservation of fats in confectionary glaze. *Eastern-European Journal of Enterprise Technologies, 6*(11 (96)), 36–43. https://doi.org/10.15587/1729-4061.2018.147760

Grevtseva, N., Gorodyska, O., Samokhvalova, O., & Bushtruk, I. (2018). Technology of confectionary glaze with the use of grape powders as an alternative to cocoa powder. *Progressive Technique and Technologies of Food Production Enterprises, Catering Business and Trade, 28*(2), 223–237. https://doi.org/10.5281/zenodo.2365648

Gubsky, S., Artamonova, M., Shmatchenko, N., Piliugina, I., & Aksenova, E. (2016). Determination of total antioxidant capacity in marmalade and marshmallow. *Eastern-European Journal of Enterprise Technologies, 4*(11(82)), 43. https://doi.org/10.15587/1729-4061.2016.73546

Gubsky, S., Labazov, M., Samokhvalova, O., Grevtseva, N., & Gorodyska, O. (2018). Optimization of extraction parameters of phenolic antioxidants from defatted grape seed flour by response surface methodology. *Ukrainian Food Journal, 7*(4), 627–639. https://doi.org/10.24263/2304-974X-2018-7-4-8

Gupta, M., Dey, S., Marbaniang, D., Pal, P., Ray, S., & Mazumder, B. (2020). Grape seed extract: Having a potential health benefits. *Journal of Food Science and Technology, 57*(4), 1205–1215. Springer. https://doi.org/10.1007/s13197-019-04113-w

Hartel, R. W., & Firoozmand, H. (2019). Emulsifiers in confectionery. In G. Hasenhuettl & R. Hartel (Eds.), *Food Emulsifiers and Their Applications* (pp. 323–346). Springer International Publishing. https://doi.org/10.1007/978-3-030-29187-7_11

Hartel, R. W., von Elbe, J. H., & Hofberger, R. (2018). Compound coatings. *Confectionery Science and Technology*, 485–499. https://doi.org/10.1007/978-3-319-61742-8_16

Iuga, M., & Mironeasa, S. (2019). Grape seeds effect on refined wheat flour dough rheology: Optimal amount and particle size. *Ukrainian Food Journal, 8*(4), 799–814. https://doi.org/10.24263/2304-974X-2019-8-4-11

Iuga, M., & Mironeasa, S. (2020). Potential of grape byproducts as functional ingredients in baked goods and pasta. *Comprehensive Reviews in Food Science and Food Safety, 19*(5), 2473–2505. https://doi.org/10.1111/1541-4337.12597

Kalli, E., Lappa, I., Bouchagier, P., Tarantilis, P. A., & Skotti, E. (2018). Novel application and industrial exploitation of winery by-products. *Bioresources and Bioprocessing, 5*(1), 46. https://doi.org/10.1186/s40643-018-0232-6

Kammerer, D., Claus, A., Carle, R., & Schieber, A. (2004). Polyphenol screening of pomace from red and white grape varieties (*Vitis vinifera* L.) by HPLC-DAD-MS/MS. *Journal of Agricultural and Food Chemistry, 52*(14), 4360–4367. https://doi.org/10.1021/jf049613b

Koca, I., Tekguler, B., Yilmaz, V. A., Hasbay, I., & Koca, A. F. (2018). The use of grape, pomegranate and rosehip seed flours in Turkish noodle (erişte) production. *Journal of Food Processing and Preservation, 42*(1), 1–12. https://doi.org/10.1111/jfpp.13343

Kuchtová, V., Kohajdová, Z., Karovičová, J., & Lauková, M. (2018). Physical, textural and sensory properties of cookies incorporated with grape skin and seed preparations. *Polish Journal of Food and Nutrition Sciences, 68*(4), 309–317. https://doi.org/10.2478/pjfns-2018-0004

Ky, I., Lorrain, B., Kolbas, N., Crozier, A., & Teissedre, P.-L. (2014). Wine by-products: Phenolic characterization and antioxidant activity evaluation of grapes and grape pomaces from six different french grape varieties. *Molecules, 19*(1), 482–506. https://doi.org/10.3390/molecules19010482

Lipp, M., & Anklam, E. (1998). Review of cocoa butter and alternative fats for use in chocolate—Part A. Compositional data. *Food Chemistry, 62*(1), 73–97. https://doi.org/10.1016/S0308-8146(97)00160-X

Lisjuk, G. M., Verezhko, N. V., & Chujko, A. N. (2011). *New Trends in the Use of Secondary Products of Grape Processing in the Production of Flour Products: Monograph.* Kharkiv State University of Food Technology and Trade.

Liu, R. H., & Felice, D. (2007). Antioxidants and whole food phytochemicals for cancer prevention. In F. Shahidi & C. T. Ho (Eds.), *Antioxidant Measurement and Applications* (pp. 15–34). https://doi.org/10.1021/bk-2007-0956.ch003

Liu, R., & Mabury, S. A. (2020). Synthetic phenolic antioxidants: A review of environmental occurrence, fate, human exposure, and toxicity. *Environmental Science & Technology, 54*(19), 11706–11719. https://doi.org/10.1021/acs.est.0c05077

Lo, S., Pilkington, L. I., Barker, D., & Fedrizzi, B. (2022). Attempts to create products with increased health-promoting potential starting with pinot noir pomace: Investigations on the process and its methods. *Foods, 11*(14), 1999. https://doi.org/10.3390/foods11141999

Lorenzo, J. M., Sineiro, J., Amado, I. R., & Franco, D. (2014). Influence of natural extracts on the shelf life of modified atmosphere-packaged pork patties. *Meat Science, 96*(1), 526–534. https://doi.org/10.1016/j.meatsci.2013.08.007

Maestre, R., Micol, V., Funes, L., & Medina, I. (2010). Incorporation and interaction of grape seed extract in membranes and relation with efficacy in muscle foods. *Journal of Agricultural and Food Chemistry, 58*(14), 8365–8374. https://doi.org/10.1021/jf100327w

Majd, M., Kordzadeh-Kermani, V., Ghalandari, V., Askari, A., & Sillanpää, M. (2022). Adsorption isotherm models: A comprehensive and systematic review (2010–2020). *Science of the Total Environment, 812*, 151334. https://doi.org/10.1016/j.scitotenv.2021.151334

Mandade, P., & Gnansounou, E. (2022). Potential value-added products from wineries residues. In G. S. Murthy, E. Gnansounou, S. K. Khanal & A. Pandey (Eds.), *Biomass, Biofuels, Biochemicals* (pp. 371–396). https://doi.org/10.1016/B978-0-12-819242-9.00008-7

Mazur, L., Gubsky, S., Dorohovych, A., & Labazov, M. (2018). Antioxidant properties of candy caramel with plant extracts. *Ukrainian Food Journal, 7*(1), 7–21. https://doi.org/10.24263/2304-974X-2018-7-1-3

Meral, R., & Doğan, İ. S. (2013). Grape seed as a functional food ingredient in bread-making. *International Journal of Food Sciences and Nutrition, 64*(3), 372–379. https://doi.org/10.3109/09637486.2012.738650

Mironeasa, S., Codina, G. G., & Mironeasa, C. (2012). The effects of wheat flour substitution with grape seed flour on the rheological parameters of the dough assessed by Mixolab. *Journal of Texture Studies, 43*(1), 40–48. https://doi.org/10.1111/j.1745-4603.2011.00315.x

Mironeasa, S., Codinə, G. G., & Mironeasa, C. (2016). Optimization of wheat-grape seed composite flour to improve alpha-amylase activity and dough rheological behavior. *International Journal of Food Properties, 19*(4), 859–872. https://doi.org/10.1080/10942912.2015.1045516

Mironeasa, S., Iuga, M., Zaharia, D., & Mironeasa, C. (2019). Optimization of white wheat flour dough rheological properties with different levels of grape peels flour addition. *Bulletin of University of Agricultural Sciences and Veterinary Medicine Cluj-Napoca. Food Science and Technology, 76*(1), 27. https://doi.org/10.15835/buasvmcn-fst:2018.0017

Montealegre, R. R., Romero Peces, R., Chacón Vozmediano, J. L., Martínez Gascueña, J., & García Romero, E. (2006). Phenolic compounds in skins and seeds of ten grape Vitis vinifera varieties grown in a warm climate. *Journal of Food Composition and Analysis, 19*(6–7), 687–693. https://doi.org/10.1016/j.jfca.2005.05.003

Naik, B., & Kumar, V. (2014). Cocoa butter and its alternatives: A reveiw. *Journal of Bioresource Engineering and Technology, 1*, 7–17.

Nassiri-Asl, M., & Hosseinzadeh, H. (2016). Review of the pharmacological effects of vitis vinifera (grape) and its bioactive constituents: An update. *Phytotherapy Research, 30*(9), 1392–1403. https://doi.org/10.1002/ptr.5644

Ortega-Heras, M., Gómez, I., de Pablos-Alcalde, S., & González-Sanjosé, M. L. (2019). Application of the just-about-right scales in the development of new healthy whole-wheat muffins by the addition of a product obtained from white and red grape pomace. *Foods, 8*(9), 419. https://doi.org/10.3390/foods8090419

Padilla-González, G. F., Grosskopf, E., Sadgrove, N. J., & Simmonds, M. S. J. (2022). Chemical diversity of flavan-3-ols in grape seeds: Modulating factors and quality requirements. *Plants, 11*(6). https://doi.org/10.3390/plants11060809

Pečivová, P. B., Kráčmar, S., Kubáň, V., Mlček, J., Jurikova, T., & Sochor, J. (2014). Effect of addition of grape seed flour on chemical, textural and sensory properties of bread dough. *Mitteilungen Klosterneuburg, 64*(3), 114–119.

Peralbo-Molina, Á., & Luque de Castro, M. D. (2013). Potential of residues from the Mediterranean agriculture and agrifood industry. *Trends in Food Science & Technology, 32*(1), 16–24. https://doi.org/10.1016/j.tifs.2013.03.007

Pérez-Jiménez, J., Arranz, S., & Saura-Calixto, F. (2009). Proanthocyanidin content in foods is largely underestimated in the literature data: An approach to quantification of the missing proanthocyanidins. *Food Research International, 42*(10), 1381–1388. https://doi.org/10.1016/j.foodres.2009.07.002

Pérez-Ramírez, I. F., Reynoso-Camacho, R., Saura-Calixto, F., & Pérez-Jiménez, J. (2018). Comprehensive characterization of extractable and nonextractable phenolic compounds by high-performance liquid chromatography–electrospray ionization–quadrupole time-of-flight of a grape/pomegranate pomace dietary supplement. *Journal of Agricultural and Food Chemistry, 66*(3), 661–673. https://doi.org/10.1021/acs.jafc.7b05901

Pinsirodom, P., & Parkin, K. L. (2005). Lipolytic enzymes. In R. E. Wrolstad, T. E. Acree, E. A. Decker, M. H. Penner, D. S. Reid, S. J. Schwartz, C. F. Shoemaker & D. Smith (Eds.), *Handbook of Food Analytical Chemistry* (pp. 369–383). https://doi.org/10.1002/0471709085.ch10

Porter, N. A. (1986). Mechanisms for the autoxidation of polyunsaturated lipids. *Accounts of Chemical Research, 19*(9), 262–268. https://doi.org/10.1021/ar00129a001

Rimbach, G., Egert, S., & de Pascual-Teresa, S. (2011). Chocolate: (Un)healthy source of polyphenols? *Genes & Nutrition, 6*(1), 1–3. https://doi.org/10.1007/s12263-010-0185-7

Sagar, N. A., Pareek, S., Sharma, S., Yahia, E. M., & Lobo, M. G. (2018). Fruit and vegetable waste: Bioactive compounds, their extraction, and possible utilization. *Comprehensive Reviews in Food Science and Food Safety, 17*(3), 512–531. https://doi.org/10.1111/1541-4337.12330

Sagdic, O., Ozturk, I., Yilmaz, M. T., & Yetim, H. (2011). Effect of grape pomace extracts obtained from different grape varieties on microbial quality of beef patty. *Journal of Food Science, 76*(7), M515–M521. https://doi.org/10.1111/j.1750-3841.2011.02323.x

Samohvalova, O., Grevtseva, N., Brykova, T., & Grigorenko, A. (2016). The effect of grape seed powder on the quality of butter biscuits. *Eastern-European Journal of Enterprise Technologies, 3*(11(81)), 61. https://doi.org/10.15587/1729-4061.2016.69838

Sánchez-Alonso, I., Jiménez-Escrig, A., Saura-Calixto, F., & Borderías, A. J. (2007). Effect of grape antioxidant dietary fibre on the prevention of lipid oxidation in minced fish: Evaluation by different methodologies. *Food Chemistry, 101*(1), 372–378. https://doi.org/10.1016/j.foodchem.2005.12.058

Santoro, H. C., Skroza, D., Dugandžić, A., Boban, M., & Šimat, V. (2020). Antimicrobial activity of selected red and white wines against Escherichia coli: In vitro inhibition using fish as food matrix. *Foods, 9*(7). https://doi.org/10.3390/foods9070936

Saura-Calixto, F. (2012). Concept and health-related properties of nonextractable polyphenols: The missing dietary polyphenols. *Journal of Agricultural and Food Chemistry, 60*(45), 11195–11200. https://doi.org/10.1021/jf303758j

Sciortino, M., Avellone, G., Scurria, A., Bertoli, L., Carnaroglio, D., Bongiorno, D., Pagliaro, M., & Ciriminna, R. (2021). Green and quick extraction of stable biophenol-rich red extracts from grape rocessing waste. *ACS Food Science & Technology, 1*(5), 937–942. https://doi.org/10.1021/acsfoodscitech.1c00123

Serina, J. J. C., & Castilho, P. C. M. F. (2022). Using polyphenols as a relevant therapy to diabetes and its complications, a review. *Critical Reviews in Food Science and Nutrition, 62*(30), 8355–8387. https://doi.org/10.1080/10408398.2021.1927977

Shaker, E. S. (2006). Antioxidative effect of extracts from red grape seed and peel on lipid oxidation in oils of sunflower. *LWT - Food Science and Technology, 39*(8), 883–892. https://doi.org/10.1016/j.lwt.2005.06.004

Shi, J., Yu, J., Pohorly, J. E., & Kakuda, Y. (2003). Polyphenolics in grape seeds—Biochemistry and functionality. *Journal of Medicinal Food, 6*(4), 291–299. https://doi.org/10.1089/109662003772519831

Soto, R. M. U., Brown, K., & Ross, C. F. (2012). Antioxidant activity and consumer acceptance of grape seed flour-containing food products. *International Journal of Food Science and Technology, 47*(3), 592–602. https://doi.org/10.1111/j.1365-2621.2011.02882.x

Spinei, M., & Oroian, M. (2021). The potential of grape pomace varieties as a dietary source of pectic substances. *Foods, 10*(4), 867. https://doi.org/10.3390/foods10040867

Stabnikova, O., Marinin, A., & Stabnikov, V. (2021). Main trends in application of novel natural additives for food production. *Ukrainian Food Journal, 10*(3), 524–551. https://doi.org/10.24263/2304-974x-2021-10-3-8

Taghvaei, M., & Jafari, S. M. (2015). Application and stability of natural antioxidants in edible oils in order to substitute synthetic additives. *Journal of Food Science and Technology, 52*(3), 1272–1282. https://doi.org/10.1007/s13197-013-1080-1

Talbot, G. (2009a). Compound coatings. In G. Talbot (Ed.), *Science and Technology of Enrobed and Filled Chocolate, Confectionery and Bakery Products* (pp. 80–100). https://doi.org/10.1533/9781845696436.1.81

Talbot, G. (2009b). Introduction. In G. Talbot (Ed.), *Science and Technology of Enrobed and Filled Chocolate, Confectionery and Bakery Products* (pp. 1–7). https://doi.org/10.1533/9781845696436.1.81

Teixeira, A., Baenas, N., Dominguez-Perles, R., Barros, A., Rosa, E., Moreno, D., & Garcia-Viguera, C. (2014). Natural bioactive compounds from winery by-products as health promoters: A review. *International Journal of Molecular Sciences, 15*(9), 15638–15678. https://doi.org/10.3390/ijms150915638

Theagarajan, R., Malur Narayanaswamy, L., Dutta, S., Moses, J. A., & Chinnaswamy, A. (2019). Valorisation of grape pomace (cv. Muscat) for development of functional cookies. *International Journal of Food Science and Technology, 54*(4), 1299–1305. https://doi.org/10.1111/ijfs.14119

Thommes, M., Kaneko, K., Neimark, A. V., Olivier, J. P., Rodriguez-Reinoso, F., Rouquerol, J., & Sing, K. S. W. (2015). Physisorption of gases, with special reference to the evaluation of surface area and pore size distribution (IUPAC Technical Report). *Pure and Applied Chemistry, 87*(9–10), 1051–1069. https://doi.org/10.1515/pac-2014-1117

Tomaz, I., Štambuk, P., Andabaka, Ž., & Stupić, D. (2017). The polyphenolic profile of grapes. In S. Thomas (Ed.), *Grapes Polyphenolic Composition, Antioxidant Characteristics and Health Benefits* (pp. 1–70). Nova Science Publishers.

Urbanski, J. (2009). Compound coatings: Handling and processing. *Manufacturing Confectioner, 79*, 75–84. https://www.gomc.com/firstpage/200912075.pdf

Vorobiev, E., & Lebovka, N. (2017). Application of pulsed electric energy for grape waste biorefinery. In D. Miklavčič (Ed.), *Handbook of Electroporation* (pp. 2781–2798). https://doi.org/10.1007/978-3-319-32886-7_152

Walker, R., Tseng, A., Cavender, G., Ross, A., & Zhao, Y. (2014). Physicochemical, nutritional, and sensory qualities of wine grape pomace fortified baked goods. *Journal of Food Science, 79*(9), S1811–S1822. https://doi.org/10.1111/1750-3841.12554

Chapter 4

Milk Whey Enriched with Magnesium and Manganese for Food Production

Oksana Kochubei-Lytvynenko, Olena Bilyk,
Viktor Stabnikov, and Anastsia Dubivko
National University of Food Technologies

CONTENTS

4.1 INTRODUCTION

Macro- and microelements enter the human body with drinking water, food, as well as mineral supplements. However, there is still a problem of deficiency of a number of minerals in the diet of the average person (Webster-Gandy et al., 2020). Among these biogenic elements, magnesium (Mg) and manganese (Mn) attract special attention (Bertinato et al., 2015). Magnesium ranks fourth in content among other cations in the human body, has extremely

DOI: 10.1201/9781003329671-4

diverse functions being a cofactor of more than 300 enzymes, and participates in energy metabolism and synthesis of carbohydrates, lipids, nucleic acids and proteins. Lack of magnesium in the diet is associated with numerous diseases including insulin resistance and type-2 diabetes mellitus, cardiovascular diseases, depression, Alzheimer's disease, obesity, bone fragility and cancer (Al Alawi et al., 2018; Barbagallo et al., 2021; Gröber et al., 2015). According to the US Food and Nutrition Board, the recommended daily allowance for magnesium is 400–420 mg for men and 310–320 mg for women (de Baaij et al., 2015); and the European Food Safety Authority considers adequate intakes for magnesium as 350 mg/day for men and 300 mg/day for women (Agostoni et al., 2015). However, it is known that populations in Europe do not consume the recommended daily intake (Al Alawi et al., 2018; de Baaij et al., 2015; Olza et al., 2017), while up to 50% of the US population is magnesium deficient (Razzaque, 2018; Uwitonze & Razzaque, 2018). Lack of magnesium affects the metabolism of vitamin D, which plays an essential role in the regulation of calcium homeostasis. This can result in different disorders such as skeleton deformities, worsen osteoporosis, vascular calcification and cardiovascular diseases. Vegetables, cereals and fruits are rich sources of magnesium (Gröber et al., 2015). Meanwhile, it is observed that the content of magnesium in these agricultural products has decreased constantly during the last 50 years, and additionally, about 80% of this microelement is lost during food preparation (Cazzola et al., 2020).

The recommended dietary allowance of manganese for adults is 2.3 mg/day for men and 1.8 mg for women, so its concentration in all diets is very low (DRI, 2001). Manganese is an essential trace element that is a cofactor for many enzymes; participates in amino acid, cholesterol and glucose metabolism in humans, bone formation and reproduction (Li & Yang, 2018; Palacios, 2006).

Milk whey, by-product from the dairy industry, and preparations derived from it are widely used in the food industry for the enrichment of various products (Ivanov et al., 2021; Kochubei-Lytvynenko et al., 2023; Królczyk et al., 2016; Kumar et al., 2018; Lange et al., 2020; Minj & Anand, 2020; Onopriichuk et al., 2023; Özer & Evrendilek, 2021). The content of magnesium and manganese in milk whey varies from 6 to 9 mg/100 g and from 0.6 to 2.8 μg/100 g, respectively (Wong et al., 1978). It should be noted that as a result of the use of baro- and electromembrane methods, which have recently been widely used in the processing of whey to remove monovalent ions of sodium, potassium, and chlorine, there is a simultaneous decrease in the content of biologically valuable magnesium and manganese (Arola et al., 2019; Kyrychuk et al., 2014; Perez et al., 1994; Soral-Śmietana et al., 2012). The presence of magnesium and manganese in raw milk improves its functional and technological properties and biological value. Therefore, an interesting area of research is the development of new methods for targeted enrichment of whey with valuable mineral elements. A possible way to compensate for the deficiency of mineral elements in milk whey with magnesium and manganese is the electrospark dispersion of conductive granules of these metals in it.

4.2 NANOBIOTECHNOLOGY FOR ENRICHMENT OF MILK WHEY WITH MINERAL ELEMENTS

4.2.1 Characteristics of Magnesium- and Manganese-Containing Particles Produced during Electric Spark Treatment

The electric spark treatment of whey was provided in the laboratory technological complex, the main parts of which were a discharge pulse generator (pulse frequency 0.2–2.0 kHz; discharge circuit inductance 1 μH) with a power section built on a thyristor element base, a capacitor used as an energy storage device and two chambers with volume 1 L each. The metal granules (magnesium, Mg, and/or manganese, Mn) were placed on the bottom of the discharge chamber between the main electrodes made of the corresponding metal. A voltage pulse passed through a layer of randomly placed magnesium or manganese granules, which caused switching of the current in the electrically conductive layer of granules with the formation of local spark discharges. As a result of electrical erosion of the surface of metal granules, the metal evaporates and condenses in the treated medium to obtain metal-containing particles. The bioavailability of mineral elements depends on the particle size and the nanosized metal fraction is a desirable result of sparking treatment. The processing of magnesium or manganese granules in an electrospark chamber at voltage of 80–100 V and a capacitor capacitance of 100 μF ensures the predominant content of particles of nano- and ultrafine fractions with sizes of 50–1,000 nm (Figure 4.1).

Images obtained with scanning electron microscopy (SEM) on a Tescan Mira 3 LMU instrument in the InBeam mode showed that magnesium-containing particles were in the form of needle-shaped crystals (Figure 4.2a); and manganese-containing particles had a spherical shape (Figure 4.2b).

Each metal had its own characteristics, which is explained by the difference in thermophysical properties: the melting points were 650°C and 1246°C, and evaporation temperatures were 1095°C and 1962°C, for magnesium and manganese, respectively. Analysis of SEM images of colloidal solutions of metal particles with energy dispersive spectroscopy (EDX) shows the formation of

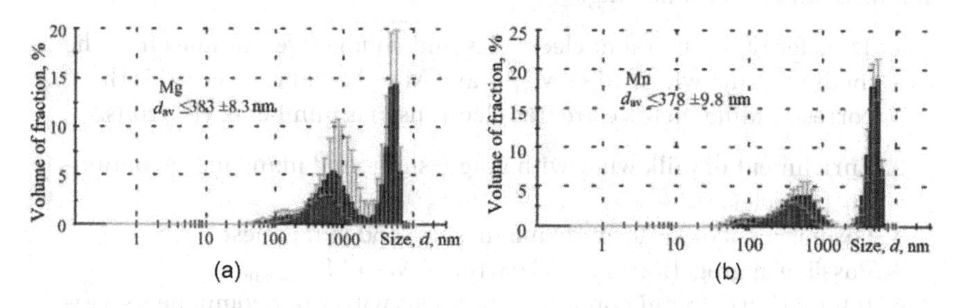

Figure 4.1 Distribution of magnesium (a) and manganese (b) particles by size.

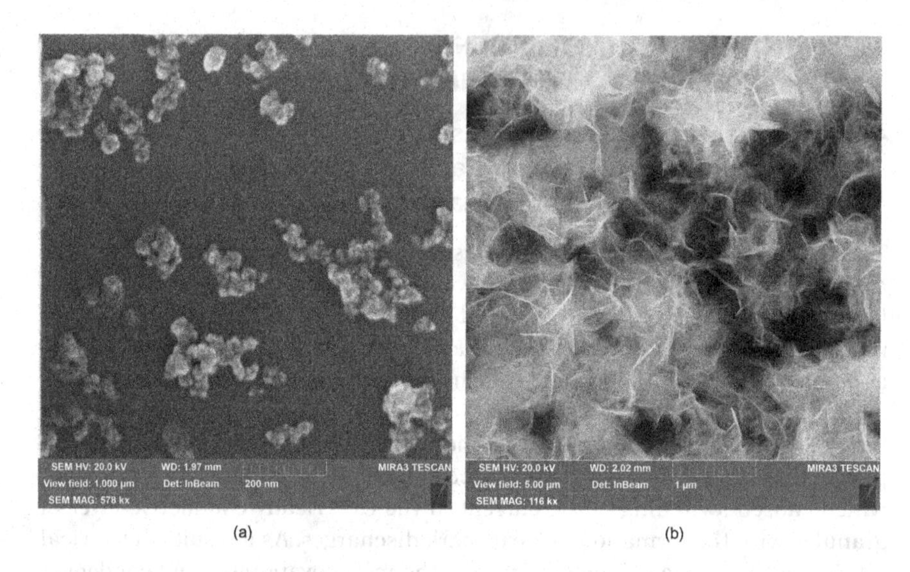

Figure 4.2 Morphology of magnesium (a) and manganese (b) particles.

both oxide and metal phases. However, the oxide phase MgO was dominant in the case of magnesium, and in the case of manganese, it was the dioxide phase MnO_2. It is known that magnesium oxide is considered to be an anticaking food additive (E 530) and prevents the formation of product lumps (Regulation EC No 1333/2008). Thus, magnesium oxide, formed as a result of the implementation of electrospark dispersion of conductive magnesium granules, can be useful for the technology of dry products.

4.2.2 Characteristics of the Composition and Properties of Milk Whey after Electric Spark Treatment

Application of electrospark dispersion of conductive metal granules in milk whey can cause a number of physicochemical and biochemical processes. Among them are the following:

1. Transfer of the metal of electrodes and conductive granules into the medium (milk whey): $M \leftrightarrow M_{(atom)}$ and $M \leftrightarrow M^{n+} + ne$, where M is the corresponding metal, e are free electrons, n is number of electrons.
2. Enrichment of milk whey with magnesium- and manganese-containing particles.
3. Formation of oxide forms of magnesium and manganese.
4. Possible aggregation of metal particles: $M \leftrightarrow M_{(colloidal)}$.
5. Interaction of metal-containing particles with whey components: carbohydrates, proteins, lactic and citric acids.

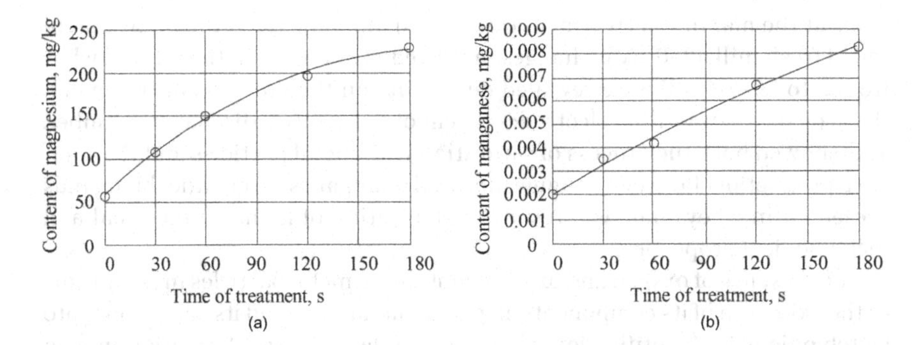

Figure 4.3 Changes in the content of magnesium (a) and manganese (b) in acid whey during its electric spark treatment.

Since milk whey, depending on the type, has different electrical conductivity values, for a reliable assessment of the nature of the process and the effect on its composition and properties, different wheys were selected—acid whey (electrical conductivity was 6.11±0.31 mSm/cm), cheese whey (electrical conductivity was 5.85±0.12 mS/cm) and demineralized whey (electrical conductivity was 3.64±0.17 mS/cm). It was found that during electrospark treatment, in all samples of whey, the content of magnesium increased by 1.8–4.1 times and manganese by 1.5–4.2 times, depending on the duration of treatment. The changes in the content of magnesium and manganese in acid whey are shown in Figure 4.3.

A similar increase in the content of magnesium and manganese with increase in processing time was observed for all studied types of whey, but at the same time, there was a difference between them in the dispersed characteristics of the obtained metal particles. It became more noticeable with increase in processing time. Thus, in acid whey and cheese whey, already after 120 s of treatment, the microfraction of dispersed metal particles, especially manganese, increased significantly. Whereas in the demineralized whey, microfraction of magnesium and manganese was insignificant. This is explained by the difference in the electrical conductivity of the whey samples and its growth over processing time. So, the duration of electric spark treatment and electrical conductivity of whey are important factors, which effect its enrichment with magnesium- and manganese-containing particles.

During electrospark treatment of milk whey, regardless of its type, the pH of the samples increases, and this effect was more pronounced when manganese electrodes were used. The explanation for this fact may be the interaction of metals with the "acidic" component of whey: lactic and citric acids and saturation with metal particles. According to the values of the redox potential, it is evident that reduction processes are taking place in the system; however, its value at the maximum treatment time (180 s) did not exceed the redox potential of fresh milk (250 mV). Changes in the redox potential in the treated whey from 5 to 70–219 mV were revealed depending on the duration of treatment, the type of whey and the electrode system of the electric discharge chamber. This showed both the process of saturation with metal particles and the complex production between magnesium and manganese ions and bioligands presented in whey, and, as a result, the formation of its new functional and technological properties.

The presence of oxygen and oxygen-containing metal particles in whey leads to the oxidation of its components, in particular lactose and its conversion into lactobionic acid. Identification of products of lactose transformation in acid whey treated with the electric spark discharges in a discharge chamber with a conductive layer of metal granules was carried out using a DionexUltiMate 3000 liquid chromatography. Acid whey was purified from ballast impurities such as fat and particles of casein dust by centrifugation. Characteristics of acid whey and lactose model solutions before and after electrospark treatment are given in Table 4.1.

The redox potential of model solution of lactose as well as of acid whey increased after the electrospark treatment. The pH of the model solution of lactose was 9.2±0.1, and after electric spark treatment for 60 s it increased to 10.3±0.1, pH value, which corresponded to one of the lactose isomerization conditions (Ganzle et al., 2008).

Lactobionic acid (LBA) is known for its prebiotic, antioxidant, antimicrobial, chelating, stabilizer, acidulant and moisturizing properties (Alonso et al., 2011; Cardoso et al., 2019; Goderska, 2019; Minal et al., 2017). LBA is present in the original (untreated) acid whey, but in a small amount. After treatment in the reaction chamber with electrodes and a conductive layer of magnesium, its amount almost doubled at a temperature of 20°C±2°C. While in the samples

TABLE 4.1 CHARACTERISTICS OF WHEY AND LACTOSE MODEL SOLUTIONS BEFORE AND AFTER ELECTRIC SPARK TREATMENT

	Acid Whey		Lactose Model Solution	
Characteristics	**Before**	**After**	**Before**	**After**
Active acidity, pH	4.2±0.2	4.8±0.2	9.2±0.1	10.3±0.1
Redox potential, mV	10.0±0.5	73.0±3.0	172.0±8.0	203.0±9.0
Electrical conductivity, mS/cm	6.5±0.3	6.98±0.3	30.5±1.0	31.2±1.5
Content of magnesium, mg/100 g	5.3±0.3	14.8±0.7	1.0±0.1	5.7±0.1

treated with electric spark discharges in the reaction chamber with a layer of manganese granules between the corresponding electrodes, the amount of lactobionic acid was close to the meaning for untreated whey. It can be assumed that nano- and ultra-sized magnesium-containing particles accumulated in the system during electrospark dispersion of metal granules due to their electrochemical activity act as catalysts for lactose oxidation. This assumption is confirmed in research (Cheng et al., 2020), in which MgO acts as a catalyst for the transformation of lactose. In the whey sequentially treated in a reaction chamber with a layer of magnesium granules between the corresponding electrodes and then in the next reaction chamber with manganese electrodes and a layer of this metal between them, the amount of lactobionic acid increased by almost four times due to the effect of double exposure to electric spark discharges with a total duration of 240 s under catalytic effect of magnesium-containing particles. The amount of lactobionic acid formed depended on the type of electrode system and the duration of treatment, as well as on its temperature. An increase in temperature enhanced the process of lactobionic acid formation both in case of using a conductive layer of magnesium granules between the corresponding electrodes or manganese, although, in the case of the latter, the amount of lactobionic acid formed was much lower. Thus, at a temperature of 60°C±2°C, in acid whey under the action of electric spark discharges in an electric discharge chamber with a conductive layer of magnesium granules, the amount of lactobionic acid increases almost 12 times, while when using a manganese electrode system with manganese granules it increases only 2-folds (Table 4.2).

TABLE 4.2 THE CONTENT OF LACTOBIONIC ACID IN ACID WHEY TREATED WITH ELECTRIC SPARK DISCHARGES

Acid Whey		Concentration of Lactobionic Acid, mg/L
Untreated		0.61±0.025
Treated in electric discharge chamber with metal granules, exposure 180 s, temperature (20±2)°C	Magnesium	1.32±0.025
	Manganese	0.57±0.025
Sequentially treated in electric discharge chambers (1) with magnesium granules (exposure 120 s, temperature 20°C±2°C) and (2) manganese (exposure 120 s, temperature 20±2°C)		2.32±0.100
Treated in electric discharge chamber with a layer of metal granules, exposure 180 s, temperature 60°C±2°C	Magnesium	7.28±0.240
	Manganese	1.19±0.030

4.3 THE USE OF ELECTRIC SPARK TREATMENT OF WHEY IN THE TECHNOLOGY OF DAIRY PRODUCTS

The electric spark treatment of whey can be used for practical application in the technology of dairy products, particularly production of dry milk whey (DMW) in the form of powder. Characteristics of milk whey treated before drying by nanofiltration (DMW_{NF}), which removes monovalent ions (K^+, Na^+, Cl^-), and with nanofiltration treatment followed with electric spark discharges, which enriches demineralized whey with valuable mineral elements (Mg and Mn) to obtain ($DMW_{NF\text{-}ESD}$), are shown in Table 4.3.

Freshly prepared dry milk wheys had no signs of particle self-compaction. However, during the year of storage under standard hermetic conditions, slight compaction was observed for both samples of whey. Freshly prepared dry wheys showed no signs of particle caking. However, after 12 months of hermetic storage, in dry milk whey treated only by nanofiltration, an almost doubling of the volume fraction of particles larger than 100 μm was found due to their coalescence (aggregation). At the same time, for dry milk whey treated by nanofiltration and electrospark discharges, caking of whey powder particles was barely noticeable after 12 months of storage.

The degree of whey caking was determined as the amount of powder appearing as lumps that cannot pass through a 500 μm sieve or 250 μm. The degree of caking more than 10% characterizes products as prone to lump formation

TABLE 4.3 CHARACTERISTICS OF DRY MILK WHEY

	Demineralized by Nanofiltration		After Nanofiltration and Electrosparking Treatment	
Parameter	Fresh	After 12 Months	Fresh	After 12 Months
Specific surface area, m^2/kg	295.9±14.5	183.6±9.1	270.4±13.1	266.3±13.0
Average particle size, μm	60.3±3.0	80.3±3.5	63.6±3.0	68.3±3.0
Volume fraction of particles with size, μm: ≤2	0.57±0.02	0.23±0.01	0.79±0.03	0.82±0.03
2.01–10	12.66±0.60	6.97±0.30	13.79±0.65	10.98±0.50
10.01–20	10.59±0.50	7.13±0.35	11.02±0.55	9.8±0.40
20.01–40	13.97±0.60	11.8±0.50	15.21±0.70	14.75±0.70
40.01–100	49.3±2.40	48.93±2.1	42.01±2.00	44.22±2.10
≥100	12.93±0.60	24.94±1.1	17.18±0.80	19.43±0.90

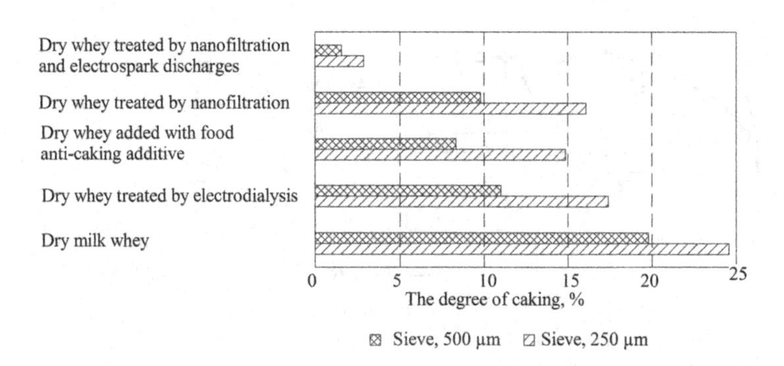

Figure 4.4 The degree of caking of dry milk whey treated by different methods.

(caking). The degree of caking of whey demineralized by nanofiltration, determined with a sieve pore size of 500 μm, approached 10%; while the dry whey, treated by nanofiltration and electrospark discharges had a degree of caking of 1.6% and 27% for the mesh size of the sieve 500 and 250 nm, respectively (Figure 4.4).

The explanation for this fact is the increase in the amount of magnesium compounds, in particular MgO, which have anticaking properties. In terms of resistance to lump formation, this new product had advantages even over milk whey, to which a food anticaking additive, silicon oxide SiO_2 (E-551), was added in an amount of 1.0% (Figure 4.4).

The dry milk whey produced by treatment with nanofiltration and electrospark discharges had improved solubility measured by the water activity index when the dry whey powder was reconstituted by adding water at a temperature of 40°C–45°C to obtain a concentration of 6% (w/v) and dispersed with ultrasound. In this case, the indicator of water activity is considered as an indirect characteristic of the amount of water bound by hydrophilic components. It was found that in the reconstructed dry whey treated by nanofiltration and electrospark discharges, stabilization of the water activity (Aw) was achieved after 45±5 min, while in reconstructed dry whey treated with nanofiltration only it took twice longer time 95±5 min (Figure 4.5a). A similar trend was also observed during reconstruction using ultrasonic cavitation, but the dissolution process for both samples was intensified (Figure 4.5b).

No signs of nonenzymatic browning were found in dry milk whey enriched with magnesium and manganese during storage for 18 months. To determine this characteristic, a whiteness index based on the reflectivity of the particles was used. The loss of whiteness indicates the occurrence of the Maillard reaction, a chemical reaction between amino acids and reducing sugars that leads to changes in food color. Signs of nonenzymatic darkening were detected by the change in the whiteness of the product during 18 months of storage under

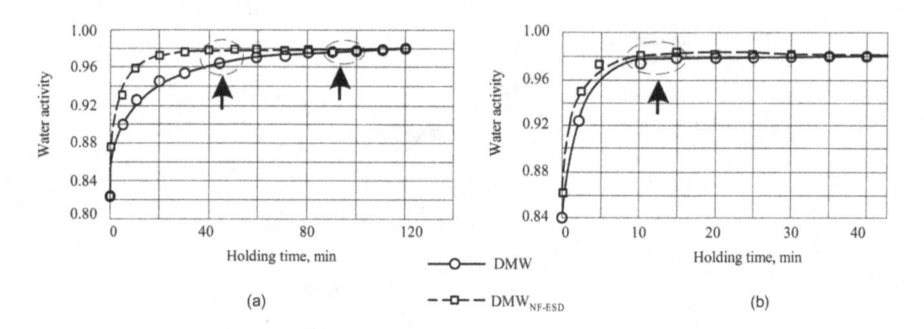

Figure 4.5 Dynamics of stabilization of the water activity index of the restored whey samples using traditional technology (a) and ultrasonic reduction (b).

TABLE 4.4 APPEARANCE OF SIGNS OF NONENZYMATIC DARKENING OF DRY WHEY DURING STORAGE UNDER STANDARD CONDITIONS

	Whiteness of Dry Milk Whey, Conventional Units, at Time of Storage, Months					
Whey	**0**	**1**	**6**	**8**	**12**	**18**
DMW	85.3±3.0	85.3±3.0	75.1±3.0	72.8±3.0	69.0±3.0	57.0±3.0
DMW$_{ACA}$	87.8±3.0	87.8±3.0	75.6±3.0	74.1±3.0	71.8±2.0	60.0±3.0
DMW$_{NF}$	90.6±2.0	90.6±2.0	83.6±2.0	80.1±2.0	79.0±2.0	72.0±2.0
DMW$_{NF-ESD}$	97.4±1.0	97.4±1.0	95.7±2.0	95.1±2.0	94.3±2.0	90.0±1.0

DMW, dry milk whey without any treatment; DMW$_{ACA}$, dry whey with the addition of food anticaking additive; DM$_{WNF}$, dry milk whey after nanofiltration; DMW$_{NF-ESD}$, dry milk whey after nanofiltration and electrospark discharge treatment.

standard conditions (in sealed packaging, temperature 18°C±2°C, relative humidity not more than 80%). The whiteness was evaluated in arbitrary units on a Blik-R3 device to measure the directional zonal reflection coefficient of the visible spectrum rays in a given wavelength range from the compacted-smoothed surface of the dry product and standard for whiteness determination. No evidence of nonenzymatic browning was observed in dry milk whey enriched with magnesium and manganese during storage for 18 months (Table 4.4).

Dry milk whey treated by nanofiltration and electrospark discharges after 18 months of storage lost whiteness by only 7.4 conventional units compared with 38.3 for dry milk whey; 27.8 for dry milk whey with addition of food anticaking additive, and 18.6 for dry milk whey treated by nanofiltration. The data obtained prove that the treatment of milk whey by nanofiltration and electrospark discharges before concentration and drying creates conditions for inhibition of reactions with the formation of carbonyl intermediates and brown pigments. This fact is explained by the blocking by magnesium-containing particles of the reactive groups of amino acids involved in reactions with reducing sugars.

Dry milk whey treated by nanofiltration and electrospark discharges differed from the whey treated by nanofiltration or dry whey with addition of food anticaking additive (DMW_{ACA}), silicon dioxide, SiO_2, in terms of glass transition temperature, Tg, and water activity, Aw, which was determined using a Hygrolab-2 device (Rotronic, Switzerland) at a temperature of 25°C. The glass transition temperature of a material characterizes the range of temperatures over which this glass transition occurs. Tg and change in heat capacity (ΔCp) were determined using a differential scanning microcalorimeter DSM-2M equipped with the ThermCap computer program for assembling and processing information written in the Delphi programming language. Microcalorimeter temperature scales were graded by two benchmark points: −95.0°C (melting temperature of chemically pure toluene) and 0°C (melting temperature of water after double distillation). Cooling of the calorimeter block was performed with the use of liquid nitrogen. Accuracy of temperature measurement was not worse than ±0.1°C. The samples were hermetically canned in aluminum containers. Weighting was performed at microanalytical scales VLM1 with an accuracy of ±0.01 mg. Overall content of water in samples was determined after measurements by dehydration till constant mass in the drying box at 104°C–105°C. To avoid artifacts connected to humidity condensation in calorimeter chambers, the measurement block was filled with dried gas-like helium, which flow was controlled during measurements. Samples were cooled at −50°C at a scanning rate of 16 K/min. Temperature intervals (ΔTg), initial temperature Tg^s and final temperatures of glass transition Tg^f were determined by DSC curves, obtained under heating of samples at scanning rate 16 K/min from −50 till +35°C. Tg was determined as ΔTg/2. The change in heat capacity under glass transition ΔCp was obtained by the difference in heat capacity at Tg^f and Tg^s. Demineralized dry milk whey (whey after nanofiltration) had the lowest Tg. Addition of silicon dioxide as an anticaking agent to milk whey increased its glass transition temperature by 10°C. The highest Tg was observed for dry whey powder treated by nanofiltration and electrospark discharges (Table 4.5).

Ability of dry product particles to stickiness and/or caking during treatment and storage was forecasted under a stickiness and caking sensitivity index (SCS). Index SCS (in diapason from 0 to 10) was determined by equation according to parameters changes diapason [T − Tg] (T—temperature of dry product) and ΔCp:

$$\text{SCSI} = \text{Number of points } [T - Tg] + \text{Number of points } [\Delta Cp],$$

where Tg is the glass transition temperature, °C; T is the dry product temperature (at the exit from drier, in the dry powder batch mixer and under storage), °C; ΔCp—change in heat capacity, J/(g·K). This index simultaneously integrates the values of [T − Tg] (ranging between 0 and 5) and ΔCp (ranging between 0 and 5). To predict the behavior of dry whey during packaging, transportation

TABLE 4.5 GLASS TRANSITION TEMPERATURES OF DRY MILK WHEY

Whey	Moisture Content, g/g of Dry Whey	Water Activity, Aw	Tg^s, °C	Tg^f, °C	Glass Transition Temperature, Tg, °C
DMW_{NF}	0.044±0.002	0.196±0.01	−18.0±0.2	+12.0±0.2	−3.0±0.2
DMW_{ACA}	0.034±0.001	0.187±0.01	−6.0±0.2	+19.0±0.2	+6.8±0.2
$DMW_{NF\text{-}ESD}$	0.031±0.001	0.130±0.01	+10.0±0.2	+27.0±0.2	+18.5±0.2

Tg^s, initial temperature of glass transition; Tg^f, final temperatures of glass transition; DM_{WNF}, dry milk whey treated with nanofiltration; DMW_{ACA}, dry whey with the addition of food anticaking additive; $DMW_{NF\text{-}ESD}$, dry whey powder treated with nanofiltration and electrospark discharge treatment.

TABLE 4.6 SOME CHARACTERISTICS OF DRY MILK WHEY DURING PACKAGING, TRANSPORTATION, AND STORAGE

		Sticking Properties		Caking Properties		
Whey	ΔCp, J/(g·K)	[T−Tg], °C	SCS (calculated)	[T−Tg], °C	SCS (calculated)	Degree of Caking, %
DMW_{NF}	0.38±0.01	33.0	8	23.0	7	16.0±0.7
DMW_{ACA}	0.32±0.01	23.2	7	13.2	5	15.0±0.4
$DMW_{NF\text{-}ESD}$	0.23±0.01	11.5	4	1.5	2	2.0±0.04

DMW_{NF}, dry milk whey treated with nanofiltration; DMW_{ACA}, dry whey with the addition of food anticaking additive; $DMW_{NF\text{-}ESD}$, dry whey powder treated with nanofiltration and electrospark discharge treatment; ΔCp, change in heat capacity, J/(g·K); SCS, caking sensitivity index.

and storage, a set of indicators is used, such as a change in heat capacity, a difference in storage/packaging and glass transition temperatures (Tg) of the product and an index of sensitivity to sticking and caking, SCS (stickiness and sensitivity to caking index). The values of SCS indicator ≤4 mean no stickiness and/or caking and values of SCSI ≥6 mean high to very high adhesiveness and/or risk of lumping.

Calculations showed that dry milk whey after nanofiltration and electrospark treatment, along with resistance to cracking, had the lowest adhesive ability among the studied samples—SCS index ≤4 (Table 4.6).

Thus, electrospark treatment of demineralized whey before drying makes it possible to improve the technological characteristics of dry whey; namely, it significantly reduces caking and stickiness, ensures the preservation of whiteness during storage and increases its solubility. Dry milk whey with elevated content of biologically important mineral elements magnesium and manganese and improved technological properties can be used for the preparation of various food products (Bilyk et al., 2021).

4.4 CONCLUSIONS

It has been shown that electric spark treatment of milk whey, a by-product of the dairy industry, which is based on the dispersion of conductive magnesium and manganese granules under the influence of electric spark discharges, leads to its enrichment with these valuable mineral elements as well as an increase in the content of lactobionic acid due to the catalytic transformation of lactose present in whey under the influence of the formed magnesium oxide. Electrospark treatment of demineralized whey improves the quality characteristics of dry whey, namely, significant reduce caking and stickiness and preserve whiteness during storage, as well as increase dry whey solubility. Thus, it is possible to improve the technological, nutritional and consumer values of dry whey for its further use in food production.

REFERENCES

Agostoni, C., Canani, R.B., Fairweather-Tait, S., Heinonen, M., Korhonen, H., La Vieille, S., Marchelli, R., Martin, A., Naska, A., Neuhäuser-Berthold, M., Nowicka, G., Sanz, Y., Siani, A., Sjödin, A., Stern, M., Strain, S., Tetens, I., Tomé, D., Turck, D., & Verhagen, H. (2015). Scientific opinion on dietary reference values for magnesium. EFSA panel on dietetic products, nutrition and allergies (NDA). *EFSA Journal*, 3(7), 4186. https://doi.org/10.2903/j.efsa.2015.4186

Al Alawi, A.M., Majoni, S.W., & Falhammar, H. (2018). Magnesium and human health: Perspectives and research directions. *International Journal of Endocrinology*, 9041694. https://doi.org/10.1155/2018/9041694

Alonso, S., Rendueles, M., & Díaz, M. (2011). Efficient lactobionic acid production from whey by *Pseudomonas taetrolens* under pH-shift conditions. *Bioresource Technology*, 102(20), 9730–9736. https://doi.org/10.1016/j.biortech.2011.07.089

Arola, K., Van der Bruggen, B., Mänttäri, M., & Kallioinen, M. (2019). Treatment options for nanofiltration and reverse osmosis concentrates from municipal wastewater treatment: A review. *Critical Reviews in Environmental Science and Technology*, 49(22), 2049–2116. https://doi.org/10.1080/10643389.2019.1594519

Barbagallo, M., Veronese, N., & Dominguez, L.J. (2021). Magnesium in aging, health and diseases. *Nutrients*, 13(2), 463. https://doi.org/10.3390/nu13020463

Bertinato, J., Xiao, C.W., Ratnayake, W.M., Fernandez, L., Lavergne, C., Wood, C., & Swist, E. (2015). Lower serum magnesium concentration is associated with diabetes, insulin resistance, and obesity in South Asian and white Canadian women but not men. *Food and Nutrition Research*, 59, 25974. https://doi.org/10.3402/fnr.v59.25974

Cardoso, T., Marques, C., Dagostin, J.L.A., & Masson, M.L. (2019). Lactobionic acid as a potential food ingredient: Recent studies and applications. *Journal of Food Science*, 84(7), 1672–1681. https://doi.org/10.1111/1750-3841.14686

de Baaij, J.H., Hoenderop, J.G., & Bindels, R.J. (2015). Magnesium in man: Implications for health and disease. *Physiological Reviews*, 95(1), 1–46. https://doi.org/10.1152/physrev.00012.2014

Bilyk, O., Vasylchenko, T., Kochubei-Lytvynenko, O., Bondarenko, Y., & Piddubnyi, V. (2021). Studying the effect of milk processing products on the structural-mechanical properties of wheat flour dough. *EUREKA: Life Sciences*, 1, 44–52. https://doi.org/10.21303/2504-5695.2021.001642

DRI. (2001). *Dietary Reference Intakes for Vitamin A, Vitamin K, Arsenic, Boron, Chromium, Copper, Iodine, Iron, Manganese, Molybdenum, Nickel, Silicon, Vanadium, and Zinc.* Washington, DC, National Academy Press.

Cazzola, R., Porta, D.M., Manoni, M, Iotti, S., Pinotti, L., & Maier, J.A. (2020). Going to the roots of reduced magnesium dietary intake: A tradeoff between climate changes and sources. *Heliyon*, 6(11), e05390. https://doi.org/10.1016/j.heliyon.2020.e05390

Cheng, S., Metzger, L.E., & Martínez-Monteagudo, S.I. (2020). One-pot synthesis of sweetening syrup from lactose. *Science Reports*, 10(1), 2730. https://doi.org/10.1038/s41598-020-59704-x

Ganzle, M.G., Haase, G., & Jelen, P. (2008). Lactose: Crystallization, hydrolysis and value-added derivatives. *International Dairy Journal*, 18(7), 685–694. https://doi.org/10.1016/j.idairyj.2008.03.003

Goderska, K. (2019). The antioxidant and prebiotic properties of lactobionic acid. *Applied Microbiology and Biotechnology*, 103, 3737–3751. https://doi.org/10.1007/s00253-019-09754-7

Gröber, U., Schmidt, J., & Kisters, K. (2015). Magnesium in prevention and therapy. *Nutrients*, 7(9), 8199–8226. https://doi.org/10.3390/nu7095388

Ivanov, V., Shevchenko, O., Marynin, A., Stabnikov, V., Gubenia, O., Stabnikova, O., Shevchenko, A., Gavva, O., & Saliuk, A. (2021). Trends and expected benefits of the breaking edge food technologies in 2021–2030. *Ukrainian Food Journal*, 10(1), 7–36. https://doi.org/10.24263/2304-974x-2021-10-1-3

Kochubei-Lytvynenko, O., Bilyk, O., Bondarenko, Y., & Stabnikov, V. (2023). Whey proteins in bakery products. In O. Paredes-López, O. Shevchenko, V. Stabnikov, & V. Ivanov, (Eds.), *Bioenhancement and Fortification of Foods for a Healthy Diet* (pp. 67–88). Boca Raton, London: CRC Press. https://doi.org/10.1201/9781003225287-6

Królczyk, J.B., Dawidziuk, T., Janiszewska-Turak, E., & Sołowiej, B. (2016). Use of whey and whey reparations in the food industry—A review. *Polish Journal of Food and Nutrition Science*, 66(3), 157–165. https://doi.org//10.1515/pjfns-2015-0052

Kumar, R., Chauhan, S.K., Shinde, G., Subramanian, V., & Nadanasabapathi, S. (2018). Whey proteins: A potential ingredient for food industry: A review. *Asian Journal of Dairy and Food Research*, 37(4), 283–290. https://doi.org/10.18805/ajdfr.DR-1389

Kyrychuk, I., Zmievskii, Y., & Myronchuk, V. (2014). Treatment of dairy effluent model solutions by nanofiltration and reverse osmosis. *Ukrainian Food Journal*, 3(2), 281–288.

Lange, I., Mleko, S., Tomczyńska-Mleko, M., Polischuk, G., Janas, P., & Ozimek, L. (2020). Technology and factors influencing Greek-style yogurt – A review. *Ukrainian Food Journal*, 9(1), 7–35. https://doi.org/10.24263/2304-974X-2020-9-1-3

Li, L., & Yang, X. (2018). The essential element manganese, oxidative stress, and metabolic diseases: Links and interactions. *Oxidative Medicine and Cellular Longevity*, 18, 7580707. https://doi.org/10.1155/2018/7580707

Minal, N., Balakrishnan, S., Chaudhary, N., & Jain, A.K. (2017). Lactobionic acid: Significance and application in food and pharmaceutical. *International Journal of Fermented Foods*, 6(1), 25–33. https://doi.org/10.5958/2321-712X.2017.00003.5

Minj, S., & Anand, S. (2020). Whey proteins and its derivatives: Bioactivity, functionality, and current applications. *Dairy*, 1(3), 233–258. https://doi.org/10.3390/dairy1030016

Olza, J., Aranceta-Bartrina, J., González-Gross, M., Ortega, R.M., Serra-Majem, L., Varela-Moreiras, G., & Gil, Á. (2017). Reported dietary intake, disparity between the reported consumption and the level needed for adequacy and food sources of calcium, phosphorus, magnesium and vitamin D in the Spanish population: Findings from the ANIBES study. *Nutrients*, 9(2), 168. https://doi.org/10.3390/nu9020168

Onopriichuk, O., Grek, O., & Tymchuk, A. (2023). Influence of malt properties on the indicators of milk-protein concentrates. In O. Paredes-López, O. Shevchenko, V. Stabnikov, & V. Ivanov, (Eds.), *Bioenhancement and Fortification of Foods for a Healthy Diet* (pp. 179–202). Boca Raton, London: CRC Press. https://doi.org/10.1201/9781003225287-6

Özer, B., & Evrendilek, G.A. (2021). Whey beverages. In A.G. da Cruz, C.S. Ranadheera, F. Nazzaro, & A.M. Mortazavian (Eds.), *Dairy Foods: Processing, Quality, and Analytical Techniques* (pp. 117–137). Woodhead Publishing. https://doi.org/10.1016/B978-0-12-820478-8.00012-2

Palacios, C. (2006). The role of nutrients in bone health, from A to Z. *Critical Reviews in Food Science and Nutrition*, 46(8), 621–628. https://doi.org/10.1080/10408390500466174

Perez, A., Andres, L.J., Alvarez, R., Coc, J., & Hill, C.G. (1994). Electrodialysis of whey permeates and retentates obtained by ultrafiltration. *Journal of Food Process Engineering*, 17, 177–190. https://doi.org/10.1111/j.1745-4530.1994.tb00334.x

Razzaque, M.S. (2018). Magnesium: Are we consuming enough? *Nutrients*, 10(12), 1863. https://doi.org/10.3390/nu1012186

Regulation (EC) No 1333/2008 of the European Parliament and of the Council of 16 December 2008 on food additives (Text with EEA relevance). https://www.legislation.gov.uk/eur/2008/1333/contents

Soral-Śmietana, M., Zduńczyk, Z., Wronkowska, M., Juśkiewicz, J., & Zander, L. (2012). Mineral composition and bioavailability of calcium and phosphorus from acid whey concentrated by various membrane processes. *Journal of Elementology*, 18, 115–125. https://doi.org/10.5601/JELEM.2013.18.1.10

Uwitonze, A.M. & Razzaque, M.S. (2018). Role of magnesium in vitamin D activation and function. *The Journal of the American Osteopathic Association*, 118(3), 181–189. https://doi.org/10.7556/jaoa.2018.037

Webster-Gandy, J., Madden, A., & Holdsworth, M. (2020). Micronutrients. In J. Webster-Gandy, A. Madden, & M. Holdsworth (Eds.), *Oxford Handbook of Nutrition and Dietetics*. Oxford Medical Handbooks. https://doi.org/10.1093/med/9780198800132.003.0006

Wong, N.P., LaCroix, D.E., & McDonough, F.E. (1978). Minerals in whey and whey fractions. *Dairy Science*, 61(12), 1700–1703. https://doi.org/10.3168/jds.S0022-0302(78)83790-4

Chapter 5

Flour from Sunflower Seed Kernels in the Production of Flour Confectionery

Iryna Tsykhanovska
Ukrainian Engineering-Pedagogics Academy

Viktoria Yevlash
State Biotechnology University

Lidiya Tovma
National Academy of the National Guard of Ukraine

Greta Adamczyk
University of Rzeszow

Aleksandr Alexandrov, Tetiana Lazarieva, and Olga Blahyi
Ukrainian Engineering-Pedagogics Academy

CONTENTS

DOI: 10.1201/9781003329671-5

5.1 INTRODUCTION

Confectionery products are high in sugar and saturated fatty acids, but at the same time, they are deficient in protein, dietary fiber, vitamins and minerals. Therefore, it is advisable to prepare flour confectionery products with additives that increase their nutritional value. In this case, the use of secondary products of vegetable raw materials processing is also a way of efficient disposal of food industry waste and ensures the implementation of the concept of a circular economy (Ancuta and Sonia, 2020; Borrello et al., 2017; Esposito et al., 2020; Stabnikova et al., 2021).

Sunflower (*Helianthus annuus* L.) is the fourth oil-producing crop in the world after palm, soybean and rapeseed (Yegorov et al., 2019). The composition and nutritional value of its seeds depend on the genotypic properties of sunflower, environmental conditions, climatic conditions and types of processing (Petraru et al., 2021). Sunflower seeds contain, %: proteins, 34 (range from 20 to 40); lipids, 53 (33–60); ash, 2.7 (2.7–4.9); dietary fiber, 8.0–20; starch, 0.4–1.3; and mono- and disaccharides, 2.1–2.7 (Akkaya, 2018; Petraru et al., 2021; Sarwar et al., 2013). The content of fatty acids in sunflower oil averages 440.6 mg/mL, of which around 10% are saturated fatty acids (Akkaya, 2018; Sarwar et al., 2013). The chemical composition of sunflower seeds includes also such antioxidants as flavonoids (quercetin, luteolin, apigenin and kaempferol), phytosterols (β-sitosterol or sitosterol, cycloartenol), chlorogenic acid, β-carotene and carotenoids; water-soluble B1, B2, B3, B4, B5, B6, B9 and fat-soluble E, D vitamins; and minerals, mainly calcium, copper, iron, magnesium, manganese, selenium, phosphorus, potassium, sodium, cesium and zinc (Adeleke & Babalola, 2020; Anjum et al., 2012; Delgado-Vargas & Paredes-Lopez, 2003; Islam et al., 2016; Petraru et al., 2021; Rosa et al., 2009; Wanjari & Waghmare, 2015).

By-products of oil production from sunflower seeds are cake (residue after pressing seeds to extract oil), meal (prepared by crushing the seed nucleus at the stage of extraction) and flour obtained from the meal (Oseyko et al., 2020). Flour produced from peeled, pressed and crushed sunflower seeds is a fine homogeneous powder of grayish-white color with a mild and delicate taste and a distinct smell of sunflower seeds (Oseyko et al., 2020). Application of flours from plant seeds as additives to food products to improve their healthy and consumer values is very popular in recent years (Stabnikova et al., 2021). There is some experience with the use of sunflower seed flour in food preparation for a partial replacement of main flour. When replacing 35% of the mass of wheat flour with sunflower samples in the cracker biscuit recipe, the nutritional and sensory properties of the product tended to increase. Similar results in relation to quality and consumer acceptance of wheat bread enriched with sunflower seed flour were observed (Man et al., 2017; Skrbic & Filipcev, 2008). The addition of defatted sunflower flour in the amount of 36% of the mass of wheat flour to the dough helps to increase the protein content by 50%, the total phenol content by 85.7%, the antioxidant activity by 350% as well as strength (viscosity) of the dough and formation of a chocolate brown color. At the same time, enriched flour products have the same calorie content as a product prepared with wheat flour only (Grasso et al., 2019).

5.2 CHARACTERISTICS OF FLOUR FROM EXTRUDED SUNFLOWER SEED KERNELS

5.2.1 Nutrient Composition of Flour from Extruded Sunflower Seed Kernels

In the present study, flour left after oil production from sunflower seed kernels in an extruder according to the following scheme was used: a husked sunflower seed was loaded into an extruder, heated to a temperature of 85°C–90°C and crushed for 18–20 min. The heat treatment ensures microbiological safety and improves the outflow of oil, reducing the fat content in the sunflower flour. During conveyor grinding, the resulting protein product, namely, meal with a particle size of 350–400 μm is used for animal feed, and a finely ground product with a particle size of 90–110 μm is the flour from extruded sunflower seed kernels (FESSK), can be used in the production of food, particularly in flour confectionery (Evlash et al., 2019). Physicochemical characteristics of FESSK are shown in Table 5.1.

FESSK compared to traditional sunflower flour has a number of advantages, namely, higher by 1.7 times content of proteins and by 1.8 times content of minerals; lower by 7.6 times content of fat. Low content of chlorogenic acid also positively characterizes FESSK because the presence of chlorogenic acid significantly contributes to the formation of undesirable green complexes due to the interaction of oxidized chlorogenic acid and its derivatives with sunflower seed proteins during baking (Wildermuth et al., 2016). Moisture content is an important factor for the long-term stability of defatted meals (Rani & Badwaik, 2021). A level below 12.0% is considered safe to store as it prevents rapid mold growth (Abdullah et al., 2011). The FESSK moisture values were 4.72%, and the total moisture and volatile content was 8.2%. It also has a fairly high content of proteins, and 76.35% of them are soluble, which is an important factor in the stabilization of foams, and emulsions; interestingly, in gel formation soluble proteins create a high homogeneous dispersion of molecules in such systems and promote intersurface interactions. The low value of the initial acid number, 0.093 mg KOH/g, and after 6 months of storage, 0.091 mg KOH/g, indicates a low level of hydrolysis and fat oxidation, and stability of FESSK during storage (the acid number increases only by 1.1%), which is explained by the action of natural

TABLE 5.1 PHYSICOCHEMICAL CHARACTERISTICS OF FLOUR FROM THE EXTRUDED SUNFLOWER SEED KERNELS (FESSK)

Characteristics	Mass fraction (%)
Moisture	4.72±0.24
Volatile substances	3.48±0.16
Dry matter	91.80±4.19
Protein on a dry matter basis	38.73±1.94
Soluble proteins in flour to the total protein content	76.35±3.11
Lipids on a dry matter basis	4.87±0.25
Fiber on a dry matter basis	11.87.±0.55
Total ash on a dry matter basis	8.0±0.35
Starch on a dry matter basis	12.53±0.59
Chlorogenic acid on a dry matter basis	0.321±0.016
Vitamin E, µg/kg	15.40±0.77
	0.091±0.004/0.093±0.004

antioxidants that contribute to the destruction of hyperoxides without the formation of free radicals (Wu, 2007).

The content of minerals in sunflower seeds ranged from 2.68% to 4.87% (Hussain et al., 2020; Nevara et al., 2021; Petraru et al., 2021; Sarwar et al., 2013), while for the FESSK this number rises up to 8.0% (Table 1.1). Mass spectroscopic studies (Agilent 7500 S, USA) according to the method (Sinkovic & Kolmanic, 2021) showed the presence of 22 elements in FESSK, namely: I<Cd<Li<Mo<Cr<Ni<Se<Fe(III)<Co<Fe(II)<Cu<Ti<Zn<Be<Na<Mn<K<Cl<Ca <S<Mg<P (Table 5.2). The results obtained were consistent with those of other authors according to the mineral composition of oilseeds by-products (Kollathova et al., 2019).

Analysis of the content of vitamins in FESSK was performed using an Agilent 1100 high-performance four-channel liquid chromatograph (Agilent Technologies, USA) in combination with a diode array detector (DAD) and mass spectrometry (MS) according to Sim et al. (2016) and Katsa et al. (2021). Totally eight water-soluble vitamins, mg/kg: C, 8.42; B1 (thiamine), 4.02; B2 (riboflavin), 1.78; PP or B3 (niacin, nicotinic acid), 47.34; B4 (choline), 28.92; B5 (pantothenic acid), 4.33; B6 (pyridoxine), 14.89; B9 (folic acid), 0.47, and four fat-soluble vitamins, mg/kg: A (retinol), 31.45; D (calciferol), 10.43; E (tocopherol), 15.40; K (phylloquinone), 0.02, were found in FESSK. The results obtained are consistent with the data of other authors (Garg et al., 2021).

TABLE 5.2 MINERAL COMPOSITION OF FLOUR FROM THE EXTRUDED SUNFLOWER SEED KERNELS (FESSK)

Macroelements, mg/100g (Wet Basis)						
P	Mg	S	Ca	Cl	K	Na
6676	4950	1540	1531	1510	1140	1062

Trace elements, 100 mg/100g (Wet Basis)														
Mn	Be	Zn	Ti	Cu	Fe(II)	Co	Fe(III)	Se	Ni	Cr	Mo	Li	Cd	I
67	33	32	19	7	5.4	4.9	4.0	3.9	3.2	0.59	0.54	0.38	0.28	0.14

Analysis of amino acid content of FESSK proteins was carried out using a liquid chromatograph (Dionex ICS-3000, USA) with an electrochemical detector (Electrochemical Detector Cell) according to the method described (Canibe et al., 1999; Oseyko, et al., 2020; Petraru et al., 2021; Rosa et al., 2009). The total content of amino acids in FESSK is 257.09 mg/g of dry protein while essential amino acids account for 112.26 mg/g of dry protein (43.67%), and nonessential 144.83 mg/g of dry protein (56.33%) (Table 5.3).

To assess the amino acid composition of FESSK proteins, the amino acid score (AC) was calculated as the ratio of the content of an essential amino acid in mg in 1 g of the FESSK protein to its amount in mg in 1 g of the reference "ideal or standard" protein (white egg). The amino acid with the lowest AC is called the first limiting amino acid, which determines the biological value and level of protein assimilation. Lysine was the first limiting amino acid with AC of 83.25%. Results of the comparative analysis of the content of crude protein and amino acids in sunflower meal and flour are shown in Table 5.4.

An increase in FESSK has been found, compared to the sunflower meal from which it was obtained, in the content of: crude protein by 1.1 times; essential amino acids by 1.2 times, and the total content of amino acids by 10.0%. Together with an increase in the content of total protein, flour from the extruded kernel of sunflower seeds is enriched with essential hydrophilic (by 1.79%) and hydrophobic (by 31.46%) amino acids; as well as nonessential hydrophilic amino acids (by 2.92%) compared to sunflower meal, from which FESSK was obtained. At the same time, the increase in the content of essential hydrophobic amino acids is 17.58 times higher than that of hydrophilic ones. This indicates that hydrophobic amino acids have a less reactive nonpolar nature, which is associated with the presence of free nonpolar hydrocarbon radicals with small dipole moments that are not involved in the formation of peptide bonds in the protein. That is why they have higher resistance to technological processing in the production of sunflower flour than hydrophilic amino acids; the more reactive nature of which is because of their polar nature due to the presence of a side

TABLE 5.3 AMINO ACID CONTENT IN FLOUR FROM THE EXTRUDED SUNFLOWER SEED KERNELS (FESSK)

Amino Acids	Content, mg/g Dry Protein	Amino Acid Score, %
Essential Amino Acids		
Hydrophilic		
Threonine	9.40±0.39	94.08
Arginine	24.33±.1.12	105.82
Histidine	6.41±0.32	92.01
Lysine	9.49±0.45	83.25
Hydrophobic		
	13.32±0.65	96.72
Leucine	16.82±0.81	94.14
Isoleucine	11.53±0.53	92.65
Methionine	5.19±0.22	86.73
Phenylalanine	11.84±0.54	93.68
Tryptophan	3.53±0.14	95.72
Nonessential amino acids		
Hydrophilic		
Serine	10.88±0.51	-
Asparagine	7.81±0.39	-
Aspartic acid	24.77±1.21	-
Glutamic acid	56.52±2.62	-
Tyrosine	6.75±0.31	-
Hydrophobic		
	11.32±0.52	-
Glycine	11.46±0.54	-
Proline	8.61±0.41	-
Cystine	6.71±0.32	-
Total content of amino acids, mg/g	257.09	-
Content of essential amino acids, mg/g	112.26	-
Content of nonessential amino acids, mg/g	144.83	-

TABLE 5.4 COMPARATIVE ANALYSIS OF THE CONTENT OF CRUDE PROTEIN AND AMINO ACIDS IN SUNFLOWER MEAL AND FLOUR MADE FROM IT

Content of	Sunflower Meal	FESSK
Crude protein, % from dry matter	36.02±1.72	38.73±1.94
Essential amino acids, % from crude protein	37.50±1.44	43.67±2.01
Nonessential amino acids, % from crude protein	62.50±3.06	56.33±2.21
Essential amino acids, % from crude protein		
Hydrophilic		
Threonine	3.54±0.16	3.66±0.18
Arginine	9.34±0.44	9.46±0.45
Histidine	2.46±0.12	2.49±0.12
Lysine	3.63±0.17	3.69±0.17
Hydrophobic		
Valine	3.98±0.17	5.18±0.24
Leucine	4.91±0.21	6.57±0.32
Isoleucine	2.96±0.13	4.48±0.19
Methionine	1.92±0.09	2.09±0.10
Phenylalanine	3.69±0.18	4.65±0.20
Tryptophan	1.07±0.05	1.39±0.02
Nonessential Amino Acid		
Hydrophilic		
Serine	4.19±0.19	4.23±0.19
Asparagine	3.01±0.14	3.04±0.14
Aspartic acid	9.59±0.45	9.63±0.47
Glutamic acid	21.94±1.01	21.98±1.01
Tyrosine	2.60±0.12	2.63±0.12
Hydrophobic		
Alanine	6.34±0.37	4.40±0.20
Glycine	7.45±0.41	4.46±0.21
Proline	5.97±0.26	3.35±0.22
Cystine	2.41±0.12	2.61±0.13
Content of essential amino acids, mg/g	37.50±1.44	43.67±2.01
Content of nonessential amino acids, mg/g	62.50±3.06	56.33±2.21

chain with hydrophilic (polar) groups and a large dipole moment (Oseyko, et al., 2020; Petraru et al., 2021). This ability of proteins to maintain functional hydrophobic properties during the technological processing of FESSK allows to use it as an ingredient for the manufacturing of food products, in particular, flour confectionery with improved consumer characteristics.

To evaluate the biological value of FESSK, the content and composition of fatty acids were studied. Determination of fatty acid content was carried out by gas chromatography–mass spectrometry (GC–MS), Clarus 600 T GC–MS vicorist mixture (PerkinElmer, USA) and the method described in Dulf (2012) and Petraru et al. (2021). FESSK contains nine saturated and four polyunsaturated higher fatty acids. For all higher fatty acids, the mass fraction was calculated in percentage values (Table 5.5).

The total content of higher fatty acids in FESSK was 49.778 mg/g of flour consisted of 65.22%±3.12% of polyunsaturated fatty acids, 19.56%±0.96%

TABLE 5.5 FATTY ACID CONTENT IN FLOUR FROM THE EXTRUDED SUNFLOWER SEED KERNELS (FESSK)

Higher Fatty Acids	Content of Fatty Acid	
	mg/g of Fat in FESSK	% of Total Amount
Lauric acid C12:0	0.006±0.0	0.01±0.0
Myristic acid C14:0	0.045±0.002	0.08±0.003
Palmitic acid C16:0	0.443±0.022	7.82±0.389
Palmitoleic acid C16:1 (ω-9)	0.113±0.005	0.20±0.010
Stearic acid C18:0	0.344±0.017	6.06±0.289
Oleic acid C18:1(ω-9)	10.954±0.545	19.32±0.913
Linoleic acid C18:2 (ω-6)	36.883±1.844	65.05±3.002
Linolenic acid C18:3 (ω-3)	0.096±0.004	0.17±0.006
Arachidic acid C20:0	0.136±0.006	0.24±0.011
Eicosenoic acid C20:1 (ω-9)	0.022±0.001	0.04±0.002
Arachidonic acid C20:4 (ω-6)	0.011±0.0	0.02±0.0
Eicosapentaenoic acid C20:5	0.153±0.007	0.27±0.012
Heneicosylic acid C21:0	0.022±0.001	0.04±0.002
Behenic acid C22:0	0.414±0.020	0.73±0.036
Tricosanoic acid C23:0	0.280±0.001	0.05±0.002
Lignoceric acid C24:0	0.108±0.003	0.19±0.007

of monounsaturated fatty acids, and 15.22%±0.75% of saturated fatty acids, among which palmitic, C16:0, 7.82%, and stearic, C18:0, 6.06%, acids were predominant. The total content of unsaturated acids in FESSK is 84.78%±4.18%, which is higher than in other oilseeds such as safflower, peanut, soybean, sesame and flax seeds (Ivanova et al., 2013; Petraru et al., 2021). The oleic, C18:1(ω-9), 19.32% and linoleic fatty, C18:2(ω-6), 65.05%, are the main components of FESSK fat. The ratio of C18:2(ω-6)/C18:1(ω-9) in FESSK fat 3.4:1 is in line with the British Nutrition Foundation's recommendations that the balanced ratio of unsaturated fatty acids of family ω -6, ω-9, especially for dietary nutrition, should be between 3:1 and 5:1 (Goiri et al., 2019). Linoleic, C18:2(ω-6), and linolenic, C18:3(ω-3), acids are polyunsaturated essential fatty acids that cannot be synthesized by the human body and play an important role in maintaining health, triglyceride and cholesterol levels, and in normalizing blood pressure. The content of linolenic acid is 0.17%. Similar results were obtained in the works by other authors (Goiri et al., 2019; Mirpoor et al., 2021; Petraru et al., 2021). Consumption of food products with FESSK contributes to an increase in the level of linoleic, linolenic and oleic acids in the human body. This makes flour from extruded sunflower seed kernels an important dietary source of unsaturated fatty acids, in particular linoleic, linolenic and oleic.

5.2.2 Functional and Technological Properties of Flour from the Extruded Sunflower Seed Kernels in the Production of Flour Confectionery

The most important functional and technological properties of FESSK include solubility, water-absorbing, water-holding, fat-absorbing, fat-holding, fat-emulsifying and foaming abilities. All experiments were performed using control wheat flour of high grade (WF).

Solubility of FESSK. The solubility of proteins, the main condition for the functionality of proteins in food raw materials, depends on the presence of hydrophilic centers, hydrophobic (Stacking interactions), electrostatic and hydrogen bonds of the surface of protein molecules with solvent molecules (Ptak-Kaczor et al., 2021). Higher solubility of proteins increases viscosity, promotes better foaming, and improves emulsification and gelation of flour systems. Reduction of flour protein solubility is usually accompanied by a change in other important functional properties of flour system, which largely affect the quality of food products, and the degree of protein digestibility in the gastrointestinal tract (Subası et al., 2021). Solubility of FESSK was determined by suspending FESSK (1:100) in aqueous solutions with different pH values, followed by centrifugation and photometric determination of the protein concentration in the supernatant by the biuret method ($\lambda = 540$ nm). The solubility of FESSK protein

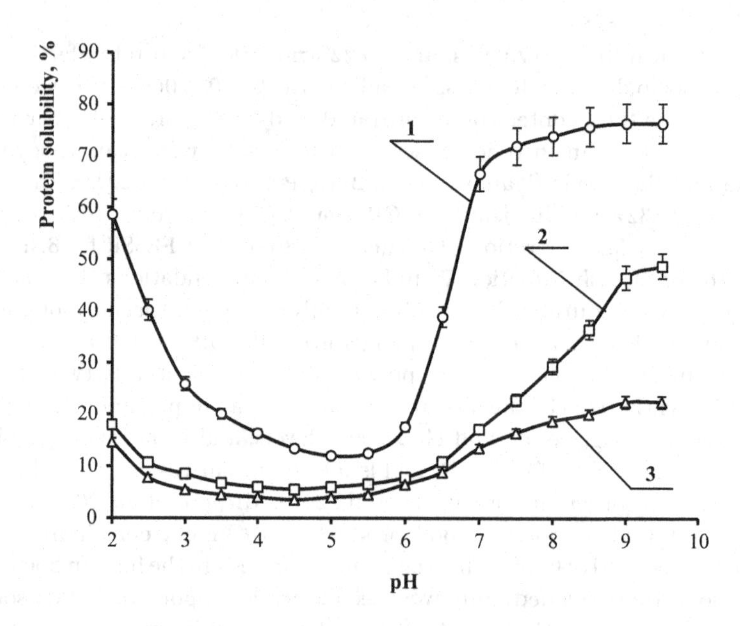

Figure 5.1 Influence of pH on the solubility of flour protein: 1, flour from the extruded kernel of sunflower seeds (FESSK); 2, mixture of FESSK with wheat flour in the ratio of 1:9; 3, wheat flour.

was calculated as the ratio of a certain protein concentration (mg/cm^3) in the supernatant to the initial one. The result was multiplied by 100 to express it as a percentage (González-Pérez et al., 2003; Ivanova et al., 2013).

The solubility of FESSK, mixture of wheat flour and FESSK, and wheat flour as control depending on the pH value was determined (Figure 5.1).

The patterns of protein solubility were the same for all flour samples at pH$\geq$3.5. In the pH range from 3.5 to 5.5, flour proteins have the lowest solubility (the pH of the medium approaches the isoelectric point). Outside this range, the solubility gradually increases, reaching maximum values in an alkaline, 9.0 because in an alkaline environment, flour proteins are negatively charged and anions have a high binding affinity for water. The solubility pattern exhibited by the proteins of the test flours as a function of pH is similar to the solubility of sunflower and wheat proteins at a low ionic strength (González-Pérez, 2003; Ivanova et al., 2013). The solubility of FESSK proteins is higher by 1.64–2.09 times than the solubility of proteins of mixtures of FESSK with wheat flour in the ratio 1:9 and by 3.29–3.35 times than the solubility of wheat flour proteins in the studied pH range from 2.0 to 9.5. Differences in protein solubility of studied flours could be explained by their chemical composition (Dabbour et al., 2018; Ivanova et al., 2013; Le Priol et al., 2019; Subası et al., 2021). FESSK has a higher

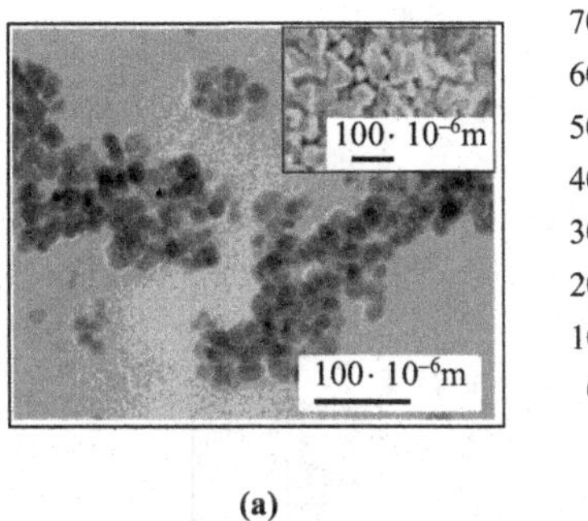

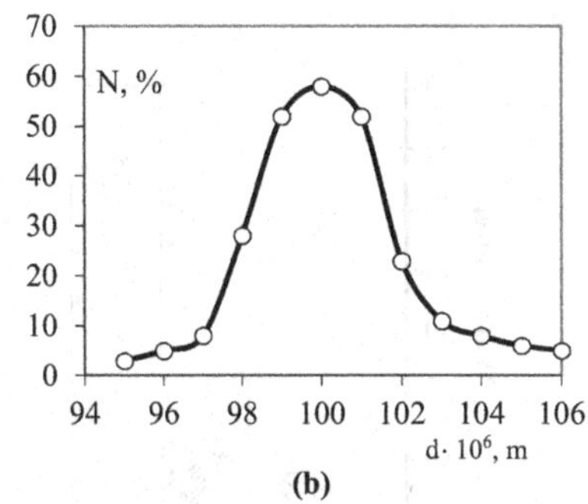

Figure 5.2 Microstructure (a) and distribution by size particles (diameter, m) of flour from the extruded sunflower seed kernels FESSK (b).

protein content (38.73%) than the mixtures of FESSK and wheat flour (33.24%), and wheat flour (15.08%).

Water-absorbing, water-binding and water-holding capacities of flour from the extruded kernel of sunflower seeds (FESSK). Water-absorbing, water-binding and water-holding capacities are associated primarily with the interaction of food raw materials proteins with water and are characterized by the adsorption of H_2O molecules with the participation of hydrophilic amino acid residues. The ability of proteins to contain water depends on the characteristics of the amino acid and fractional and structural protein composition; method of technological processing; medium pH; temperature; the presence of carbohydrates, lipids, and proteins in raw materials (Subası et al., 2021).

Particles of FESSK are characterized by a fairly uniform diameter distribution (<d100 nm), close to monodispersity. Ultrafine dispersion leads to the ability of FESSK to sorption and hydration, while there is a shift toward chemically bound moisture compared to the free form of H_2O (Figure 5.2).

By determination of the water-binding capacity by indicator and thermogravimetric methods, after swelling of FESSK, the content of bound water was determined: physicochemical, 69.5%–70.3%; physicomechanical, 22.1%–23.1%, and free, 6.9%–7.8% (Figure 5.3).

In flour systems containing proteins, fats, polysaccharides (carbohydrates), and water, their water- and fat-holding abilities are adjusted due to the structure-forming and stabilizing properties of the main raw ingredients, in particular flour (Subası et al., 2021). It has been found that in technological media such as milk, whey, kefir, NaCl, acetic acid and sucrose solutions, FESSK has a high hydration capacity, which increases with increase in ambient temperature (Table 5.6).

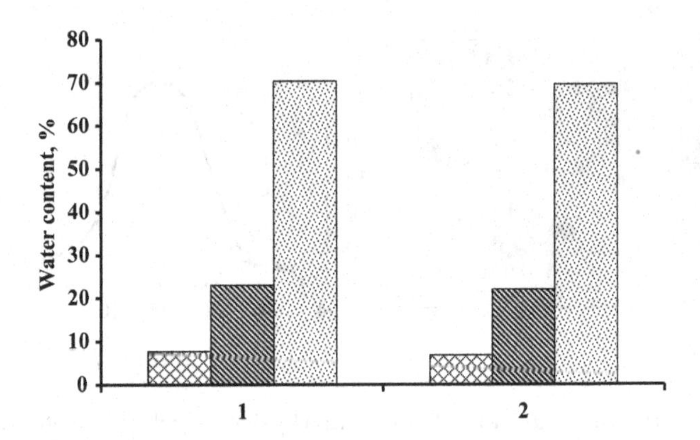

Figure 5.3 Distribution of water in the FESSK after swelling determined by: 1, thermogravimetric method; 2, chemogravimetric method: [dotted], the content of physicochemically bound water, %; [hatched], the content of physically and mechanically bound water, %; [cross-hatched], the content of free water, %.

TABLE 5.6 WATER ABSORPTION COEFFICIENTS OF FESSK IN TECHNOLOGICAL MEDIA

Polar Medium	Water Absorption Coefficients, at Temperature	
	18 ± 2°C	40 ± 2°C
Sodium hydrocarbonate, pH 6.0	11.4±0.4	11.8±0.4
Acetic acid solution, pH 4.5	10.8±0.4	11.1±0.4
Sodium chloride solution, 0.5%	12.1±0.5	12.4±0.5
Sodium chloride solution, 1.7%	12.4±0.5	12.5±0.5
Sucrose solution, 1.1%	12.6±0.5	12.7±0.5
Sucrose solution, 5.0%	12.8±0.5	13.0±0.5
Milk	11.9±0.4	12.2±0.4
Kefir	11.8±0.4	12.0±0.4
Whey	12.1±0.4	12.4±0.4

The pronounced hydrophilic properties and the tendency of FESSK to form aqua associates are due to the high dispersion and active specific surface of its particles with polarized regions of biopolymer molecules (proteins, polysaccharides) and with their charged intramolecular complexes (Dabbour et al.,

2018; Ivanova et al., 2013; Oseyko et al., 2020). These factors increase the intensity of hydration (swelling coefficient, K_w) and water absorption capacity (WAC) of FESSK (Figure 5.4).

During 40×60^{-1}s, the weight of all "FESSK+H_2O" systems increases to the maximum by 1.9–2.3 times (Figure 5.4a). The highest WAC value is determined for "FESSK+H_2O" system with the ratio of 1:1.8, and the lowest for the system with the ratio of 1:0.6. A high swelling rate is characteristic of the area at $t<10\times60^{-1}$s (Figure 5.4b). The swelling rate of FESSK is higher by 1.16–1.19 times compared to flour mixture FESSK with wheat flour in the ratio 1:9 and by 1.21–23 times compared to wheat flour. This can be explained by different: (a) chemical composition (amino acids, proteins and hydrocarbons), and (b) the number of hydrophilic sites (including hydrophilic amino acid residues) capable of adsorbing H_2O molecules and forming aqua associates with protein–carbohydrate complexes of flour systems. Due to low moisture content, the hydrophilic groups of flour interact with water molecules, forming a monomolecular layer; at high moisture content, a multilayer structure (aqua-associate) is formed around the protein globule with simultaneous penetration of water into interior sites. After $(40\pm5)\times60^{-1}$s, the swelling process ends and reaches its maximum value $K_{sw}\sim178\%$ after 40×60^{-1}s for FESSK; $K_{sw}\sim165\%$ after 42×60^{-1}s for FESSK with wheat flour in the ratio 1:9; $K_{sw}\sim154\%$ after 45×60^{-1} for wheat flour. That is, the swelling process of FESSK is reduced by $(2–5)\times60^{-1}$s compared to the mixture of FESSK with wheat flour in the ratio 1:9 and wheat flour due to the greater aquaphilic effect of FESSK particles. As a result, the hydration capacity of FESSK increases due to an increase in the number of hydrophilic centers (Subası et al., 2021). The flour from the extruded kernel of sunflower seeds (FESSK) shows ability to bind and hold water by 1.25–1.27 times higher than the mixture of FESSK with wheat flour in the ratio of 1:9 times and higher by 1.45–1.47 times than wheat flour (Figure 5.5).

The evaluation of the structure-forming properties of flour from the extruded sunflower seed kernels was carried out by its ability to reduce the surface tension of water at different amounts of FESSK from 0.1% to 1.0% in flour suspension (Figure 5.6).

A decrease in surface tension at the phase distribution boundary shows an increase in the concentration of surfactants in the monolayer, which is due to an increase in the content of proteins in the solution. The ability of FESSK to reduce surface tension during adsorption at the phase boundary indicates its high surface activity, which confirms the expediency of using FESSK in the composition of emulsions and foams. The processes of structure formation, in particular foaming and emulsification, are determined by the interfacial tension index and dispersion density. The effective viscosity increases by 3.33 times with an increase in the FESSK concentration from 1.0% to 3.0% (Figure 5.7).

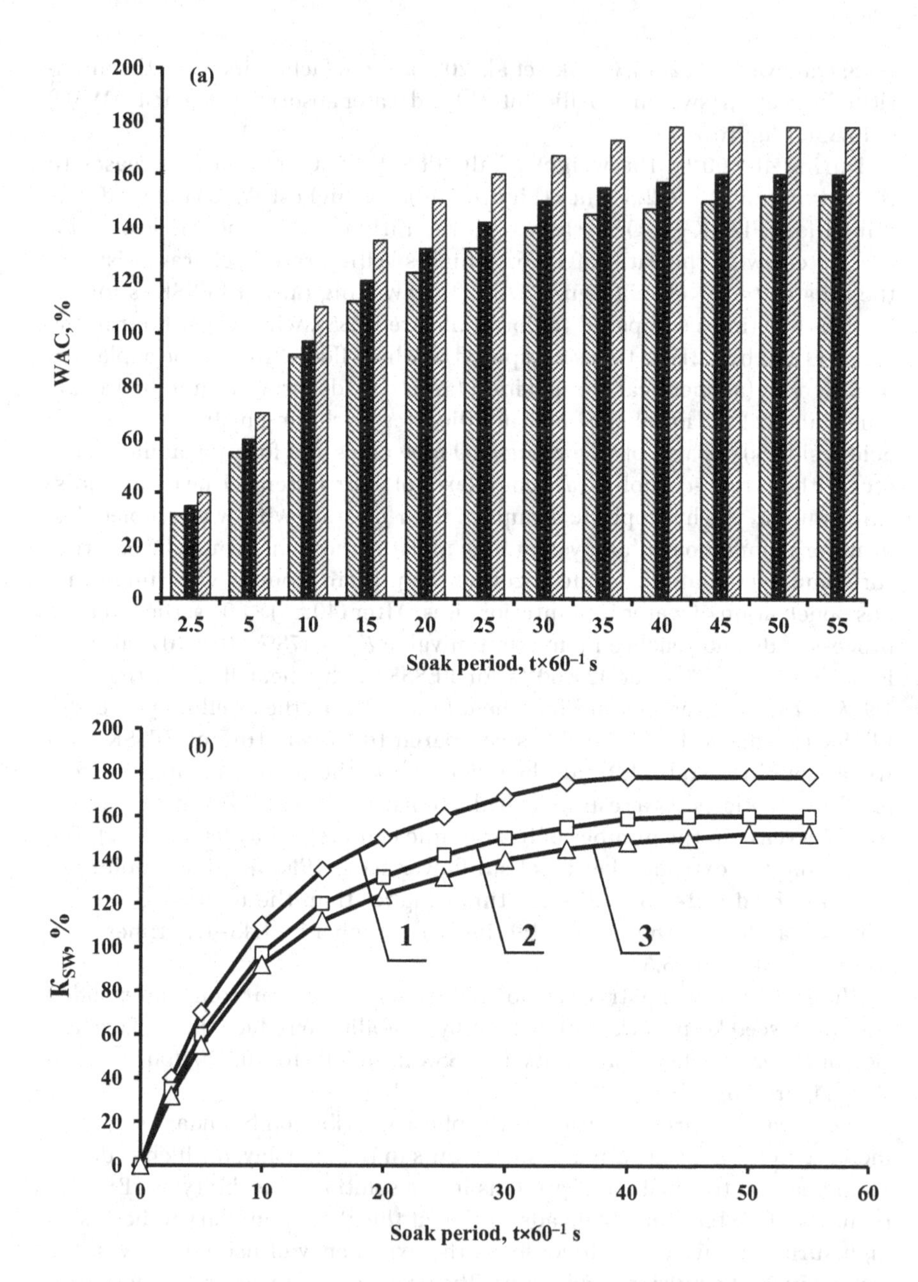

Figure 5.4 Intensity of FESSK hydration (a): water absorption capacity (WAC) of "FESSK+H_2O" systems in the ratio of FESSK:H_2O: ▧, 1:0.6; ■, 1:1.2; ▨, 1:1.8; (b), swelling kinetics: 1, FESSK; 2, FESSK with wheat flour in the ratio of 1:9; 3, wheat flour.

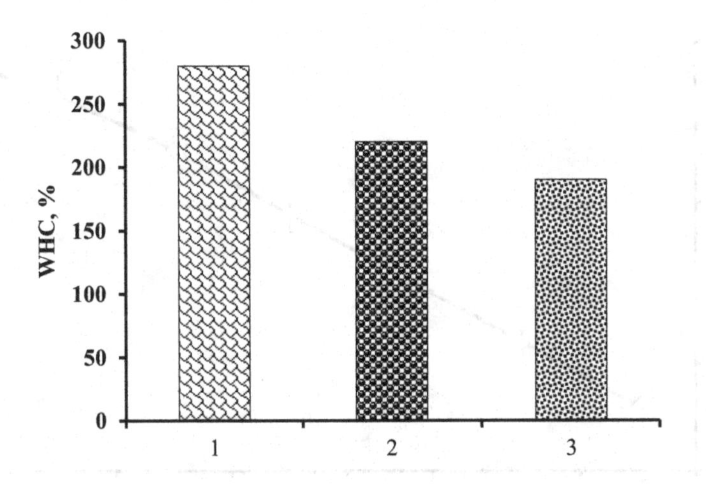

Figure 5.5 Water-holding capacity (WHC) of flour samples: ▨, FESSK; ▩, FESSK/WF in the ratio of 1:9; ▦, wheat flour.

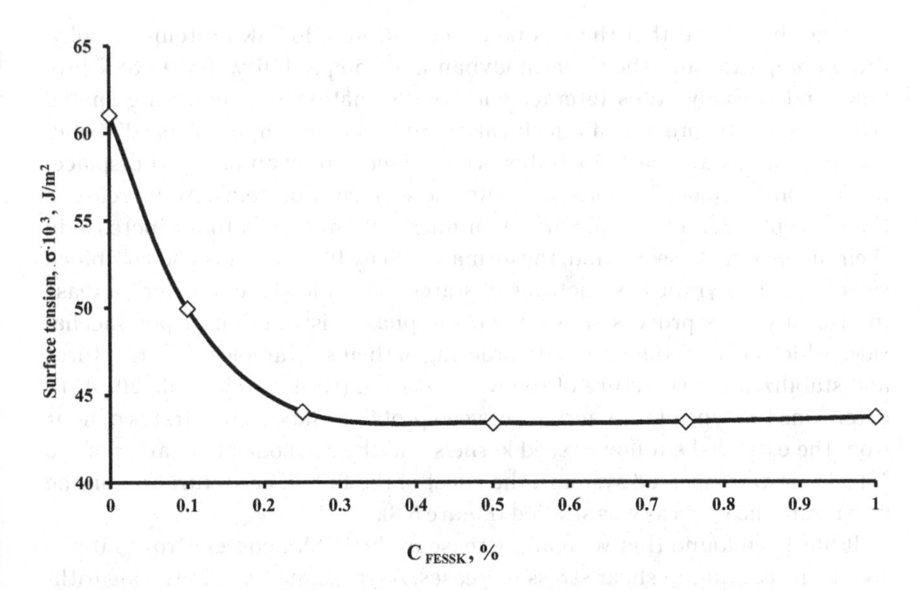

Figure 5.6 Surface tension isotherm of aqueous suspensions of flour from the extruded sunflower seed kernels (FESSK) with different concentration (C_{FESSK}).

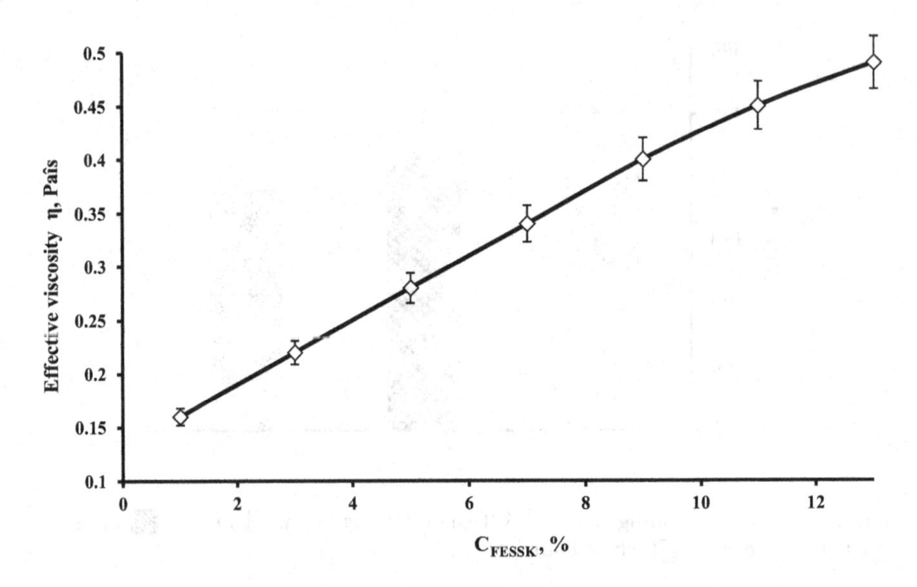

Figure 5.7 Effective viscosity of aqueous suspensions of flour from the extruded sunflower seed kernels (FESSK) at the shear rate $100\,s^{-1}$.

It can be argued that there is no coacervation of FESSK protein–carbohydrate complexes and their thermodynamic incompatibility, and FESSK proteins and carbohydrates interact with the formation of penetrating spatial structures in the process of micellization. In this case, a microphase distribution of proteins and carbohydrates occurs, followed by an oriented displacement of polysaccharide molecules onto the surface of proteins. An increase in the concentration of carbohydrates in microvolumes leads to an increase in their independent association, the formation of hydrogen bonds and combined sites from the pyranose structures of starch, which leads to a faster increase in viscosity. This process slows down the phase distribution of polysaccharides which ensures the necessary ordering of their supramolecular structures and stabilizes the structure of the whole system (Hnitsevych et al., 2019). To determine the type of interaction between proteins and carbohydrates of flour from the extruded sunflower seed kernels and the rheological behavior of the "FESSK+H_2O" dispersed systems, the effect of the FESSK concentration on the maximum shear stress was studied (Figure 5.8).

It has been found that with an increase in the FESSK content from 1.0% to 15.0%, the maximum shear stress increases. At the same time, in the logarithmic coordinate system dependence of the maximum shear stress on the content of flour from extruded sunflower seed kernels, there is a breakpoint in the curve

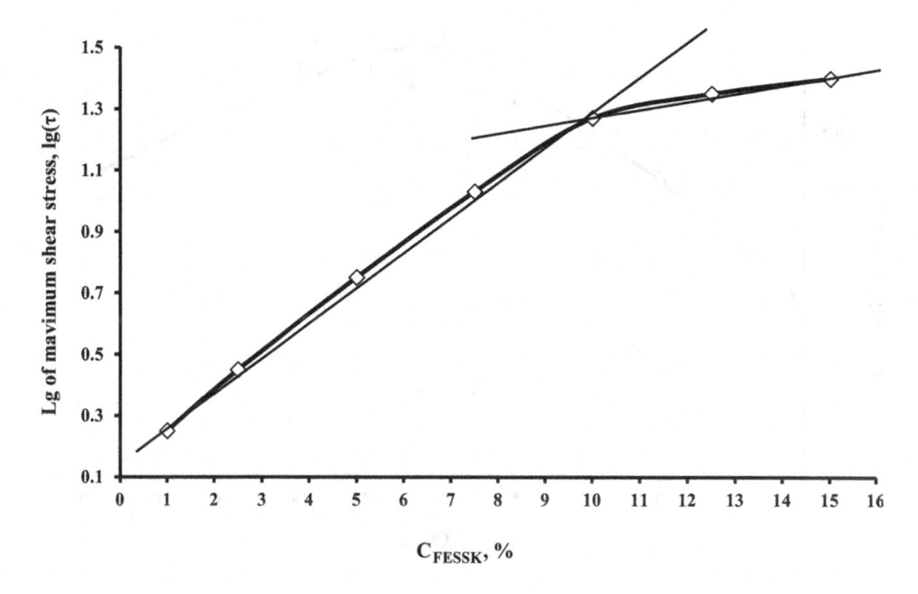

Figure 5.8 Dependence of the logarithm of the maximum shear stresses of the dispersed system "FESSK+H_2O" on the concentration of flour from the extruded sunflower seed kernels (FESSK).

for the FESSK content of 10.0%. This is due to a change in the type of interaction between proteins and carbohydrates, in particular, pentosans: starch, soluble fractions of hemicellulose and pectin of FESSK. The structure-forming ability of "FESSK+H_2O" aqueous dispersions increases by 3.8±0.2 times up to a FESSK concentration of 10.0% and by 2.3±0.1 times at FESSK concentration more than 10.0%. That is, with a FESSK content of 10.0%, the maximum implementation of the structure-forming properties of flour from the extruded sunflower seed kernels (FESSK) is achieved, and the dispersed systems are characterized as viscoplastic. A further increase in the maximum shear stress is a consequence of a change in the interaction of proteins and carbohydrates of FESSK. This is evidenced by the rate of increase in the maximum shear stress, determined by the tangent of the slope of the curve, which decreases by 39.5%±0.2% that is associated with a change in the solubility of protein–carbohydrate complexes, their molecular weight and diffusion coefficient. This is consistent with the studies of other authors, which proved that with an increase in the content of pentosans, the particle size of the protein-pentosan increases (Hnitsevych et al., 2019). Thus, the introduction of FESSK into food dispersion systems increases their nutritional and biological value; improves the ability to hold liquid in the structure of the product, increases the stability during storage, and increases

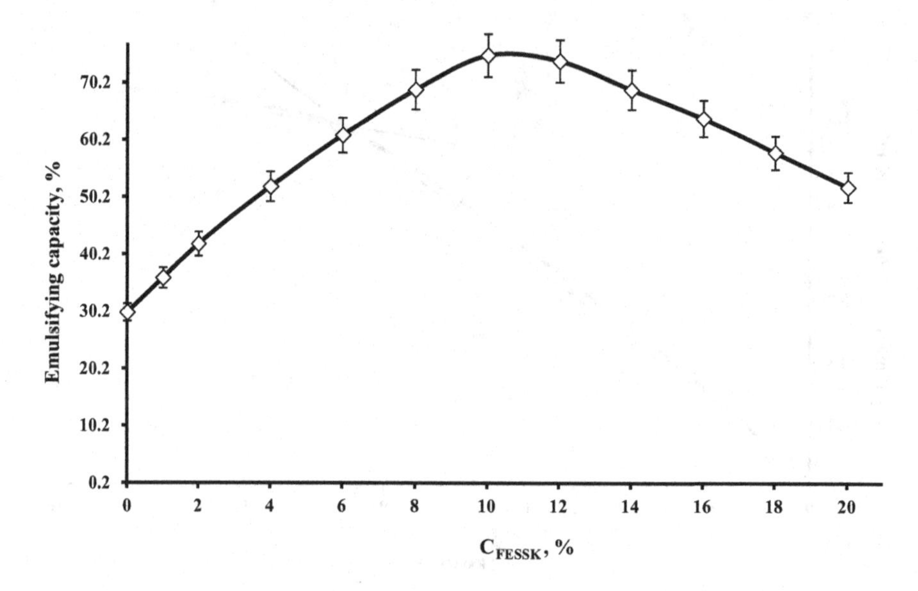

Figure 5.9 Dependence of the phase inversion point of the "Oil+H_2O+FESSK" emulsion system on the mass fraction of flour from the extruded sunflower seed kernels (FESSK).

the viscosity of polyphasic food systems. It is shown that the introduction of different amounts of FESSK into the composition of dispersed systems makes it possible to control the viscosity as a factor of the system stability. Since the counteracting factor of the oil emulsification is the value of effective viscosity, which leads to significant energy consumption, it is necessary to evaluate the emulsifying ability of FESSK.

Emulsifying properties of FESSK. Gel-forming properties are characterized by the ability of a colloidal solution of proteins to pass from a freely dispersed state into a bound dispersed state due to the creation of a three-dimensional network of intermolecular interactions by protein molecules, thereby affecting the structural, mechanical and rheological properties of food products (Subası et al., 2021). The emulsifying capacity of the "Oil+H_2O+FESSK" systems was evaluated by the phase inversion point (Figure 5.9).

Emulsions were prepared by dispersing refined sunflower oil in the aqueous phase of the emulsifier. An aqueous suspension of FESSK with concentration from 1.0% to 20.0% was used as an emulsifier, to which oil was added under constant stirring. No FEESK was added to control. The dependence of the inversion point on the content of FESSK is extreme, and with an increase in the FESSK content from 1.0% to 10.0%, the emulsifying capacity increases by 3.7–3.9 times

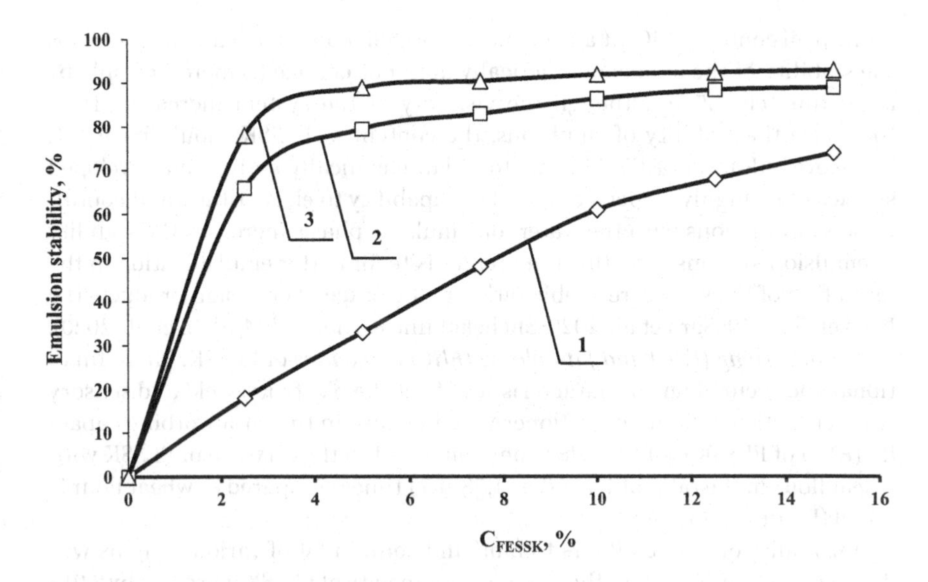

Figure 5.10 Dependence of the emulsion stability of the "Oil+H_2O+FESSK" emulsion on oil content: 1, 20.0%; 2, 40.0%; 3, 60.0%.

compared with control; a subsequent increase in the FESSK content to 20.0% leads to its decrease by 1.62–1.64 times. The maximum emulsifying capacity of FESSK in a water–oil emulsion is 62.0%±1.0% which testifies the correlation of the maximum shear stress of the dispersed systems "FESSK+H_2O" and the point of inversion of the phases of the emulsion "Oil+H_2O+FESSK." This is due to the formation of complexes, the maximum hydrophobicity of which is formed in dispersed systems with a FESSK content from 10.0% to 12.0%. When FESSK content is more than 12.0%, the hydrophilic–lipophilic balance probably changes, the dimensional characteristics increase and, as a result, the diffusion coefficient decreases. This negatively affects the emulsifying capacity; however, a positive effect on emulsion stability can be predicted (Dabbour et al., 2018; Hnitsevych et al., 2019; Salgado et al., 2012b). Studies of the stability of the emulsion "Oil+H_2O+FESSK" with an oil content of 20.0%, 40.0% and 60.0% found that with an increase in its oil content, emulsion stability increases (Figure 5.10).

Thus, with a FESSK content of 10.0% and an oil content of 60.0%, the maximum emulsion stability of 90%±2% is achieved. Discrete variation of the FESSK content from 2.5% to 15.0% increases the stability of emulsions by 4.70–4.72 times with an oil content of 20.0%; by 1.35–1.37 times with an oil content of 40.0%, and by 1.08–1.10 times with an oil content of 60.0%. In emulsion systems with an oil content of 40.0% and 60.0% with FESSK content from 10.0% to 15.0%, the stability of the emulsion practically does not change (it increases only by approximately 1.5%), although the viscosity of the system increases. Thus, to ensure the stability of emulsions, the content of FESSK should be 10.0%. Therefore, addition of FESSK, due to polar, chemically active and developed surface of its highly dispersed particles, capability to electrostatic and coordination interactions with the water–oil emulsion phase, increases the stability of emulsion systems. And this is a prerequisite for better emulsification in the formation of the structure, stabilization and storage of emulsion products (Le Priol et al., 2019; Sara et al., 2020; Shchekoldina & Aider, 2014; Slabi et al., 2020).

Fat-absorbing (FAC) and fat-holding (FHC) capacities of FESSK. These functional and technological characteristics affect the plasticity, yield and sensory characteristics of flour confectionery. An increase in the fat-absorbing capacity (FAC) of FESSK by 1.24–1.28 times compared to the mixture of FESSK with wheat flour in the ratio of 1:9 and by 1.58–1.74 times compared to wheat flour is found (Figure 5.11a).

The ability of studied flours to bind and contain fat of various origins was determined (Figure 1.16b). The fat-holding capacity of FESSK increases by 2.01–2.42 times compared to the mixture FESSK with wheat flour in the ratio of 1:9 and by 3.17–3.48 times compared to wheat flour. This is explained by the high lipophilic properties of FESSK due to the formation of protein–carbohydrate complexes with higher hydrophobicity. An upward shift in fat-absorbing and fat-holding capacities was found for oils compared to solid fats due to hydrophobic interactions of the "«π- π-Stacking" type (Lin et al., 1974; Shchekoldina &Aider, 2014). Thus, FESSK due to its specific nutrient composition and high fat-holding capacity is a promising ingredient for inclusion in a variety of food products, including dietary samples.

Foaming capacity and foam stability (foam resistance) FESSK. The foam ability (the foam resistance) of the model systems "H_2O+flour" showed that it increases with FESSK by 4.9%±0.2% compared with the mixture of FESSK and wheat flour with ratio 1:9 and by 10.5%±0.5% compared to wheat flour (Figure 5.12).

This is explained by a large decrease in the surface tension of FESSK proteins and an acceleration of the coagulation of protein molecules, which leads to an increase in the volume concentration of the air phase in the protein solution and a decrease in the size of air bubbles. As a result, the process of air dispersion in the heterogeneous "H_2O+FESSK" system is simplified. It has been found that the foam resistance in the "H_2O+FESSK" system is higher by 3.7%±0.2% compared to the mixture of FESSK and wheat flour in the ratio 1:9 and by 8.7%±0.4% compared to wheat flour because of the better stabilizing effect of FESSK proteins. This is due to the presence of highly dispersed FESSK particles with a chemically active surface on the phase interface in

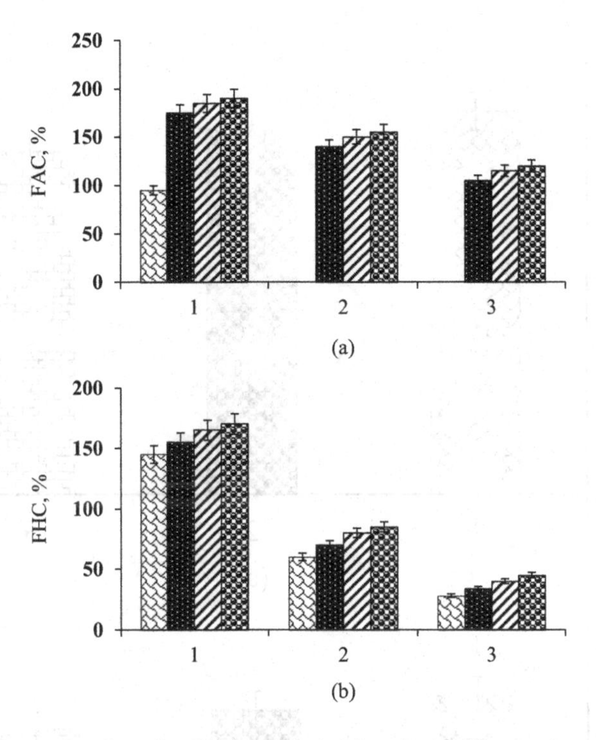

Figure 5.11 Fat-absorbing (FAC) (a) and fat-holding (FHC) (b) capacities: 1, flour from the extruded sunflower seed kernels (FESSK); 2, mixture of FESSK and wheat flour in the ratio of 1:9; 3, wheat flour: ▨, milk fat; ■, a mixture of milk fat: sunflower oil in the ratio of 5:1; ▨, sunflower oil; ▩, corn oil.

the adsorption layer, which increases the adhesion force between protein molecules. As a result, the mobility of the liquid decreases, and its flow in the film slows down, thereby preventing the coalescence of foam bubbles and increasing their dispersion. The viscosity of the liquid in the foam films also increases, which slows down their destruction and stabilizes the foam resistance at a higher level (Arntfield, 2018; Dabbour et al., 2018; Hakkinen et al., 2018; Pickardt et al., 2015; Salgado et al., 2012a, b; Sara et al., 2020; Shchekoldina & Aider, 2014; Slabi et al., 2020).

Gelling properties of FESSK. The lowest gelation concentration (LGC) is the lowest level at which the test sample does not fall or slide when the sample tube is inverted (Adebowale & Lawal, 2003; Hnitsevych et al., 2019). LGC was determined with minor modifications according to the method reported by (Adebowale & Lawal, 2003). Model systems of aqueous suspensions

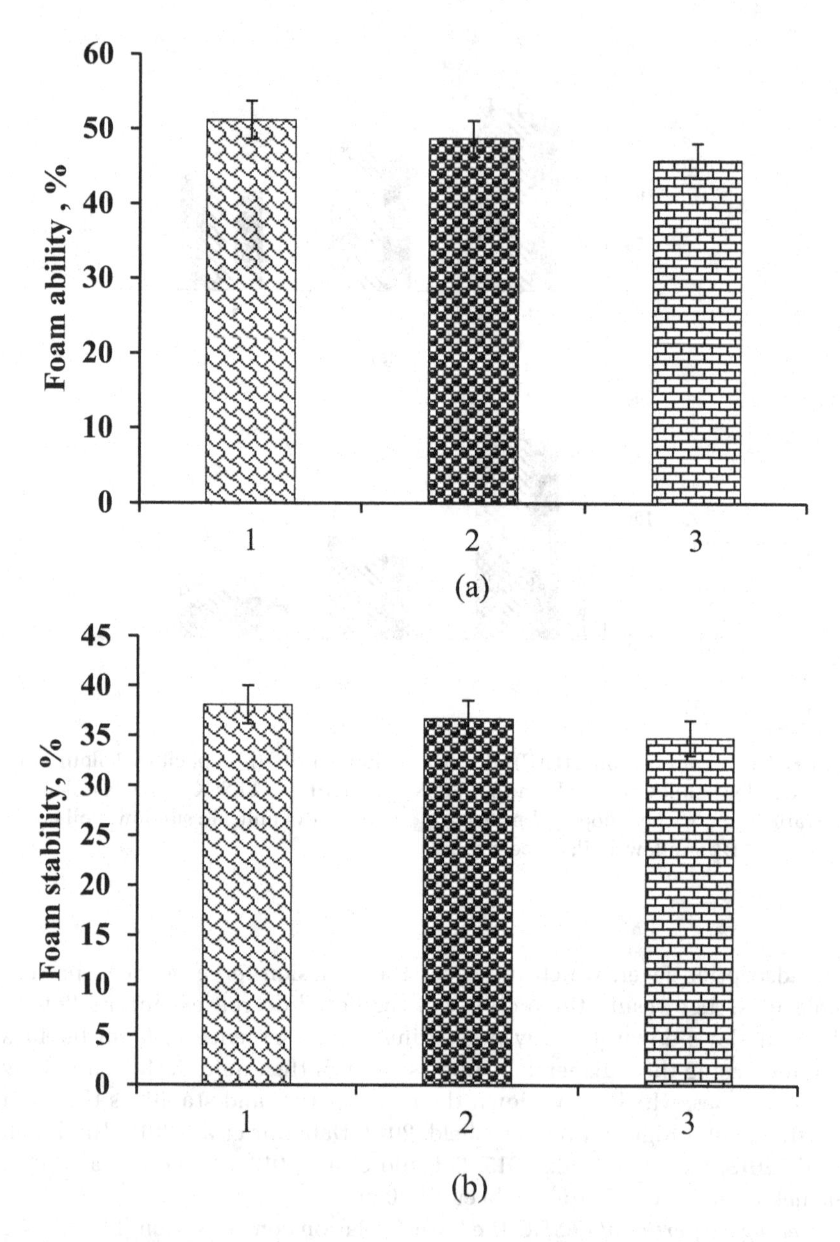

Figure 5.12 Foam ability (a) and foam stability (b) of aqueous suspension "H_2O + flour": ▨, flour from extruded sunflower seed kernels (FESSK); ▩, flour mixture "FESSK: wheat flour" in the ratio of 1:9; ▤, wheat flour.

"H_2O+flour" were used for analysis at different concentrations of flour: 6.0%; 8.0%; 10.0%; 12.0%; 14.0%; 16.0% (w/v) at temperature 80°C (on a water bath) for 5, 10, 15, 20 and 25 min, and then cooled to 20°C. The viscosity and maximum shear stress of the studied water–flour system, which forms a gel-like mass, increase with an increase in the mass fraction of FESSK (Table 5.7; Figure 5.13). This was observed on similar dispersed systems in other works (Hnitsevych et al., 2019).

It is determined that the viscosity of gel-like systems subjected to thermal induction at a temperature of 80°C for 5–25 min increases with an increase in flour concentration from 6.0% to 16.0% (w/v): by 3.91±0.19 times for FESSK; by 3.79%±0.18% times or the flour mixture FESSK with wheat flour in the ratio 1:9; in 3.67±0.17 times for wheat flour. At the same time, the strength of the gel-like mass with FESSK is greater by 13.2%±0.6% compared to FESSK/WF in the ratio 1:9 and by 26.4%±0.9% compared to wheat flour (WF). This is due to the different composition of the biopolymer fractions that form the spatial structure of the FESSK, FESSK/WF in the ratio 1:9, and WF gels. The lowest concentration of aqueous flour dispersions required for the formation of gels at a temperature of 80°C is 8.0% (w/v) for FESSK (after 10 min); 8.0% (w/v) for FESSK/WF (after 15 min); 8.0% (w/v) for WF (after 20 min). However, only the FESSK dispersed system at a concentration of 8.0% (w/v) forms a gel of medium strength after 20 min; aqueous dispersion of FESSK forms a hard and very strong gel at a concentration of 10.0% (w/v) after 20 and 25 min (Table 5.7). That is, an increase in the concentration of water–flour suspensions above 8.0% (w/v) and/or an extension of the heating time contributes to the formation of more stable and solid gels than those obtained at suspension concentrations below this value and/or a decrease in the heating time (Table 5.5). This is explained by the fact that a high concentration of biopolymers (proteins and polysaccharides) simplifies the interaction between polypeptide and polysaccharide chains. In addition, an increase in the heating time of aqueous flour dispersions contributes to the increase in biopolymer structures availability, which has a positive effect on the degree of interaction between polymer chains. An inverse relationship has also been established between gel time formation and LGC (Figure 5.18). The obtained data are in line with the studies of other authors (Adebowale & Lawal, 2003; Gonzalez-Perez & Vereijken, 2008; Salgado et al., 2012a, b.

5.2.3 Flour Confectionery Products Enriched with Flour from the Extruded Sunflower Seed Kernels

The basis for the development of a new type of gingerbread is the recipe for gingerbread using traditional technology with cocoa powder. The recipes for gingerbread made according to traditional technology from wheat flour and

TABLE 5.7 GEL-FORMING PROPERTIES OF AQUEOUS FLOUR SUSPENSIONS (THERMAL INDUCTION AT 80°C)

Flour	Concentration of Flour, % (w/w)	The Consistency of the Gel Mass					Maximum Shear Stress (kPa)				
		Heating Time (min)									
		5	10	15	20	25	5	10	15	20	25
FESSK	6.0	a*	a	a	a	a	–	–	–	–	–
	8.0	a	b	c	d	d	–	0.1618	0.2434	0.3285	0.3322
	10.0	b	c	d	e	f	0.1648	0.2469	0.3325	0.5118	0.6085
	12.0	b	c	d	f	f	0.1675	0.2482	0.3347	0.6104	0.6125
	14.0	c	d	e	f	f	0.2492	0.3351	0.5163	0.6192	0.6237
	16.0	c	e	f	f	f	0.2528	0.5239	0.6259	0.6293	0.6312
Mixture of FESSK and wheat flour in the ratio 1:9	6.0	a	a	a	a	a	–	–	–	–	–
	8.0	a	a	b	c	d	–	–	0.1613	0.2321	0.3182
	10.0	a	b	c	d	e	–	0.1624	0.2326	0.3237	0.5026
	12.0	b	c	d	e	f	0.1646	0.2353	0.3272	0.5102	0.6048
	14.0	c	d	e	e	f	0.2398	0.3299	0.5044	0.5108	0.6084
	16.0	c	d	e	f	f	0.2452	0.3319	0.5099	0.6089	0.6167
Wheat flour	6.0	a	a	a	a	a	–	–	–	–	–
	8.0	a	a	a	b	c	–	–	–	0.1611	0.2218
	10.0	a	a	b	c	c	–	–	0.1618	0.2248	0.2276
	12.0	b	c	c	d	d	0.1626	0.2259	0.2299	0.3142	0.3188
	14.0	b	c	d	e	f	0.1649	0.2286	0.3196	0.5026	0.6028
	16.0	c	d	e	e	f	0.2386	0.3276	0.5065	0.5098	0.5917

* a, no gel formation; b, very weak gel (very low viscosity mass); c, weak gel (low viscosity mass); d, medium gel (medium viscosity mass); e, hard gel (viscous mass); f, very strong gel (very viscous mass).

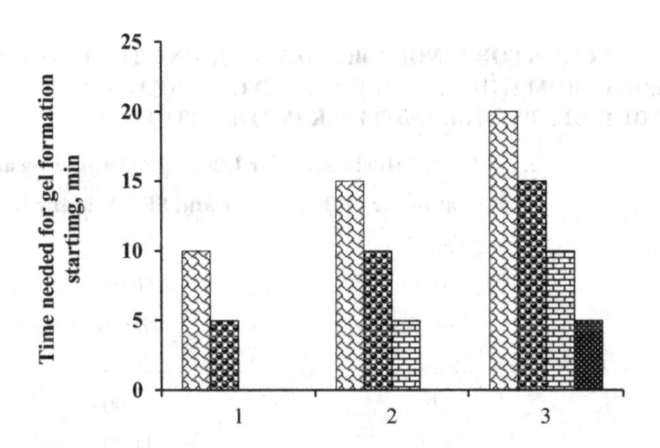

Figure 5.13 Gel-forming properties of aqueous suspension "H_2O+flour" (thermoinduction at 80°C): 1, flour from extruded sunflower seed kernels (FESSK); 2, FESSK mixed with the wheat flour in the ratio of 1:9; 3, wheat flour. Flour concentration, % (w/v): , 8.0; , 10.0; , 12.0; , 14.0.

gingerbread with partial replacement of wheat flour (10%) with flour from the extruded sunflower seed kernels (FESSK) are shown in Table 5.8.

Based on the functional and technological potential of flour from the extruded sunflower seed kernels regarding water and fat-holding, fat-emulsifying and stabilizing abilities, FESSK affects the time of formation, physical and chemical parameters, structural and mechanical properties of gingerbread dough. That is, using FESSK, it is possible to obtain a more resistant flour product with different shapes. The indicator of the "falling number" of flour is responsible for the taste, volume and appearance of flour products. It evaluates the state of the carbohydrate–amylase complex of flour, which is characterized by the gas-forming ability of flour and the activity of amylase in the dough. Determination of the "falling number" of flour using the Hagberg–Perten method (Petrenko et al., 2015) showed that when mixing wheat flour with flour from extruded sunflower seed kernels, rarefaction of the flour system does not occur (Figure 5.14).

That is, the autolytic enzymes of FESSK do not have a significant effect on the hydrolysis of starch in the "H_2O+wheat flour+FESSK" system. An increase in the "falling number" by 1.11–1.15 times along with an increase in the FESSK content is due to swelling of proteins and starch of flour from the extruded sunflower seed kernels and the formation of protein–carbohydrate complexes (Shchekoldina & Aider, 2014; Slabi et al., 2020). To predict the formation of the structure of flour confectionery products during baking, the process of gelatinization of water–flour suspensions at a constantly increasing temperature was studied using a Brabender amylograph (Pojic et al., 2013). It was found that the partial replacement of wheat flour with FESSK in the amount of 5.0%, 10.0%

TABLE 5.8 RECIPES FOR GINGERBREAD ACCORDING TO TRADITIONAL TECHNOLOGY FROM WHEAT FLOUR AND GINGERBREAD FROM THE MIXTURE OF WHEAT FLOUR AND FESSK IN THE RATIO 90:10

	Raw Materials, kg for 1,000 kg of Gingerbread from	
Raw Materials	**Wheat Flour**	**Wheat Flour and FESSK in the Ratio 90:10**
Wheat flour	509.07	458.16
FESSK	–	50.91
Sugar	230.89	230.89
Natural honey	221.95	221.95
Butter	56.00	56.00
Melange	11.70	11.70
Baking soda	1.54	1.54
Ammonium carbonate	7.28	7.28
Cocoa powder	11.19	11.19
Cinnamon	3.05	3.05
Palenque	10.18	10.18
Total	1062.85	1,062.85
Yield	1000.00	1,048.00

and 15.0% reduces the temperature of the gelatinization starting of the flour mixture by 2.0°C–3.0°C, reduces the time before the gelatinization starting by 2.0–2.5 min, but reduces the value of the maximum suspension viscosity by 5.1%–5.9% (Table 5.9).

Significant changes in the temperature of the maximum viscosity of the flour paste in the dispersed systems "H_2O+wheat flour+FESSK" are not observed in comparison with wheat flour. The maximum viscosity of the flour paste in the studied dispersed systems is in the temperature range of 96°C–97°C, regardless of their composition. This is due to the complexity and multicomponent nature of the studied water–flour suspensions, in which, in addition to starch, there are other polysaccharides, in particular pentosans and hexosans, soluble in water affecting the viscosity of the gelatinized suspension (Figure 5.15).

The presence of water-soluble proteins, in particular those present in FESSK, also indicates a significant effect on the duration of the gelatinization process, and the initial and final viscosity of the aqueous dispersion. As a result, the recommended ratio of the wheat flour to flour from the extruded sunflower seed kernels (FESSK) was determined as 9:1 or as a percentage 90:10.

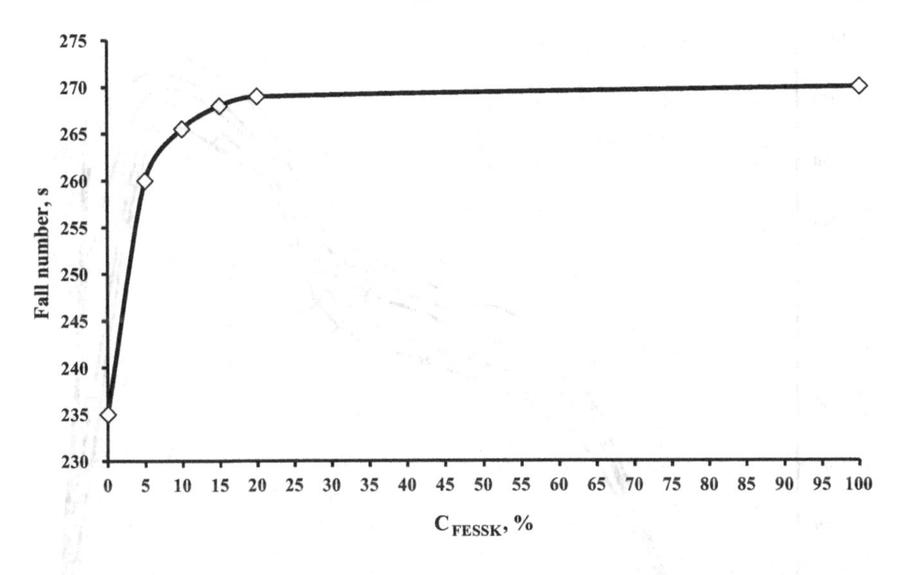

Figure 5.14 The influence of the flour from an extruded sunflower seed kernels (FESSK) content on the "fall number" of "H_2O+wheat flour+FESSK" dispersions.

TABLE 5.9
RHEOLOGICAL AND TECHNOLOGICAL PROPERTIES OF WATER–FLOUR SUSPENSIONS

Parameter	The Composition of Water–Flour Suspensions				
	Wheat Flour	Wheat Flour with Partial Replacement with FESSK, %			FESSK
		5	10	15	
Flour gelatinization start temperature, °C	75.0	73.0	72.5	72.0	71.0
Time before gelatinization, min	19.5	17.5	17.3	17.0	16.8
Temperature of maximum viscosity of flour paste, °C	96.0	96.5	97.0	97.0	97.0
Maximum viscosity of flour paste, units of device	1000.0	949.2	944.3	942.4	939.5

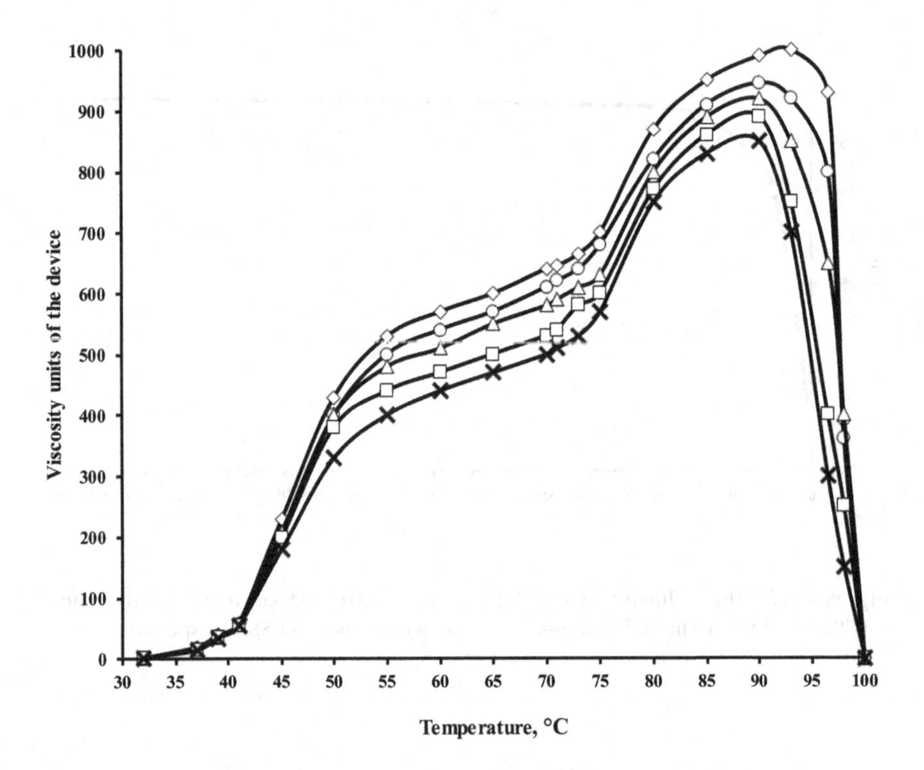

Figure 5.15 Dependence of the viscosity of water–flour suspension on temperature. Contents flour from the extruded sunflower seed kernels (FESSK) in flour taken for suspension preparation, %: ▨, 0; ▩, 5; ⊟, 10; ■, 15; ◆, 100.

Studies of the physicochemical parameters and structural and mechanical properties of gingerbread dough found that partial replacement of wheat flour (10%) with flour from the extruded sunflower seed kernels increases the maximum shear stress by 13.8%–14%; plastic viscosity by 28.9%–31.8%, and decreases dough formation time by 1.8±0.3 min and dough density by 7.5%–7.7% compared to control (Table 5.10).

These changes are due to the ability of FESSK to a stronger holding of moisture and fat in the structure of the dough because of the complexing action of FESSK biopolymers with H_2O dipoles and lipids, and also due to the activation of biochemical processes. In addition, FESSK contributes to a decrease in adhesive strength by 1.16–1.18 times compared to control to the steel surface due to the formation of an aqua complex and a decrease in the free moisture content on the dough surface.

TABLE 5.10 PHYSICOCHEMICAL PARAMETERS, STRUCTURAL AND MECHANICAL PROPERTIES OF GINGERBREAD DOUGH

Parameters	Gingerbread Dough from	
	Wheat Flour	Wheat Flour and FESSK in the Ratio 90:10
Content of moisture, %	24.6±1.2	26.2±1.3
Density, g/cm³	1.31±0.06	1.21±0.04
Maximum shear stress, Pa	504.9±20.0	576.8±22.0
Adhesion strength (steel), kPa	2.8±0.1	2.3±0.1
Plastic viscosity, kPa·s (at $\gamma = 0.02\ s^{-1}$)	7.8±0.4	10.3±0.5
Dough formation time, min	6.8±0.3	5.0±0.2

TABLE 5.11 A SENSORY ASSESSMENT OF CUSTARD GINGERBREAD QUALITY

Parameters	Custard Gingerbread from Wheat Flour with Partial Replacement with FESSK (%)			
	0	5	10	15
Taste, smell	5.00±0.02	5.00±0.02	5.00±0.02	5.00±0.02
Color	5.00±0.02	5.00±0.02	5.00±0.02	5.00±0.02
View at the break	4.90±0.02	4.92±0.02	5.00±0.02	4.98±0.02
Consistency	4.90±0.01	4.92±0.01	5.00±0.01	4.97±0.01
Surface	4.98±0.01	4.99±0.01	5.00±0.01	4.98±0.01
Form	4.98±0.01	4.99±0.01	5.00±0.01	4.99±0.01
Overall score	29.76	29.82	30.00	29.92

Sensory analysis found an increase in the overall quality index due to a partial replacement of wheat flour with FESSK in the amount of 5.0%; 10.0% and 15.0% of the total amount of flour: 0.06±0.01; 0.24±0.01 and 0.16±0.01 points, respectively (Table 5.11).

In the production of gingerbread with a partial replacement of wheat flour with 5.0% FESSK, the quality indicators of the product did not change much compared to control from wheat flour only, but when wheat flour was replaced with 15.0% FESSK, the texture of the product became dry and brittle. Products have the best sensory characteristics by replacing of wheat flour (10%) with FESSK (Figure 5.16).

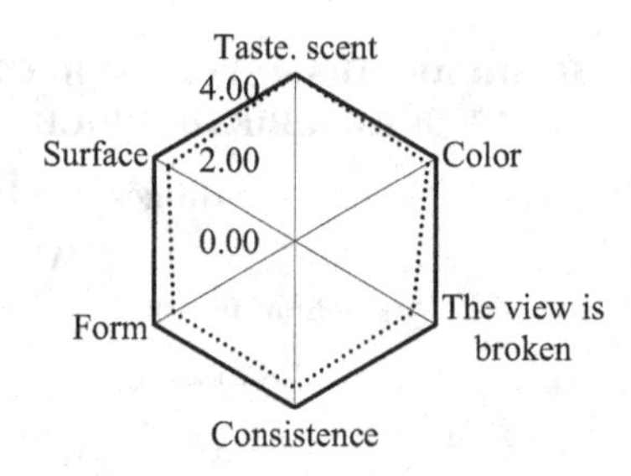

Figure 5.16 Sensory analysis of gingerbread samples: ....., gingerbread from wheat flour; ——, gingerbread with replacing of wheat flour (10%) with FESSK.

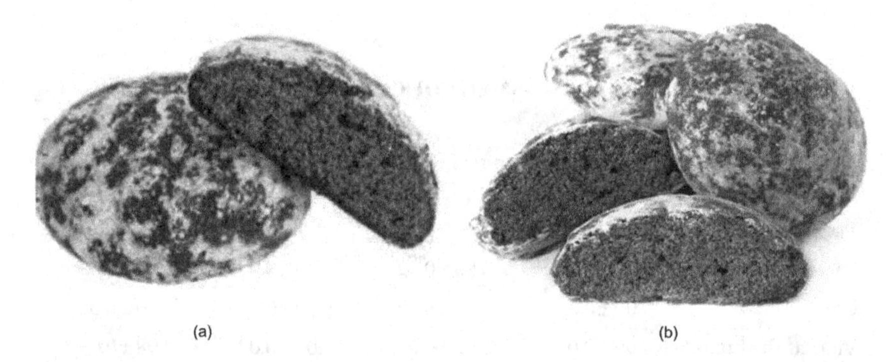

Figure 5.17 Gingerbreads from: (a) wheat flour; (b) wheat flour with replacement (10%) with flour from the extruded sunflower seed kernels (FESSK).

This preserves the texture; products have a pleasant nutty flavor; appearance is correct, with a convex glazed surface shape; smooth, nonsticky, glossy, glazed, crack-free; consistency is finely porous, elastic, tender and fractured, and a structure has small evenly distributed pores (Figure 5.17).

An improvement in the quality indicators of gingerbread with a partial replacement of wheat flour with FESSK in the amount of 5.0%; 10.0% and 15.0% has been shown: strength increases by 1.89–2.61 times, and the loss in weight during baking decreases by 1.09–1.21 times in comparison with wheat flour. In addition, the yield of the finished product increases by 5.6%–6.1% (Figure 5.18).

This is due to the structure-forming and stabilizing effect of FESSK. Moreover, changes in strength, the loss of weight during baking and yield when replacing 15.0% of wheat flour with FESSK are lower. That is, the preferable amount of wheat flour replacement with FESSK is 10.0%.

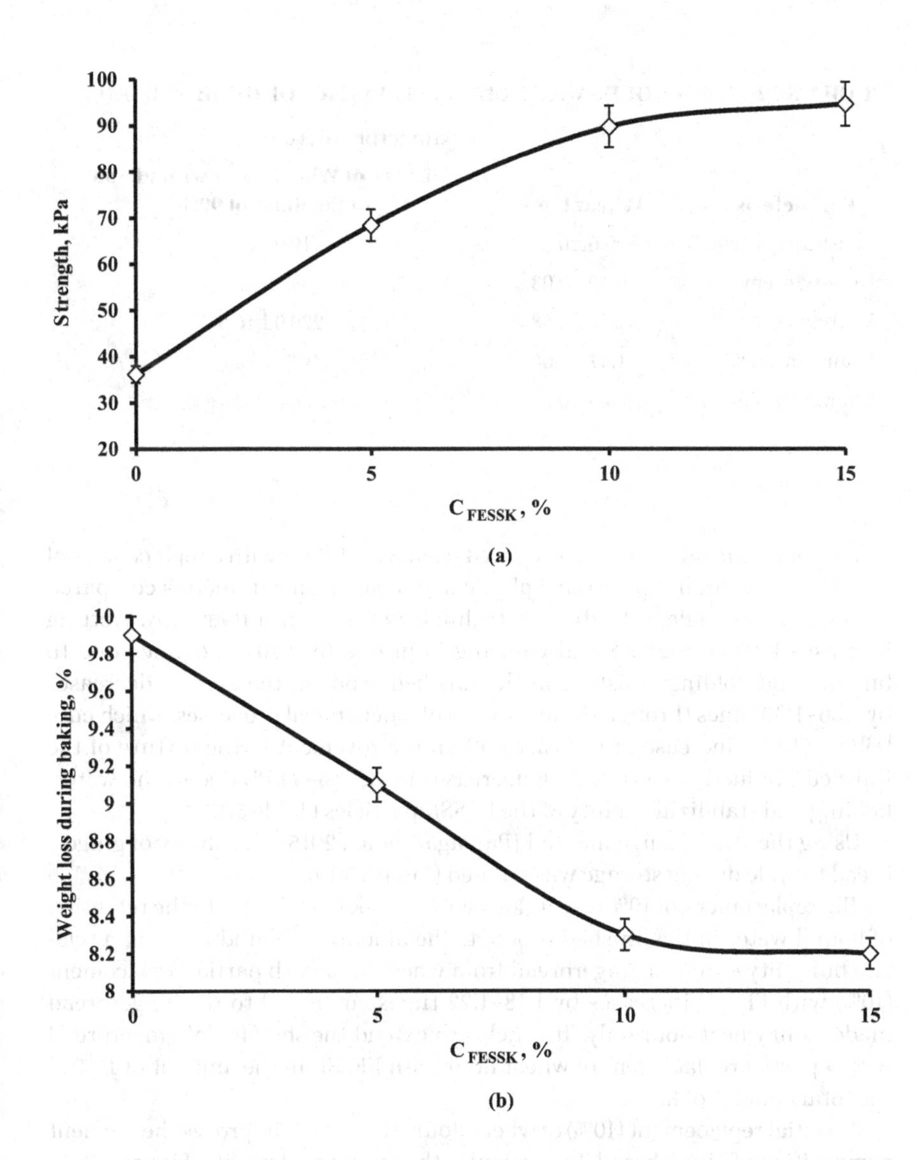

Figure 5.18 Dependence of the quality indicators of gingerbread on the amount of wheat flour replaced with FESSK: (a) strength, kPa; (b) the loss in weight during baking, %.

TABLE 5.12 PHYSICOCHEMICAL CHARACTERISTICS OF GINGERBREAD

Characteristics	Gingerbread from	
	Wheat Flour	Mixture of Wheat Flour with FESSK in the Ratio of 90:10
Moisture content, %	9.6±0.5	10.0±0.5
Density, g/cm^3	0.50±0.03	0.39±0.02
Wetting, %	190.0±9.8	229.0±10.9
Crunchiness, %	1.47±0.06	0.97±0.03
Chewing, points	4.3±0.2	4.8±0.2

It is found that gingerbread prepared from wheat flour with a replacement of 10.0% with FESSK has preferable physical and chemical parameters compared to gingerbread made by traditional technology from wheat flour only. Wetting increases by 19.4%–21.4% and chewing improves by 1.10–1.12 times due to binding and holding moisture in the finished product; the density decreases by 1.26–1.30 times through the activation of biochemical processes, which contributes to an increase in porosity and an improvement in the texture of the finished product. The crumbling decreases by 32.3%–33.7% due to the water-holding and stabilizing ability of the FESSK particles (Table 5.12).

Using the strain gauge method (Panjagari et al., 2015), the ability of gingerbread to stale during storage was studied (Figure 5.19).

The replacement of 10% wheat flour with FESSK contributes to the retention of bound water in the finished product. The amount of bound water at a relative humidity $\varphi=0.7$ in gingerbread from wheat flour with partial replacement (10%) with FESSK increases by 1.18–1.22 times compared to the gingerbread made from wheat flour only. This helps to extend the shelf life of gingerbread with a partial replacement of wheat flour with FESSK in the amount of 10% of the total amount of flour.

A partial replacement (10%) of wheat flour with FESSK improves the nutrient composition of gingerbread, in particular the amino acid profile (Figure 5.20).

This is consistent with the results obtained by a number of authors for flour products made with partial replacement of wheat flour with sunflower cake, meal and flour (Bhise et al., 2014; Martins et al., 2017; Mohammed et al., 2018; Shchekoldina & Aider, 2014).

It was shown that the partial replacement of wheat flour with FESSK in the amount of 10.0% of the total amount of flour improves the biological and nutritional value of the finished product compared to the control (Table 5.13).

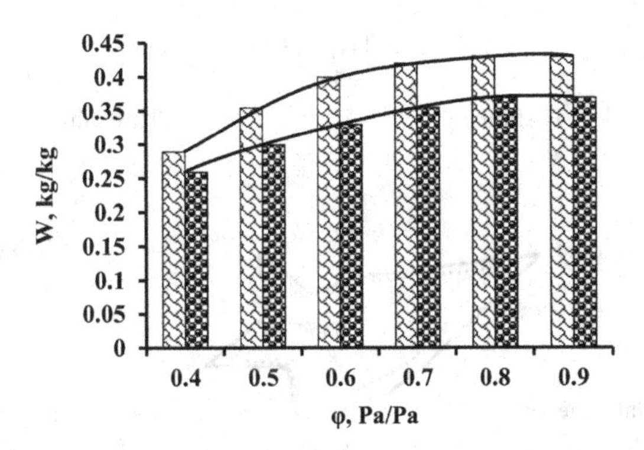

Figure 5.19 Water absorption capacity of flour (W, кг/кг) at different relative air humidity (φ, Pa/Pa) at temperature 20°C±2°C. Gingerbread, made from: ▧, wheat flour; ▩, wheat flour with replacement of 10% with flour from the extruded sunflower seed kernels (FESSK).

The content of protein, fat and ash in gingerbread from the mixture of wheat flour with FESSK in the ratio 90:10 was higher by 1.83 times, 1.03 times and 1.40 times, respectively, compared with gingerbread made from wheat flour only, while the content of carbohydrates was lower by 1.11 times. In addition, the vitamin and mineral contents in finished products are enriched. Therefore, the incorporation of FESSK as a raw ingredient in flour confectionery improves the nutritional value and consumer properties of the finished products.

5.3 CONCLUSIONS

The expediency of using a secondary product of the fat-and-oil industry, flour from extruded sunflower seed kernels (FESSK) as a raw ingredient in flour confectionery on the example of gingerbread is shown. A sufficiently high protein content, 38.73%, in FESSK with a well-balanced amino acid composition and a significant amount of soluble proteins, 76.35% of total protein value, are an important factor in the stabilization of food systems. The content of fat, 4.87% rich in unsaturated fatty acids (16 fatty acids) and the content of different vitamins and minerals (12 vitamins and 22 chemical elements) increase the biological value of FESSK. A sufficiently high solubility of FESSK proteins, by 3.29–3.35 times higher than the solubility of wheat flour proteins, is the basis for the formation of the viscosity of the flour product, better foaming, increased emulsification and gelation capabilities. High surface activity; sorption, aqua- and lipophilic abilities of FESSK contribute to the improvement of the functional

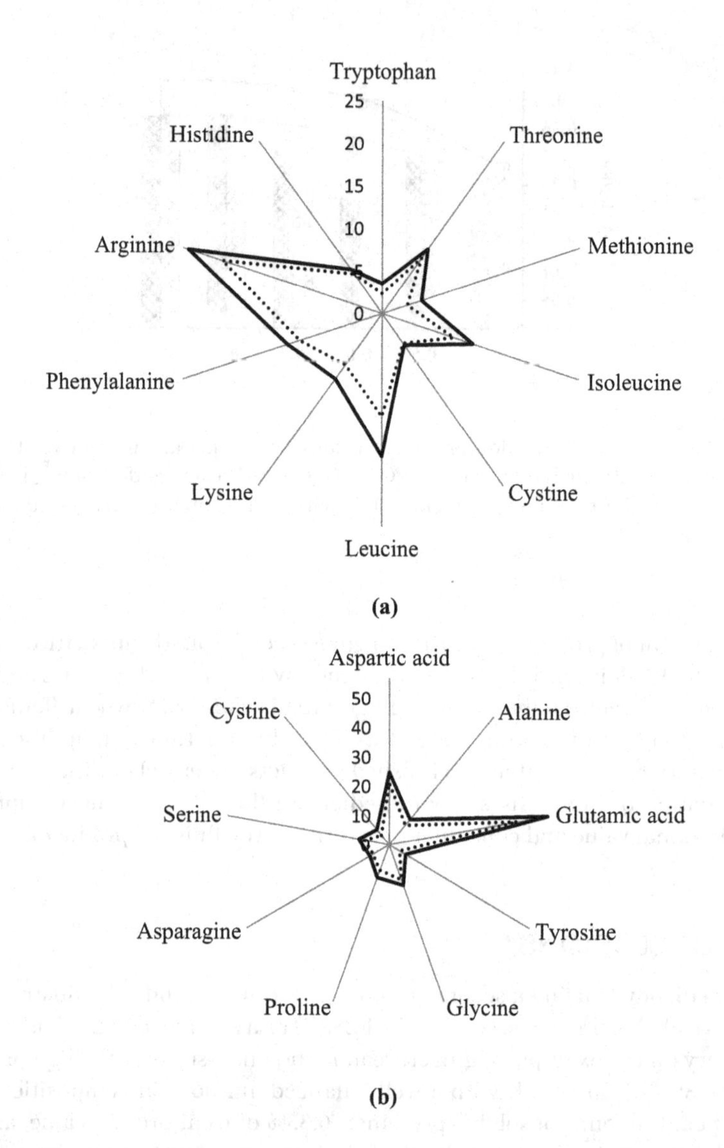

Figure 5.20 Amino acid profile of gingerbread samples: ····· gingerbread made from wheat flour; ▬▬ gingerbread wheat flour with replacement of 10% with flour from the extruded sunflower seed kernels: (a) essential amino acids and (b) nonessential amino acids.

TABLE 5.13 NUTRIENT COMPOSITION OF GINGERBREAD

Components	Gingerbread from	
	Wheat Flour	Mixture of Wheat Flour with FESSK in the Ratio of 90:10
Carbohydrates, %	71.52±0.75	64.06±0.64
Fats, %	16.28±0.51	17.23±0.41
Proteins, %	4.26±0.08	7.85±0.12
Ash, %	2.12±0.02	2.98±0.03
Content, Kcal/100 g	432.94	442.71
Calories from protein, %	3.94	7.09

and technological properties of heterogeneous flour systems. The structure-forming properties of FESSK are confirmed by an increase in the viscosity of water–flour suspensions by 3.32–3.34 times. Addition of FESSK in wheat flour increased the stability of emulsion oil-water–flour systems. The recommended amount of replacement of wheat flour with FESSK in the technology of gingerbread is 10.0%. It is shown that addition of FESSK improves the biological and nutritional value, and quality properties of flour confectionery on the example of gingerbread. Partial replacement, 10.0%, of wheat flour by flour from extruded sunflower seed kernels reduces the time of formation of gingerbread dough by 1.8 min, heat treatment losses by 1.15 times, improves the nutrient composition and helps to stabilize the consumer properties of products during storage. As it is indicated above, the development of technological procedures for the use of secondary products from the oilseed industry, as it is the case of sunflower, allows the manufacture of products with high functional, sensorial and nutritional properties; at the same time, this general strategy is more environment-friendly.

REFERENCES

Abdullah, M.H.R.O., Ch'ng, P.E., & Lim, T.H. (2011). Some physical properties of Parkia speciosa seeds. *International Conference on Food Engineering and Biotechnology*, 9(1), 43–47. http://ipcbee.com/vol9/9-B021.pdf.

Adebowale, K.O. & Lawal, O.S. (2003). Foaming, gelation and electrophoretic characteristics of mucuna bean (Mucuna pruriens) protein concentrates. *Food Chemistry*, 83(2), 237–246. https://doi.org/10.1016/S0308-8146(03)00086-4.

Adeleke, B.S. & Babalola, O.O. (2020). Oilseed crop sunflower (Helianthus annuus) as a source of food: Nutritional and health benefits. *Food Science and Nutrition*, 8, 4666–4684. https://doi.org/10.1002/fsn3.1783.

Akkaya, M.R. (2018). Fatty acid compositions of sunflowers (Helianthus annuus L.) grown in east Mediterranea region. *Rivista Italiana Delle Sostanze Grasse*, XCV(4), 239–247.

Ancuta, P. & Sonia, A. (2020). Oil press-cakes and meals valorization through circular economy approaches: A review. *Applied Sciences*, 10(21), 7432. http://doi.org/10.3390/app10217432.

Anjum, F.M., Nadeem, M., Khan, M.I, & Hussain, S. (2012). Nutritional and therapeutic potential of sunflower seeds: A review. *British Food Journal*, 114(4), 544–552. https://doi.org/10.1108/00070701211219559.

Arntfield, S.D. (2018). Proteins from oil-producing plants. In: R.Y. Yada (Ed.), *Proteins in food processing*, 2nd ed., 187–221. London: Woodhead Publishing, Elsevier Ltd.

Bhise, S., Kaur, A., Ahluwali, P., & Thind, S. (2014). Texturization of deoiled cake of sunflower, soybean and flaxseed into food grade meal and its utilization in preparation of cookies. *Nutrition Food Science*, 44(6), 576–85. https://doi.org/10.1108/NFS-01-2014-0002.

Borrello, M., Caracciolo, F., Lombardi, A., Pascucci, S., & Cembalo, L. (2017). Consumers' perspective on circular economy strategy for reducing food waste. *Sustainability*, 9(1), 141. https://doi.org/10.3390/su9010141.

Canibe, N., Martin-Pedrosa, M., Robredo, L.M., & Bach-Knudsen, K.E. (1999). Chemical composition, digestibility and protein quality of 12 sunflower (Helianthus annuus L.) cultivars. *Journal of the Science of Food and Agriculture*, 79(13), 1775–1782.

Dabbour, M., He, R., Ma, H., & Musa, A. (2018). Optimization of ultrasound assisted extraction of protein from sunflower meal and its physicochemical and functional properties. *Journal of Food Process Engineering*, 41(5), 12799. https://doi.org/10.1111/jfpe.12799.

Delgado-Vargas, F., & Paredes-Lopez, O. (2003). *Natural colorants for food and nutraceutical uses*. Boca Raton, FL: CRC Press. https://doi.org/10.1201/9781420031713.

Dulf, F.V. (2012). Fatty acids in berry lipids of six sea buckthorn (Hippophae rhamnoides L., subspecies carpatica) cultivars grown in Romania. *Chemistry Central Journal*, 6(1), 106–118. https://doi.org/10.1186/1752-153X-6-106.

Esposito, B., Sessa, M.R., Sica, D., & Malandrino, O. (2020). Towards circular economy in the agri-food sector. A systematic literature review. *Sustainability*, 12(18), 7401. https://doi.org/10.3390/su12187401.

Evlash, V., Tovma, L., Tsykhanovska, I., & Gaprindashvili, N. (2019). Innovative technology of the scoured core of the sunflower seeds after oil expression for the bread quality increasing. In: V. Nadykto (Ed.), *Modern development paths of agricultural production*. Cham: Springer. https://doi.org/10.1007/978-3-030-14918-5_65.

Garg, M., Sharma, A., Vats, S., Tiwari, V., Kumari, A., Mishra, V., & Krishania, M. (2021). Vitamins in cereals: A critical review of content, health effects, processing losses, bioaccessibility, fortification, and biofortification strategies for their improvement. *Frontiers in Nutrition*, 16, 1–15. https://doi.org/10.3389/fnut.2021.586815.

Goiri, I., Zubiria, I., Benhissi, H., Atxaerandio, R., Ruiz, R., Mandaluniz, N., & Garcia-Rodriguez, A. (2019). Use of cold-pressed sunflower cake in the concentrate as a low-input local strategy to modify the milk fatty acid profile of dairy cows. *Animals*, 9(10), 803. http://doi.org/10.3390/ani9100803.

González-Pérez, S. (2003). Physico-chemical and functional properties of sunflower proteins. Wageningen, The Netherlands: Wageningen University, PhD dissertation.

Gonzalez-Perez, S., & Vereijken, J.M. (2008). Sunflower proteins: overview of their physicochemical, structural and functional properties. *Journal of Science Food and Agriculture*, 87(12), 2173–2191. https://doi.org/10.1002/jsfa.2971

Grasso, S., Omoarukhe, E., Wen, X., Papoutsis, K., & Methven, L. (2019). The use of upcycled defatted sunflower seed flour as a functional ingredient in biscuits. *Foods*, 8(8), 305–316. https://doi.org/10.3390/foods8080305.

Hakkinen, S.T., Nuutila, A.M., & Ritala, A. (2018). Modifying seeds to produce proteins. In: R.Y. Yada (Ed.), *Proteins in food processing*, 2nd ed., 413–441. London: Woodhead Publishing, Elsevier Ltd.

Hnitsevych, V., Honchar, Y., & Yevdomakha, T. (2019). The structure-forming properties of semi-product on the basis of concentrated low lactos milk whey. *Equipment and Technologies of Food Production*, 2(39), 20–27. https://doi.org/10.33274/2079-4827-2019-39-2-20-27.

Hussain, S., Joudu, I., & Bhat, R. (2020). Dietary fiber from underutilized plant resources-a positive approach for valorization of fruit and vegetable wastes. *Sustainability*, 12(13), 5401. https://doi.org/10.3390/su12135401.

Islam, R.T., Hossain, M.M., Majumder, K., & Tipu, A.H. (2016). In vitro phytochemical investigation of helianthus annuus seeds. *Bangladesh Pharmaceutical Journal*, 19(1), 100–105. https://doi.org/10.3329/bpj.v19i1.29245.

Ivanova, P., Chalova, V., Koleva, L., & Pishtiyski, I. (2013). Amino acid composition and solubility of proteins isolated from sunflower meal produced in Bulgaria. *The International Food Research Journal*, 20(6), 2995–3000.

Katsa, M., Papalouka, N., Mavrogianni, T., Papagiannopoulou, I., Kostakis, M., Proestos, C., & Thomaidis, N.S. (2021). Comparative study for the determination of fat-soluble vitamins in rice cereal baby foods using HPLC-DAD and UHPLC-APCI-MS/MS. *Foods*, 10(3), 648. https://doi.org/10.3390/foods10030648.

Kollathova, R., Varga, B., Ivanisova, E., Biro, D., Rolinec, M., Juracek, M., Simko, M., & Galik, B. (2019). Mineral profile analysis of oilseeds and their by-products as feeding sources for animal nutrition. *Slovak Journal of Animal Science*, 52(1), 9–15.

Le Priol, L., Dagmey, A., Morandat, S., Saleh, K., Kirat, K.E., & Nesterenko, A. (2019). Comparative study of plant protein extracts as wall materials for the improvement of the oxidative stability of sunflower oil by microencapsulation. *Food Hydrocolloids*, 95, 105–120. https://doi.org/10.1016/j.foodhyd.2019.04.026.

Lin, M. J.Y., Humbert, E.S., & Sosulski, F.W. (1974). Certain functional properties of sunflower meal products. *Journal of Food Science*, 39(2), 368–370. https://doi.org/10.1111/j.1365-2621.1974.tb02896.x.

Man, S., Păucean, A., Muste, S., Pop, A., Sturza, A., Mureșan, V., & Salanță, L.C. (2017). Effect of incorporation of sunflower seed flour on the chemical and sensory characteristics of cracker biscuits. *Bulletin UASVM Food Science and Technology*, 74(2), 95–98. https://doi.org/10.15835/buasvmcn-fst:0018.

Martins, Z.E., Pinho, O., & Ferreira, I.M.P.L.V.O. (2017). Food industry by-products used as functional ingredients of bakery products. *Trends in Food Science & Technology*, 67, 106–128. https://doi.org/10.1016/j.tifs.2017.07.003.

Mirpoor, S.F., Giosafatto, C.V.L., & Porta, R. (2021), Biorefining of seed oil cakes as industrial co-streams for production of innovative bioplastics. A review. *Trends in Food Science & Technology*, 109, 259–270. http://doi.org/10.1016/j.tifs.2021.01.014.

Mohammed, K.M., Obadi, J.O. Omedi, K.S., Letsididi, M., Koko, Zaaboul, F., Siddeeg, A., & Li, Y. (2018). Effect of sunflower meal protein isolate (SMPI) addition on wheat bread quality. *Journal of Academia and Industrial Research*, 6(9), 159–164.

Nevara, G.A., Kharidah, S., Muhammad, S., Zawawi, N., Mustapha, N.A., & Karim, R. (2021). Dietary fiber: Fractionation, characterization and potential sources from defatted oilseeds. *Foods*, 10(4), 754. https://doi.org/10.3390/foods10040754.

Oseyko, M., Romanovska, T., & Shevchyk, V. (2020). Justification of the amino acid composition of sunflower proteins for dietary and functional products. *Ukrainian Food Journal*, 9(2), 394–403. https://doi.org/10.24263/2304-974X-2020-9-2-11.

Panjagari, N.R., Singh, A.K., Ganguly, S., & Indumati, K.P. (2015). Beta-glucan rich composite flour biscuits: Modelling of moisture sorption isotherms and determination of sorption heat. *Journal of Food Science and Technology*, 52(9), 5497–5509. https://doi.org/10.1007/s13197-014-1658-2.

Petraru, A., Ursachi, F., & Amariei, S. (2021). Nutritional characteristics assessment of sunflower seeds, oil and cake. Perspective of using sunflower oilcakes as a functional ingredient. *Plants*, 10, 2487. https://doi.org/10.3390/plants10112487.

Petrenko, V., Osipova, T., Lyubich, V., & Homenko, L. (2015). Relation between Hagberg-Perten falling number and acidity of wheat flour according to storage and agricultural systems. *Ratarstvo i Povrtarstvo*, 52(3), 120–124. https://doi.org/10.5937/ratpov52-8485.

Pickardt, C., Eisner, P., Kammerer, D.R., & Carle, R. (2015). Pilot plant preparation of light-coloured protein isolates from de-oiled sunflower (*Helianthus annuus* L.) press cake by mild-acidic protein extraction and polyphenol adsorption. *Food Hydrocolloids*, 44, 208–227. https://doi.org/10.1016/j.foodhyd.2014.09.020.

Pojic, M., Hadnadev, M., & Dapcevic-Hadnadev, T. (2013). Gelatinization properties of wheat flour as determined by empirical and fundamental rheometric method. *European Food Research and Technology*, 237, 299–307. https://doi.org/10.1007/s00217-013-1991-0.

Ptak-Kaczor, M., Banach, M., Stapor, K., Fabian, P., Konieczny, L., & Roterman, I. (2021). Solubility and aggregation of selected proteins interpreted on the basis of hydrophobicity distribution. *Internation Journal of Molecular Sciences*, 22(9), 5002. https://doi.org/10.3390/ijms22095002.

Rani, R., & Badwaik, L.S. (2021). Functional properties of oilseed cakes and defatted meals of mustard, soybean and flaxseed. *Waste Biomass Valorization*, 12(10), 5639–5647. https://doi.org/10.1007/s12649-021-01407-z.

Rosa, P.M., Antoniassi, R., Freitas, S.C., Bizzo, H.R., Zanotto, D.L., Oliveira, M.F., & Castiglioni, V.B.R. (2009). Chemical composition of Brazilian sunflower varieties. *HELIA*, 32(50), 145–156. https://doi.org/10.2298/HEL0950145R.

Salgado, P.R., Ortiz, S.M.E., Petruccelli, S., & Mauri, A.N. (2012a). Functional food ingredients based on sunflower protein concentrates naturally enriched with antioxidant phenolic compounds. *Journal of the American Oil Chemists' Society*, 89(5), 825–836. https://doi.org/10.1007/s11746-011-1982-x.

Salgado, P.R., Lopez-Caballero, M.E., Gomez-Guillen, M.C., Mauri, A.N., & Montero, M.P. (2012b). Exploration of the antioxidant and antimicrobial capacity of two sunflower protein concentrate films with naturally present phenolic compounds. *Food Hydrocolloids*, 29(2), 374–381. https://doi.org/10.1016/j.foodhyd.2012.03.006.

Sara, A.S., Mathe, C., Basselin, M., Fournier, F., Aymes, A., Bianeis, M., Galet, O., & Kapel, R. (2020). Optimization of sunflower albumin extraction from oleaginous meal and characterization of their structure and properties. *Food Hydrocolloids*, 99, 105335. https://doi.org/10.1016/j.foodhyd.2019.105335.

Sarwar, M.F., Sarwar, M.H., Sarwar, M., Qadri, N.A., & Moghal, S. (2013).The role of oil-seeds nutrition in human health: A critical review. *Journal of Cereals and Oilseeds*, 4(8), 97–100. https://doi.org/10.5897/JCO12.024.

Shchekoldina, T., & Aider, M. (2014). Production of low chlorogenic and caffeic acid containing sunflower meal protein isolate and its use in functional wheat bread making. *Journal of Food Science and Technology*, 51(10), 2331–2343. https://doi.org/10.1007/s13197-012-0780-2.

Sim, H.J., Kim, B., & Lee, J. (2016). A Systematic approach for the determination of B-group vitamins in multivitamin dietary supplements by high-performance liquid chromatography with diode-array detection and mass spectrometry. *Journal of AOAC International*, 99(5), 1223–1232. https://doi.org/10.5740/jaoacint.16-0093.

Sinkovic, L., & Kolmanic, A. (2021). Elemental composition and nutritional characteristics of cucurbita pepo subsp. Pepo seeds, oil cake and pumpkin oil. *Journal of Elementology*, 26(1), 97–107. https://doi.org/10.5601/jelem.2020.25.4.2072.

Skrbic, B., & Filipcev, B. (2008). Nutritional and sensory evaluation of wheat breads supplemented with oleic-rich sunflower seed. *Food Chemistry*, 108(1), 119–129. https://doi.org/10.1016/j.foodchem.2007.10.052.

Slabi, S.A., Mathe, C., Basselin, M., Framboisier, X., Ndiaye, M., Galet, O., & Kapel, R. (2020). Multi-objective optimization of solid/liquid extraction of total sunflower proteins from cold press meal. *Food Chemistry*, 317, 126423. https://doi.org/10.1016/j.foodchem.2020.126423.

Stabnikova, O., Marinin, A., & Stabnikov, V. (2021). Main trends in application of novel natural additives for food production. *Ukrainian Food Journal*, 10(3), 524–551. https://doi.org/10.24263/2304-974X-2021-10-3-8

Subası, B.G., Vahapoglu, B., Capanoglu, E., & Mohammadifar, M.A. (2021). A review on protein extracts from sunflower cake: Techno-functional properties and promising modification methods. *Critical Reviews in Food Science and Nutrition*, 62(24), 6682–6697. https://doi.org/10.1080/10408398.2021.1904821.

Wanjari, N., & Waghmare, J. (2015). Phenolic and antioxidant potential of sunflower meal. *Advances in Applied Science Research*, 6(4), 221–229.

Wildermuth, S.R., Young, E.E., & Were, L.M. (2016). Chlorogenic acid oxidation and its reaction with sunflower proteins to form green-colored complexes. *Comprehensive Reviews in Food Science and Food Safety*, 15(5), 829–843. https://doi.org/10.1111/1541-4337.12213.

Wu, L. (2007). Effect of chlorogenic acid on antioxidant activity of Flos Lonicerae extracts. *Journal of Zhejiang University Science B*, 8(9), 673–679. https://doi.org/10.1631/jzus.2007.B0673.

Yegorov, B., Turpurova, T., Sharabaeva, E., & Bondar, Y. (2019). Prospects of using by-products of sunflower oil production in compound feed industry. *Journal of Food Science and Technology*, 13(1), 106–113. https://doi.org/10.15673/fst.v13i1.1337.

Chapter 6

Microbial Reduction and Oxidation of Iron for Wastewater Treatment

Olena Stabnikova, Viktor Stabnikov, and Viktoriia Krasinko
National University of Food Technologies

Zubair Ahmed
Mehran University of Engineering and Technology

CONTENTS

6.1 INTRODUCTION

Iron is the fourth most abundant element after oxygen, silicon and aluminum in the biosphere with a mass content of more than 5% in the Earth's crust. Iron (Fe) plays an essential role in the ecosystem's functioning being involved in diverse interactions between biochemical cycles of iron, carbon (C), nitrogen (N) and phosphorus (P) (Ivanov, 2020; Liu et al., 2022). An important place in biogeochemical cycles is occupied by the chemical or biological oxidation of ferrous

DOI: 10.1201/9781003329671-6

iron, Fe(II), to ferric iron, Fe(III), and bioreduction of Fe(III) to Fe(II) (Ivanov & Stabnikov, 2017; Song et al., 2021; Weber et al., 2006). The compounds of Fe(II) are not stable under aerobic conditions and are almost absent in natural deposits. Meanwhile, Fe(III) can be found in naturally abundant iron-rich minerals such as iron ores, iron-bearing clay minerals and iron hydroxides in wetland ecosystems (Ivanov et al., 2014; Liu et al., 2017; Yu et al., 2021). Bioreduction of Fe(III) is performed anaerobically as a nonspecific reduction by hydrogen produced by fermenting bacteria or as a specific reduction by iron-reducing bacteria (IRB) using different organic substances, for example, organic acids, as electron donors. Activity of IRB is very important in iron cycling in the environment. Ferrous, Fe(II), at neutral pH is soluble and stable under anaerobic conditions. It can be produced from Fe(III) by anaerobic dissimilatory iron-reducing bacteria which use the oxidation of H_2 or organic substrates for the reduction of ferric iron (Lovley, 2000). IRB can produce ferrous ions from undissolved Fe(III) oxides, particularly goethite (α-FeOOH) and hematite (α-Fe_2O_3), hydroxides, low-grade iron ore, iron (III) of clay minerals and is capable of dissolution and reduction of magnetite (Fe_3O_4) (Kostka et al., 2002; Zhang F. et al., 2020). Fe(III) which is bound with the colloids of humic acids can be also used as an electron acceptor in dissimilatory Fe(III) reduction; meanwhile, humic substances could serve as redox mediators (Wu et al., 2013). The dissolution of produced Fe(II) is accompanied by important geochemical reactions, such as the sorption and desorption of phosphate and heavy metals on the surface of iron oxides and hydroxides (Lovley, 2000).

IRB is considered an important group of microorganisms for ecology and environment, which along with the reduction of Fe(III) participate in the degradation of organic matter in the natural anaerobic zones. IRB is widespread in bottom sediments of natural aquatic systems. Thus, seven strains of IRB belonging to the genera *Clostridium*, *Bacillus*, *Aeromonas*, *Geobacter* and *Pantoea* were isolated from river sediment slurry (Scala et al., 2006). Iron-reducing bacteria from the genera *Pseudomonas*, *Thiobacillus*, *Geobacter*, *Rhodoferax* and *Clostridium* were present in the sediments of the Yellow River (Zhang et al., 2020). The strain *Desulfuromonas* sp. with iron(III)-reducing activity was isolated from sea sediment (Guo et al., 2020). IRB is also present in anaerobic digesters of organic wastes. The strain of iron-reducing bacteria *Stenotrophomonas maltophilia* was isolated from anaerobic sludge from a wastewater treatment plant (Ivanov et al., 2005). The representatives of the *Desulfuromonadales* order, mainly *Geobacter* species, which are known as iron-reducing bacteria, were found in sewage sludge during the anaerobic process (Mustapha et al., 2020). Microbial community of anaerobic digester-treated industrial food waste included bacteria from the genus *Shewanella* from the order *Alteromonadales* which along with *Geobacter* is the most often used for iron reduction (Ike et al., 2010).

Figure 6.1 The coarse iron ore (particle size 1 mm) (a) and fine iron ore (particle size less than 0.5 mm) (b). Magnification × 40.

Production of dissolved ferrous Fe(II) by iron-reducing bacteria could be performed for different engineering systems (Tay et al., 2008). For bioreduction of Fe(III), organic substances can be used as electron donors. As the source of ferric, it is possible to use Fe(III) hydroxide and natural materials such as iron-containing clay or hematite-containing iron ore (Fe_2O_3) (Figure 6.1).

Application of iron-reducing bacteria in environmental engineering for the intensification of treatment of different wastewaters and production of value-added products from them will be described in the present chapter.

6.2 APPLICATION OF IRON-REDUCING BACTERIA (IRB) FOR THE TREATMENT OF SULFATE-CONTAINING WASTEWATER

Seawater contaminated with organic matter, wastewater from seafood-processing, tanneries, leather processing, coal-burning power plants, paper and pulp industry, as well as from biotechnology companies, especially those that use molasses, contain significant amounts of sulfate (Colleran & Pender, 2002; van Lier et al., 2001). Under anaerobic treatment of these wastewaters, with a concentration of sulfate above 1,000 mg/L, the production of toxic hydrogen sulfide (H_2S) occurs. H_2S is generated due to the activity of sulfate-reducing bacteria (SRB) in anaerobic bioreactors where they compete for electron donors (acetate and hydrogen) with methanogens, and precipitation of trace elements by sulfide is performed (Muyzer & Stams, 2008). There is no methane generation in an anaerobic bioreactor at a concentration of free sulfide above 250

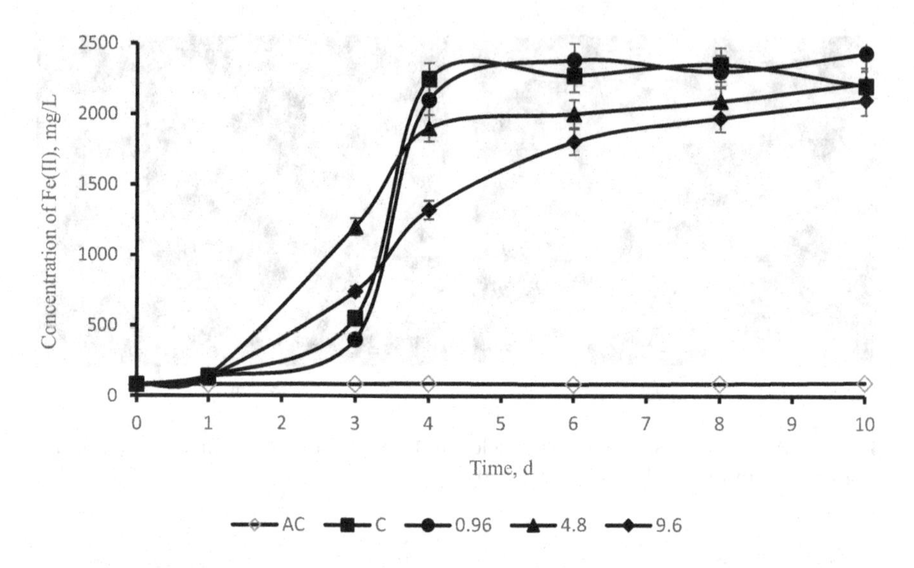

Figure 6.2 Quantity of Fe(II) produced by iron-reducing bacteria in the medium containing sodium sulfate: medium with 0 (control, C), 0.96 g/L, 4.80 g/L, 9.60 g/L of $NaSO_4$ and abiotic control without IRB (AC).

mg/L, and inhibition of methanogenesis is observed at COD (chemical oxygen demand)/SO_4 ratio below 10 (Stabnikov & Ivanov, 2006). The presence of sulfate, even at high concentrations, had no negative effect on the final amount of Fe(II) produced from ferric hydroxide by IRB (Figure 6.2).

However, the maximum Fe(III) reduction rate at high sulfate concentrations, 60 g of Na_2SO_4/L, was 40%–50% of the one in control, which may be due to the competition for an electron donor between IRB and sulfate-reducing bacteria.

Addition of ferric hydroxide and iron-reducing bacteria in anaerobic reactor-treated sulfate-containing wastewater could inhibit sulfate bioreduction and production of sulfide. The study of the influence of the Fe/SO_4^{2-} ratio was carried out on model wastewater with a sulfate concentration of 2.94 g/L with the addition of an enrichment culture of iron-reducing bacteria. Iron was added to the anaerobic reactor as ferric hydroxide. The molar ratio of Fe/SO_4^{2-} was 0.06 in the control; 0.5, 1 and 2 in the experiments. The concentration of sulfide constantly increased in the control and experiment with the ratio of Fe/SO_4^{2-}0.5; however, the formation of sulfide was not observed at the ratio of Fe/SO_4^{2-}1 and 2. At the molar ratio of Fe(III)/SO_4^{2-}0.06, 0.5, 1 and 2, the content of methane in biogas was 25±2% vol., 41±3% vol., 55±4% vol. and 62±3% vol., respectively. The addition of iron likely improves methanogenesis because the reduction of Fe(III) decreases the reduction of sulfate and forms Fe(II), which precipitates

TABLE 6.1 CHARACTERISTICS OF TWO ANAEROBIC TREATMENTS OF SULFATE-CONTAINING MODEL WASTEWATER WITHOUT (I) AND WITH (II) ADDITION OF IRB AND FE(III)

	Biotechnology	
Parameters	**I**	**II**
Initial molar ratio Fe(III)/SO_4^{2-}	0	1.75:1
Maximum rate of COD removal, kg/m³·day	0.037	0.068
The rate of fermentation, g TOC/g VSS·h	0.23	0.69
Average rate of biogas production, m³/m³·day	0.045	0.082
Maximum rate of biogas production, m³/m³·day	0.067	0.093
Content of methane in biogas, %	29	55
Concentration of H_2S in biogas, g/m³	72	0
Ratio of cells of sulfate-reducing bacteria and methanogens*	35	6

**Note:* It was determined by fluorescent *in situ* hybridization using 16S rRNA-targeted probes: Dsv687 for sulfate-reducing bacteria; Arc915 for methanogens.

COD, chemical oxygen demand; TOC, total organic carbon; VSS, volatile suspended solids.

sulfide. A comparison of the effectiveness of applying biotechnologies without (I) and with (II) the addition of IRB and Fe(III) hydroxide for anaerobic treatment of sulfate-containing wastewater was made (Table 6.1).

As can be seen from the data presented, the introduction of Fe(III) and IRB during anaerobic treatment of wastewater with a high content of organic matter and sulfate intensifies the treatment process due to the activity of IRB increasing the removal of total organic carbon and methane production. Results described above were confirmed by the research of other authors (Ahmed & Lin, 2017; Ahmed et al., 2021).

6.3 APPLICATION OF IRB FOR THE REMOVAL OF PHOSPHATE FROM REJECT WATER

High content of phosphorus in aquatic environments can cause their eutrophication characterized by the exuberant growth of algae and aquatic higher plants and a decrease in concentration of dissolved oxygen. The limits of phosphorus in the treated wastewater discharged to aquatic systems in many countries are from 0.5 mg/L to 2.0 mg/L; meanwhile, its concentration in municipal wastewater varies from 5 to 20 mg/L (Ivanov et al., 2005). From 10% to 50% of the phosphorus entering, the aerobic reactor of municipal wastewater treatment plants (WWTP) is phosphorus from the liquid fraction remaining after dewatering of anaerobic sludge, so-called reject water, or return liquor or digested supernatant, which contains 50 to 200 mg P/L. The volume of reject water usually consists of 2%–5% of the incoming wastewater volume, but loads of total phosphorus could reach almost 50% of the loads in raw wastewater (Mucha and Mikosz, 2021). So, it will be useful to treat reject water additionally to remove phosphorus before it will be returned to an aerobic bioreactor. Source of carbon in the reject water is represented mainly by branched volatile fatty acids such as isobutyric, isocaproic, valeric and acetic acids (Ivanov et al., 2007; Reyhanitash et al., 2017; Zacharof & Lovitt, 2017). These compounds serve as a carbon source for IRB for the reduction of added ferric iron to dissolved ferrous ions, which could precipitate phosphate.

A suspension of ferric hydroxide was added as a source of iron to the reject water to ensure the ratio of added Fe(III) to initial phosphate 4. Prepared medium was inoculated with iron-reducing bacteria *Stenotrophomonas maltophilia* BK and the initial content of volatile suspended solids (VSS) in the experiment with IRB was 0.37 g/L. It was shown that the concentration of total ferrous in experiment was 227 mg/L on day 10 (Figure 6.3).

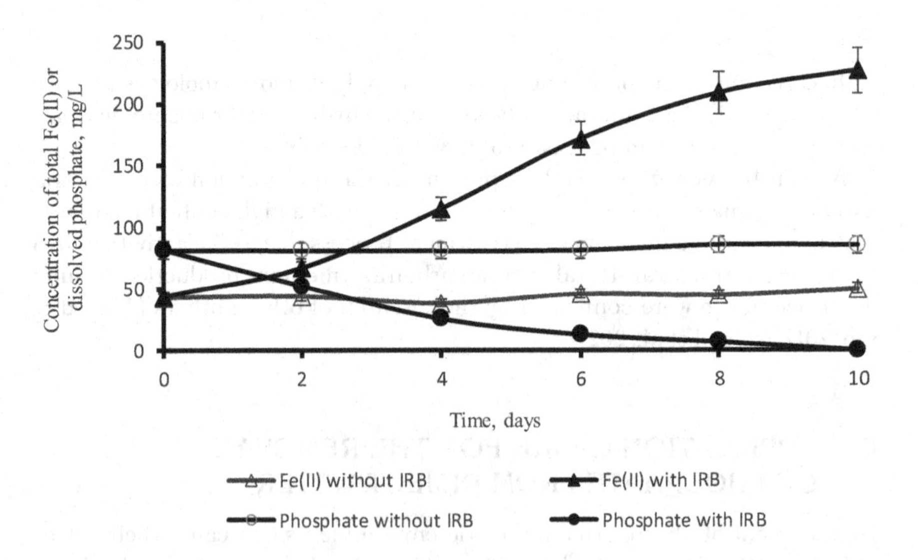

Figure 6.3 Changes of total ferrous and phosphate concentrations in control without the addition of IRB and in an experiment with the addition of IRB *Stenotrophomonas maltophilia* BK.

TABLE 6.2 EFFECT OF PIP ADDITION TO SANDY LOAM ON GROWTH OF SWAMP CABBAGE

Characteristics	C	C1 (with N)	C2 (with P)	E
Length of stem, cm	$19.2\pm0.5^{a*}$	16.2 ± 0.8^{b}	11.6 ± 1.3^{b}	33.6 ± 3.7^{c}
Length of root, cm	15.1 ± 2.0^{b}	10.4 ± 1.3^{a}	7.8 ± 1.1^{a}	15.3 ± 1.6^{b}
Dry mass of stem, g	0.32 ± 0.03^{a}	0.23 ± 0.01^{a}	0.16 ± 0.02^{a}	1.17 ± 0.27^{b}
Dry mass of root, g	0.15 ± 0.03^{ab}	0.12 ± 0.02^{ab}	0.10 ± 0.02^{a}	0.18 ± 0.03^{b}

Note: The values in one line with the same letter are not statistically significant at $P<0.05$ (calculated with the help of ANOVA and Tukey statistical procedures).

The pH was in the range from 7.2 to 8 during all 10 days of cultivation in the control and experiment. Concentration of Fe(II) remained unchanged in control without IRB but the concentration of phosphate increased slightly, probably due to the release of phosphate after the partial lysis of anaerobic biomass. 98% of added Fe(III) was reduced with iron reduction rate 2.96 mg Fe^{3+}/g VSS·h and the concentration of phosphate decreased from 70 mg/L to 0.4 mg/L on day 10 in experiment. The average rate of phosphate removal calculated for the 10 days of the process was 33 mg/g VSS·d (Figure 6.3).

Described upper principals were used for the development of technology for phosphate removal from wastewater using anoxic reduction of iron ore in the rotating reactor (Guo et al., 2009), and were also confirmed by the application of iron to remove phosphorus in anaerobic sequencing batch reactor (Zhang, 2012), as well as in the study for removal of phosphorus in activated sludge system using Fe(III) reduction (Wang et al., 2015). Phosphate–iron precipitate (PIP) obtained due to the anaerobic treatment of the reject water from municipal WWTP with the addition of IRB and iron (III) can be used as a phosphorus-containing fertilizer. The introduction of PIP as a phosphorus fertilizer into the sandy loam increased the dry mass of the ground part of *Ipomoea aquatica* (swamp cabbage) in the experiment (E, soil with the addition of ammonium sulfate and PIP as phosphorus fertilizer in ratio of N/P=2:1) by 4–7 times compared with control (C, soil without any additions), control 1 (C1, soil with addition of phosphorus-containing fertilizer) and control 2 (C2, soil with addition of ammonium sulfate as nitrogen-containing fertilizer) (Table 6.2) and (Figure 6.4).

Figure 6.4 Growth of swamp cabbage in controls and experiment: C (control, soil without any additions); C1 (control 1, soil with addition of phosphorus-containing fertilizer); C2 (control 2, soil with addition of ammonium sulfate as nitrogen-containing fertilizer); E (experiment, soil with addition of ammonium sulfate and PIP as phosphorus fertilizer in ratio of N/P = 2:1).

6.4 APPLICATION OF IRB FOR THE TREATMENT OF LIPID-CONTAINING WASTEWATER

Lipids are one of the major types of organic matter presented in municipal and food-processing wastewater, produced by slaughterhouses, meat and dairy product factories, fish meal and oil-processing plants, restaurants and wool scouring process. It is possible to remove up to 90% of lipids using physicochemical methods, but to remove the remaining lipids, biological cleanup should be used. During anaerobic treatment, fats are hydrolyzed to glycerol and long-chain fatty acids (LCFA) followed by their subsequent β-oxidation (Matikevičienė et al., 2012). The presence of LCFA even in very low concentrations inhibits both acidogenic fermentation and methanogenesis due to absorption on the cells causing damage to cell membranes. Inhibitory effect of LCFA positively correlated with the length of carbon chains and the number of double bonds in the fatty acids (Czatzkowska et al., 2020). Addition of dissolved ferrous/ferric salts diminished the inhibitory effect of LCFA because of the precipitation of LCFA as iron salt.

It was shown that the presence of Fe(II) produced from iron-containing clay due to activity of IRB decreased the inhibition effect of the model compound of

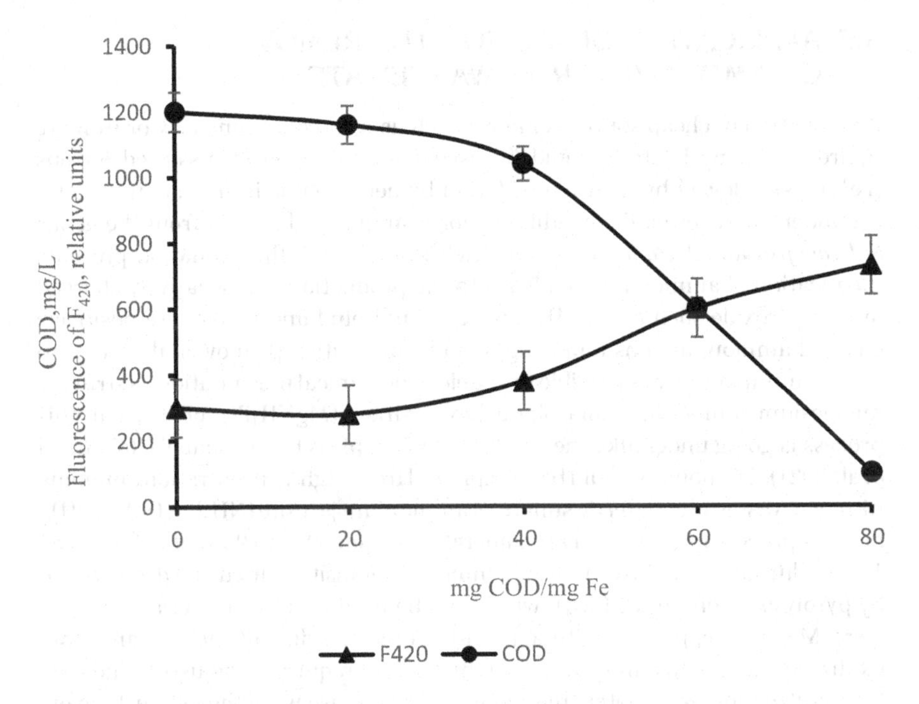

Figure 6.5 The effect of iron addition on the anaerobic fermentation of vegetable corn oil.

LCFA, stearic acid, added in concentration 0.5 g COD/L to model wastewater; its methanogenic fermentation resulted in an increase of production of total biogas by 37.2% and methane by 63% (Ivanov et al., 2002). Effectiveness of the anaerobic fermentation of vegetable corn oil (1.38 g/L or 4 COD/L) depended on the ratio of COD to mg of iron (Figure 6.5).

The efficiency of anaerobic digestion at low loading of iron in the bioreactor practically did not differ significantly from the control (without the introduction of iron); however, it increased sharply at the ratio of 80 mg COD/mg Fe. The methane production increased 1.5 times as compared to control at the ratio of 80 mg COD/mg Fe. COD removal efficiency increased by elevating the amount of added iron and was 77%, 80% and 98% when iron was added to ensure the ratio of 20, 40 and 80 mg COD/mg Fe, respectively. Activity of methanogens was estimated by autofluorescence of coenzyme F_{420} present in methanogenic bacteria (Grinter & Greening, 2021). The activity of methanogens changed similarly and was more active at 80 mg COD/mg Fe. Results described here were confirmed by the research of other authors (Li et al., 2006).

6.5 APPLICATION OF IRB FOR THE REMOVAL OF AMMONIUM FROM WASTEWATER

Bioreduction of cheap sources of ferric such as iron-containing clay or iron ore by iron-reducing bacteria could be used for production of dissolved ferrous (Fe^{2+} ions) followed by their biooxidation by neutrophilic iron-oxidizing bacteria under microaerophilic conditions, for example, by bacteria from the genus *Siderocapsa* and *Metallogenium* (Hedrich et al., 2011). This oxidation prevents nitrification of ammonium and leads to the production of a negatively charged ferric hydroxide such as $Fe(OH)_4^-$, which could bind and precipitate positively charged ammonium ions removing them from solution (Ivanov et al., 2004). To remove ammonium, it is possible to use electrochemical precipitation as struvite, magnesium ammonium phosphate hexahydrate ($MgNH_4PO_4{\cdot}6H_2O$), but this process is going under alkaline conditions when pH is higher than 9 (Rajaniemi et al., 2021). Ammonium could be precipitated from high-concentration ammonia nitrogen wastewater by ferric sulfate as ammonium jarosite $(NH_4)Fe_3(SO_4)_2(OH)_6$, but this process may proceed at temperatures from 50°C to 95°C at pH between 1 and 4 during 3–12 h (Li & Li, 2012). Ammonium jarosite is dried and decomposed by pyrolysis to obtain $Fe_2(SO_4)_3$ which can be used as a raw material for fertilizers. Meanwhile, process with application of iron-reducing bacteria and iron-oxidizing bacteria to remove ammonium does not require pH adjusting and high temperature for wastewater treatment. As in the case with phosphate, to avoid overloading of aerobic reactors with nitrogen, separate treatment of reject water after dewatering of anaerobically digested sludge will be a reasonable solution. The ammonium concentration in the reject water varies between 500 and 1500 mg NH_4^+-N/L, and reject water contributes 15-20% of the total nitrogen load at a municipal WWTP (Gustavsson, 2011).

It was shown the possibility of the use of the combined microbial iron reduction—iron oxidation to remove nitrogen from reject water using the sequential process consisting from

a. anaerobic bioreduction of a cheap source of iron like iron ore by iron-reducing bacteria to produce dissolved ferrous ions. Addition of acetate as the electron donor increased the rate of ferric bioreduction

$$8Fe^{3+}\left(\text{undissolved}\right) + CH_3COO^-\left(\text{acetate}\right)$$

$$+ 4H_2O \rightarrow 8Fe^{2+}\left(\text{dissolved}\right) + 2HCO_3^- + 9H^+$$

b. Fe(II), produced in an anaerobic bioreactor, being chelated with organic acids, oxidases by iron-oxidizing bacteria that degrade the organic “envelope” of the iron atom, which is then chemically oxidized under microaerophilic conditions:

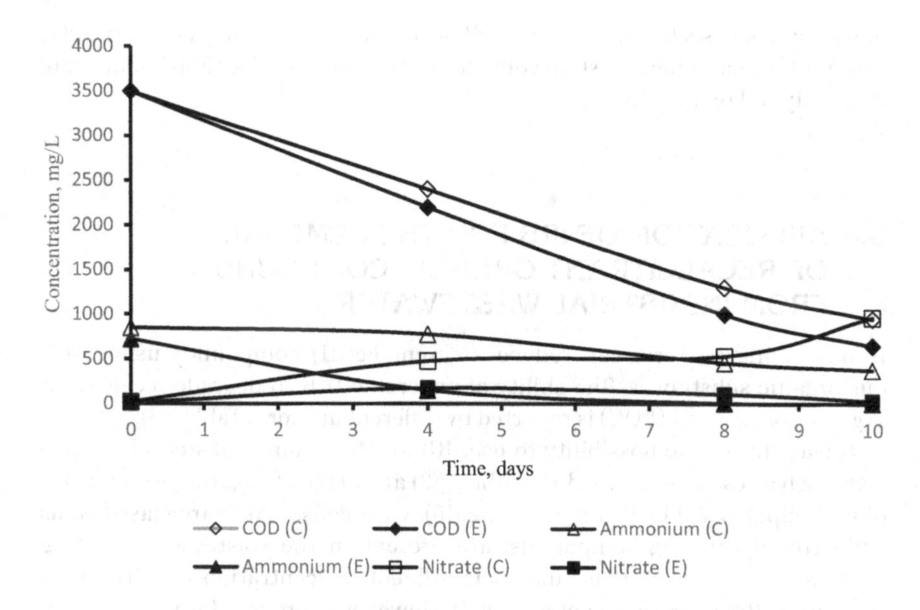

Figure 6.6 Effect of Fe(II) and iron-oxidizing bacteria addition on aerobic treatment of wastewater with high COD content. C—control, E—experiment.

$$4Fe^{2+}(\text{dissolved}) + O_2 + 4H^+ \rightarrow Fe^{3+} + 2H_2O$$

The main product of ferrous oxidation under the neutral pH is iron hydroxide $Fe(OH)_4^-$ (Cornell & Schwertmann, 2003). This oxidation prevents nitrification of ammonium and leads to the production of a negatively charged ferric hydroxide such as $Fe(OH)_4^-$, which could bind and precipitate positively charged ammonium ions removing them from the solution (Figure 6.6): The equation should be written

$$4Fe^{3+} + O_2 + 12H_2O + 4NH_4^+ \rightarrow Fe(NH_4)(OH)_4 \downarrow + 12H^+$$

The precipitate could be used as a slowly released ammonium fertilizer.

Microbial reduction of Fe(III) could enhance liquid-phase anaerobic digestion of organic-rich waste, for example, chicken or pig manure. The addition of clay containing a sufficient amount of Fe(III) and iron-reducing bacteria in the chicken manure slurry allows to conduct anaerobic digestion at an increased solids content of up to 15% (w/v) without stirring, which leads to a decrease in energy consumption and the volume of generated wastewater, saving up to 13.3 liters of water per kilogram of dry organic matter. At the same time, the production of biogas is intensified and increased due to the elimination of inhibition of methanogenesis by ammonium, sulfide and long-chain fatty acids. The clay slurry after anaerobic digestion could be used as soil fertilizer and enhancer of the sandy soil texture (Ivanov et al., 2019).

6.6 APPLICATION OF IRB FOR THE REMOVAL OF RECALCITRANT ORGANIC COMPOUNDS FROM INDUSTRIAL WASTEWATER

Iron-reducing bacteria can reduce different Fe(III) compounds using various organic substances. The ability of anaerobic IRB to degrade recalcitrant organic compounds (ROC) is reported by different authors (Table 6.3).

It was shown the possibility to use IRB for the removal of steroidal estrogens such as estrone (E1), 17β-estradiol (E2) and estriol (E3), from reject water of municipal WWTPs (Ivanov et al., 2010). Estrogens, which are classified as endocrine-disrupting compounds, are present in the wastewater entering in it *via* human excretions, and their influent concentration usually varies from 10 to 100 ng/L (Braga et al., 2005). However, there is information about much higher amounts of estrogens, for example, concentrations of E1 and E2 in influents for WWTPs in Brazil were 3050 ng/L and 776 ng/L, respectively (Pessoa et al., 2014); total concentration of 468 ± 27 ng/L of E1, E2, E3, 17α-ethinylestradiol (EE2) and diethylstilbestrol (DES) in treated wastewater was reported for the Beijing–Tianjin–Hebei region, China (Lei et al., 2020); the amount of E1 discharged daily with the effluent of WWTPs in Zeekoegat, South Africa, was 8002.3 ± 6416.3 mg/d (Kibambe et al., 2020). It is considered that the removal efficiency of estrogens from raw sewage ranges from 50% to 95% depending on the treatment technologies applied (Pessoa et al., 2014). So, relatively large portions of hormones remained in reject water after anaerobic digestion of aerobic sludge.

Iron ore as a source of ferric, cell suspension of IRB *Shewanella baltica*; estrone, 17β-estradiol and estriol in concentrations of 100 μg/L for each estrogen were added to reject water (Ivanov et al., 2010). Anaerobic process of reject water treatment was conducted at ambient temperature under rotation at 120 rpm for 15 days. The removal of E2 was 92% after 15 days of treatment, while the concentrations of E1 and E3 in the aqueous phase were reduced by 27% and 60%, respectively. Maxima of the biodegradation rates in batch culture of iron-reducing bacteria were 0.5, 1.0 and 2.0 mg/L for E1, E2 and E3, respectively. However, synthetic 17a-ethinylestradiol was resistant and was not degraded by IRB. The removal of recalcitrant compounds from wastewater could be due to the activity of iron-reducing bacteria, namely (a) anoxic oxidation; (b) formation of Fe^{2+} and insoluble ferrous and ferric salts and chelates of organic acids

TABLE 6.3 DEGRADATION OF RECALCITRANT ORGANIC COMPOUNDS BY IRON-REDUCING BACTERIA

Iron-Reducing Bacteria	Recalcitrant Compounds	References
Geobacter metallireducens GS-15	Phenol	Schleinitz et al., 2009
Geobacter metallireducens	Benzoate	Wischgoll et al., 2005
Geobacter metallireducens GS-15	Aniline, o-, m-, and p-cresol	Kazumi et al., 1995
Pure and mixed cultures of iron-reducing bacteria	Phenol/Benzoate	Lu et al., 2008
Stenotrophomonas maltophilia BK	Diphenylamine, m-cresol, 2,4-dichlorphenol and p-phenylphenol	Ivanov et al., 2004
Enriched iron-reducing bacterial culture	Naphthalene	Kleemann & Meckenstock, 2011
Enriched iron-reducing bacterial culture BF (mainly *Peptococcaceae*-related bacteria)	Benzene	Laban et al., 2010
Enriched cultures containing iron-reducing bacteria	Benzene, toluene, meta- and para-xylene (BTX)	Botton & Parsons, 2007
Chelatobacter heintzii	Atrazine	Rousseaux et al., 2001
Dehalobacterium	4-chlorophenol	Li et al., 2014
Dehalococcoides spp.	Tetrachloroethene	Yoshikawa et al., 2021
Shewanella decolorationis	Recalcitrant textile dye	Xu et al., 2005
Shewanella baltica	17β-estradiol, estriol and estrone	Ivanov et al., 2010

and phenols or (c) sedimentation of polar recalcitrant molecules during coagulation of ferrous produced from iron ore (Ivanov et al., 2020). Due to the activity of iron-reducing bacteria, positively or negatively charged ferrous or ferric hydroxides are formed such as $Fe(OH)^+$, $Fe(OH)_2$, $Fe(OH)_3^-$, $Fe(OH)^{2+}$, $Fe(OH)_3$, $Fe(OH)_4^-$, which can adsorb and precipitate recalcitrant organic compounds.

The process of iron reduction using IRB could find application in the treatment of wastewater from the production of sucralose. Artificial sweetener sucralose, $C_{12}H_{19}Cl_3O_8$, produced by the attachment of three chlorine atoms to a molecule of saccharose, is widely used in the food industry as a zero-calorie sugar substitute, and its production is permanently increased. Sucralose is hardly a biodegradable compound; it does not degrade in aerobic or anaerobic biological reactors, and it may be present in untreated and treated municipal wastewaters as well as in surface waters due to the discharge of effluents from WWTPs. Thus, the average sucralose concentration of seven WWTP effluents in Arizona was 2,800±1,000 ng/L, and surface waters contained sucralose at concentrations up to 300±30 ng/L (Torres et al., 2011). Industrial wastewater of the plants producing sucralose contains a lot of nonbiodegradable compounds, which cannot be removed at present in an economically and reasonable way by any known chemical or physicochemical method. Therefore, these substances are discharged into the environment, where they are accumulated with a high probability of dangerous consequences.

The ratio of chemical oxygen demand (COD) to biological oxygen demand (BOD) for wastewater of sucralose production after conventional aerobic biotreatment is about 300. This parameter shows that there are almost no biodegradable organics. The technology of sucralose production is described in US patent 5298611 "Sucralose pentaester production" (Navia et al., 1994). According to the data described in the patent, different organic solvents can be the major contaminants of wastewater. These solvents include methyl acetate, ethyl acetate, methyl ethyl ketone, methyl iso-butyl ketone, methyl iso-amyl ketone, methylene chloride, chloroform, diethyl ether and methyl tert-butyl ether. A preferred solvent is ethyl acetate. Another solvent, which can be found in high concentration of wastewater, is toluene, which is used in the production of sucralose in ratio of 5–10 mL toluene/g sucralose-6-ester. Meanwhile, toluene (C_7H_8) is an aromatic hydrocarbon belonging to the BTEX components (benzene, toluene, ethylbenzene and xylene), and the maximum level of toluene according to the US Environmental Protection Agency drinking water regulations for permissible drinking water is 2 mg/L (American Water Works Association, 1990).

It is known that iron ore with the major component of hematite has a high surface area and can serve as an efficient adsorbent for such heavy metals as Pb(II) and Hg(II) (Sarkar et al., 2017); bog iron ores were proposed to be used to remove lead, arsenic and zinc from contaminated waters (Debiec et al., 2018); natural iron oxide can be used as an adsorbent for dyes removal and is a promising adsorbent for the removal of toxic pollutants from water (Bien & Ha, 2019; Puiatti et al., 2021).

The positively charged particles of cheap iron ore can be used for the purification of wastewater of sucralose production, but in this case, surface of the iron ore particles will be quickly saturated and adsorption capacity of iron ore in terms of g of adsorbed COD/g of iron ore will be low (Ivanov et al., 2018). The wastewater of the industrial production of artificial sweetener sucralose contained an average 1,100 mg/L of total organic carbon with 2,100 mg/L of COD and 10 mg/L of BOD. Biodegradability of the wastewater components was low due to chlorinated organic substances. The mechanism of recalcitrant organics removal from the wastewater of sucralose production is adsorption of the recalcitrant compounds on the positively charged particles of iron (hydr)oxides produced after the dissolution of the iron ore surface. The main role of IRB is the regeneration of surface of iron ore used for adsorption of recalcitrant organic compounds from wastewater from sucralose production, and production of the positively charged particles of iron (hydr)oxides to be used as adsorbents in the purification of wastewater process.

The iron ore with major mineral hematite (Fe_2O_3), iron content of 60% (w/w), and particles sizes between 0.5 and 7 mm was used in the experiment. The combined chemical and biological treatment of this wastewater in the bioreactors with hematite iron ore removed up to 70% of TOC. About 20% of TOC was removed quickly by adsorption on iron ore particles, but adsorption/precipitation of others up to 50% of TOC was due to ferrous/ferric ions and hydroxides produced during microbial reduction and dissolution of iron ore. The calculated dosage of iron ore with 150 regeneration cycles could be 46.7 g/L of wastewater (Ivanov et al., 2014; 2018). Thus, the treatment of wastewater with iron ore and iron-reducing bacteria diminished the quantity of granulated activated carbon that is used in the treatment of sucralose production wastewater by up to 70%.

6.7 APPLICATION OF IRB FOR REMOVAL OF BACTERIAL PATHOGENS IN ENGINEERING PROCESSES

There are some new applications of iron-reducing bacteria in environmental engineering. For example, it was shown that ions of Fe^{2+} produced by IRB could serve as a bactericidal agent under anoxic conditions (Auffan et al., 2008; Kim et al., 2021). It was found that inactivation of *Escherichia coli* cells occurred in the presence of ions Fe^{2+}, and *E. coli* inactivation rates had a linear correlation with Fe^{2+} concentrations. Meanwhile, the inactivated cells were further used by iron-reducing bacteria as an electron source for ferric reduction. According to the cited authors, this knowledge could be used for the enhancement of pathogen removal in many engineering processes.

6.8 CONCLUSIONS

Application of iron-reducing bacteria (IRB) for bioreduction of Fe(III) from the relatively cheap iron-containing minerals such as iron ore or clay should be used in anaerobic environmental technologies to improve the process of wastewaters treatment from different industrial food enterprises; and also for recovery of essential nutrients such as ammonium and phosphate for their further use as value-added products. The major advantage of this process is that iron ore could be used for large-scale water or wastewater treatments and will be much more economical technologies than those requiring a supply of expensive chemical reagents or adsorbents. Instead of pure cultures of IRB, enrichment or mixed cultures could be used which are more suitable because of the diversity of temperature, pH, salinity, type of electron donors, type of ferrous and ferric chelating substances and redox potential of certain applications.

REFERENCES

Ahmed, M., Anwar, R., Deng, D., Garner, E. & Lin, L.S. (2021). Functional interrelationships of microorganisms in iron-based anaerobic wastewater treatment. *Microorganisms, 9*(5), 1039. https://doi.org/10.3390/microorganisms9051039

Ahmed, M. & Lin, L.S. (2017). Ferric reduction in organic matter oxidation and its applicability for anaerobic wastewater treatment: a review and future aspects. *Reviews in Environmental Science and Bio/Technology, 16,* 273–287 https://doi.org/10.1007/s11157-017-9424-3

American Water Works Association. (1990). *Water Quality and Treatment: A Handbook of Community Water Supplies,* 4th ed. McGraw-Hill, Inc.

Auffan, M., Achouak, W., Rose, J., Roncato, M.A., Chanéac, C., Waite, D.T., Masion, A., Woicik, J.C., Wiesner, M.R. & Bottero, J.Y. (2008). Relation between the redox state of iron-based nanoparticles and their cytotoxicity toward *Escherichia coli. Environmental Science and Technology, 42*(17), 6730–6735. https://doi.org/10.1021/es800086f.010.02248.x

Bien, N. & Ha, N.T.H. (2019). Adsorption of arsenic and heavy metals from solutions by modified iron ore sludge. *Vietnam Journal of Earth Sciences, 41*(3), 259–271. https://doi.org/10.15625/0866-7187/41/3/13947

Botton, S. & Parsons, J.R. (2007). Degradation of BTX by dissimilatory iron-reducing cultures. *Biodegradation, 18,* 371–381. https://doi.org/10.1007/s10532-006-9071-9

Braga, O., Smythe, G.A., Schafer, A.I. & Feitz, A.J. (2005). Steroid estrogens in primary and tertiary wastewater treatment plants. *Water Science and Technology, 52*(8), 273–278. https://www.oieau.org/eaudoc/system/files/documents/40/204491/204491_doc.pdf

Colleran, E. & Pender, S. (2002). Mesophilic and thermophilic anaerobic digestion of sulphate-containing wastewaters. *Water Science and Technology, 45*(10), 231–235. https://doi.org/10.2166/wst.2002.0339

Cornell, R.M. & Schwertmann, U. (2003). *The Iron Oxides: Structure, Properties, Reactions, Occurrences and Uses.* 2nd ed. Wiley-VCH. https://doi.org/10.1002/3527602097

Czatzkowska, M., Harnisz, M., Korzeniewska, E. & Koniuszewska, I. (2020). Inhibitors of the methane fermentation process with particular emphasis on the microbiological aspect: A review. *Energy Science and Engineering, 8*(5), 1880–1897. https://doi.org/10.1002/ese3.609

Debiec, K., Rzepa, G., Bajda, T., Uhrynowski, W., Sklodowska, A., Krzysztoforski, J. & Drewniak, L. (2018). Granulated bog iron ores as sorbents in passive (bio)remediation systems for arsenic removal. *Frontiers in Chemistry, 6*, 54. https://doi.org/10.3389/fchem.2018.00054

Grinter, R. & Greening, C. (2021). Cofactor F420: an expanded view of its distribution, biosynthesis and roles in bacteria and archaea. *FEMS Microbiology Reviews, 45*(5), fuab021. https://doi.org/10.1093/femsre/fuab021

Guo, C., Stabnikov, V., Kuang, S. & Ivanov, V. (2009). The removal of phosphate from wastewater using anoxic reduction of iron ore in the rotating reactor. *Biochemical Engineering Journal, 46*(2), 223–226. https://doi.org/10.1016/j.bej.2009.05.011

Guo, Y., Aoyagi, T., Inaba, T., Sato, Y., Habe, H. & Hori, T. (2020). Complete genome sequence of *Desulfuromonas* sp. strain AOP6, an iron(III) reducer isolated from subseafloor sediment. *ASM Journals Microbiology Resource Announcements, 9*(12), e01325–19. https://doi.org/10.1128/MRA.01325-19

Gustavsson, D.J. (2011). Biological sludge liquor treatment at municipal wastewater treatment plants – a review. *Vatten, 66*, 179–192. https://www.tidskriftenvatten.se/wp-content/uploads/2017/04/48_article_4264.pdf

Hedrich, S., Schomann, M. & Johnson, D.B. (2011). The iron-oxidizing proteobacteria. *Microbiology, 157*, 1551–1564. https://doi.org/10.1099/mic.0.045344-0

Ike, M., Inoue, D., Miyano, T., Liu, T.T., Sei, K., Soda, S. & Kadoshin, S. (2010). Microbial population dynamics during startup of a full-scale anaerobic digester treating industrial food waste in Kyoto eco-energy project. *Bioresource Technology, 101*(11), 3952–3957. https://doi.org/10.1016/j.biortech.2010.01.028

Ivanov, V. (2020). *Environmental Microbiology for Engineers*, 3rd ed. CRC Press, Taylor & Francis Group.

Ivanov, V. & Stabnikov, V. (2017). *Construction Biotechnology: Biogeochemistry, Microbiology and Biotechnology of Construction Materials and Processes*. Springer Science+Business Media. https://www.springer.com/gp/book/9789811014444

Ivanov, V.N., Stabnikova, E.V., Stabnikov, V.P., Kim, I.S. & Ahmed, Z. (2002). Effects of iron compounds on the treatment of fat-containing wastewater. *Applied Biochemistry and Microbiology, 38*(3), 255–258. https://doi.org/10.1023/A:1015475425566

Ivanov, V., Wang, J.Y., Stabnikova, O., Krasinko, V., Stabnikov, V., Tay, S.T.L. & Tay, J.H. (2004). Iron-mediated removal of ammonia from strong nitrogenous wastewater of food processing. *Water Science and Technology, 49*(5–6), 421–431. https://doi.org/10.2166/wst.2004.0783

Ivanov, V., Stabnikov, V., Zhuang, W.Q., Tay, J.H. & Tay, S.T.L. (2005). Phosphate removal from the returned liquor using iron-reducing bacteria. *Journal of Applied Microbiology, 98*(5), 1152–1161. https://doi.org/10.1111/j.1365-2672.2005.02567.x

Ivanov, V., Lim, J.J.W., Stabnikova, O. & Gin, K.Y.H. (2010). Biodegradation of estrogens by facultative anaerobic iron-reducing bacteria. *Process Biochemistry, 45*(2), 284–287. https://doi.org/10.1016/j.procbio.2009.09.017

Ivanov, V., Stabnikov, V., Guo, C.H., Stabnikova, O., Ahmed, Z., Kim, I.S. & Shuy, E.B. (2014). Wastewater engineering applications of BioIronTech process based on the biogeochemical cycle of iron bioreduction and (bio)oxidation. *AIMS Environmental Journal, 1*(2), 53–66. http://doi.org/10.3934/environsci.2014.2.53

Ivanov, V., Stabnikov, V. & Tay, J.H. (2018). Removal of the recalcitrant artificial sweetener sucralose and its by-products from industrial wastewater using microbial reduction/oxidation of iron. *ChemEngineering, 2*(3), 37. http://doi.org/10.3390/chemengineering2030037

Ivanov, V., Stabnikov, V., Stabnikova, O., Salyuk, A., Shapovalov, E., Ahmed, Z. & Tay, J.H. (2019). Iron-containing clay and hematite iron ore in slurry-phase anaerobic digestion of chicken manure. *AIMS Materials Science, 6*(5), 821–832. https://doi.org/10.3934/matersci.2019.5.821

Kazumi, J., Haggblom, M.M. & Young, L.Y. (1995). Degradation of monochlorinated and nonchlorinated aromatic compounds under iron-reducing conditions. *Applied and Environmental Microbiology, 61*(11), 4069–4073. https://doi.org/10.1128/AEM.61.11.4069-4073.1995

Kim, L., Yan, T. & Pham, V.T. (2021). Inactivation of *Escherichia coli* enhanced by anaerobic microbial iron reduction. *Environmental Science and Pollution Research, 28*, 63614–63622. https://doi.org/10.1007/s11356-020-11209-w

Kibambe, M.G., Momba, M.N.B., Daso, A.P., Zijl, C.V. & Coetzee, M. (2020). Efficiency of selected wastewater treatment processes in removing estrogen compounds and reducing estrogenic activity using the T47D-KBLUC reporter gene assay. *Journal of Environmental Management, 260*, 110135. https://doi.org/10.1016/j.jenvman.2020.110135

Kleemann, R. & Meckenstock, R.U. (2011). Anaerobic naphthalene degradation by Gram-positive, iron-reducing bacteria. *FEMS Microbiology Ecology, 78*(3), 488–496. https://doi.org/10.1111/j.1574-6941.2011.01193.x

Khelifi, S., Choukchou-Braham, A., Oueslati, M., Sbih, H. & Ayari, F. (2020). Identification and use of local-ores deposit as adsorbent: adsorption study and photochemical regeneration. *Desalination and Water Treatment, 206*, 429–438. https://doi.org/10.5004/dwt.2020.26307

Kostka, J.E., Dalton, D.D., Skelton, H., Dollhopf, S. & Stucki, J.W. (2002). Growth of iron(III)-reducing bacteria on clay minerals as the sole electron acceptor and comparison of growth yields on a variety of oxidized iron forms. *Applied and Environmental Microbiology, 68*(12), 6256–6262. https://doi.org/10.1128/AEM.68.12.6256-6262.2002

Laban, N.A, Selesi, D., Rattei, T., Tischler, P. & Meckenstock, R.U. (2010). Identification of enzymes involved in anaerobic benzene degradation by a strictly anaerobic iron-reducing enrichment culture. *Environmental Microbiology*, 12(10), 2783–2796. https://doi.org/10.1111/j.1462-2920.2

Lei, K., Lin, C.Y., Zhu, Y., Chen, W., Pan, H.Y., Sun, Z., Sweetman, A., Zhang, Q. & He, M.C. (2020). Estrogens in municipal wastewater and receiving waters in the Beijing-Tianjin-Hebei region, China: Occurrence and risk assessment of mixtures. *Journal of Hazardous Materials, 389*, 121891. https://doi.org/10.1016/j.jhazmat.2019.121891

Li, R. & Li, X. (2012) Method for treating high-concentration ammonia nitrogen waste water by utilizing ferric sulfate. Patent CN102139944B. https://patents.google.com/patent/CN104028098A/en

Li, Z., Wrenn, B.A. & Venosa, A.D. (2006). Effects of ferric hydroxide on methanogenesis from lipids and long-chain fatty acids in anaerobic digestion. *Water Environment Research, 78*(5), 522–530. http://doi.org/10.2175/106143005x73064

Li, Z., Suzuki, D., Zhang, C., Yang, S., Nan, J., Yoshida, N., Wang, A. & Katayama, A. (2014). Anaerobic 4-chlorophenol mineralization in an enriched culture under iron-reducing conditions. *Journal of Bioscience and Bioengineering, 18*(5), 529–532. http://doi.org/10.1016/j.jbiosc.2014.04.007

Liu, G., Qiu, S., Liu, B., Pu, Y., Gao, Z., Wang, J., Jin, R. & Zhou, J. (2017). Microbial reduction of Fe(III)-bearing clay minerals in the presence of humic acids. *Scientific Reports, 7,* 45354. https://doi.org/10.1038/srep45354

Liu, Y., Wang, Y., Wang, Z. & Gao, T. (2022). Characteristics of iron cycle and its driving mechanism during the development of biological soil crusts associated with desert revegetation. *Soil Biology and Biochemistry, 164,* 108487. https://doi.org/10.1016/j.soilbio.2021.108487

Lovley, D.R. (2000). Fe(III) and Mn(IV) reduction. In D.R. Lovley (Ed.), *Environmental Microbe-Metal Interactions* (pp. 3–30). ASM Press.

Lu, W., Wang, H., Huang, C. & Reichardt, W. (2008). Aromatic compound degradation by iron reducing bacteria isolated from irrigated tropical paddy soils. *Journal of Environmental Science (China), 20*(12), 1487–1493. https://doi.org/10.1016/s1001-0742(08)62554-1

Matikevičienė, V., Grigiškis, S., Levišauskas, D., Kinderyt, O. & Baškys, E. (2012). Technology for treatment of lipid-rich wastewater and pipelines clogged by lipids using bacterial preparation. *Journal of Environmental Engineering and Landscape Management, 20*(1), 49–57. https://doi.org/10.3846/16486897.2012.662747

Mucha, Z. & Mikosz, J. (2021). Technological characteristics of reject waters from aerobic sludge stabilization in small and medium-sized wastewater treatment plants with biological nutrient removal. *International Journal of Energy and Environmental Engineering, 12,* 69–76. https://doi.org/10.1007/s40095-020-00358-w

Mustapha, N.A., Toya, S. & Maeda, T. (2020). Effect of Aso limonite on anaerobic digestion of waste sewage sludge. *AMB Express, 10,* 74 https://doi.org/10.1186/s13568-020-01010-w

Muyzer, G. & Stams, A.J.M. (2008). The ecology and biotechnology of sulphate-reducing bacteria. *Nature Reviews Microbiology, 6,* 441–454. https://doi.org/10.1038/nrmicro1892

Navia, J.L., Walkup, R.E., Vernon, N.M., Wingard, R.E. & Jr. (1994). Sucralose pentaester production. US patent 5298611. https://patents.google.com/patent/US5298611A/en

Pessoa, G.P., de Souza, N.C., Vidal, C.B., Alves, J.A., Firmino, P.I., Nascimento, R.F. & dos Santos, A.B. (2014). Occurrence and removal of estrogens in Brazilian wastewater treatment plants. *Science of the Total Environment, 490,* 288–295. https://doi.org/10.1016/j.scitotenv.2014.05.008

Puiatti, G.A., Elerate, E.M., de Carvalho, J.P., Luciano, V.A., de Carvalho Teixeira, A.P., Lopes, R.P. & Teixeira de Matos, A. (2021). Reuse of iron ore tailings as an efficient adsorbent to remove dyes from aqueous solution. *Environmental Technology,* 1–12. https://doi.org/10.1080/09593330.2021.2011427

Rajaniemi, K., Hu, T., Nurmesniemi, E.T., Tuomikoski, S. & Lassi, U. (2021). Phosphate and ammonium removal from water through electrochemical and chemical precipitation of struvite. *Processes, 9*(1), 150. https://doi.org/10.3390/pr9010150

Reyhanitash, E., Kersten, S.R. & Schuur, B. (2017). Recovery of volatile fatty acids from fermented wastewater by adsorption. *ACS Sustainable Chemistry & Engineering, 5*, 9176–9184. https://doi.org/10.1021/acssuschemeng.7b02095

Rousseaux, S., Hartmann, A. & Soulas, G. (2001). Isolation and characterisation of new Gram-negative and Gram-positive atrazine-degrading bacteria from different French soils. *FEMS Microbiology Ecology, 36*(2–3), 211–222. https://doi.org/10.1111/j.1574-6941.2001.tb00842.x

Sarkar, S., Sarkar, S. & Biswas, P. (2017). Effective utilization of iron ore slime, a mining waste as adsorbent for removal of Pb(II) and Hg(II). *Journal of Environmental Chemical Engineering, 5*(1), 38–44. https://doi.org/10.1016/j.jece.2016.11.015

Scala, D.J., Hacherl, E.L., Cowan, R., Young, L.Y. & Kosson, D.S. (2006). Characterization of Fe(III)-reducing enrichment cultures and isolation of Fe(III)-reducing bacteria from the Savannah River site, South Carolina. *Research in Microbiology, 157*(8), 772–783. https://doi.org/10.1016/j.jbiosc.2015.12.014

Schleinitz, K.M., Schmeling, S., Jehmlich, N., von Bergen, M., Harms, H., Kleinsteuber, S., Vogt, C. & Fuchs, G. (2009). Phenol degradation in the strictly anaerobic iron-reducing bacterium *Geobacter metallireducens GS-15. Applied and Environmental Microbiology, 75*(12), 3912–3919. https://doi.org/10.1128/AEM.01525-08

Song, Y., Yang, L., Wang, H., Sun, X., Bai, S., Wang, N., Liang, J. & Zhou, L. (2021). The coupling reaction of Fe^{2+} bio-oxidation and resulting Fe^{3+} hydrolysis drastically improve the formation of iron hydroxysulfate minerals in AMD, *Environmental Technology, 42*(15), 2325–2334. https://doi.org/10.1080/09593330.2019.1701564

Stabnikov, V.P. & Ivanov, V.N. (2006). The effect of various iron hydroxide concentrations on the anaerobic fermentation of sulfate-containing model wastewater. *Applied Biochemistry & Microbiology, 42*, 284–288. https://doi.org/10.1134/S0003683806030112

Tay, J.H., Tay, S.T.L., Ivanov, V., Stabnikova, O. & Wang, J.Y. (2008). Compositions and methods for the treatment of wastewater and other waste. US Patent 7393452. https://patents.google.com/patent/US7393452B2/en

Torres, C.I., Ramakrishna, S., Chiu, C.A., Nelson, K.G., Westerhoff, P. & Krajmalnik-Brown, R. (2011). Fate of sucralose during wastewater treatment. *Environmental Engineering Science, 28*, 325–331. https://doi.org/10.1089/ees.2010.0227

van Lier, J.B., Lens, P.N. & Pol, L.W. (2001). Anaerobic treatment for C and S removal in "zero-discharge" paper mills: effects of process design on S removal efficiencies. *Water Science and Technology, 44*(4), 189–195. https://doi.org/10.2166/wst.2001.0218

Wang, Y., Zhai, S., Geng, M. & Li, J. (2015). Effect of pH and phosphorus concentration on phosphorus removal by dissimilatory Fe(III) reduction in activated sludge. *Chinese Journal of Environmental Engineering, 9*(8), 4002–4008. https://doi.org/10.12030/j.cjee.20150866

Weber, K., Achenbach, L. & Coates, J. (2006). Microorganisms pumping iron: Anaerobic microbial iron oxidation and reduction. *Nature Reviews Microbiology, 4*, 752–764 https://doi.org/10.1038/nrmicro1490

Wischgoll, S., Heintz, D., Peters, F., Erxleben, A., Sarnighausen, E., Reski, R., Van Dorsselaer, A. & Boll, M. (2005). Gene clusters involved in anaerobic benzoate degradation of *Geobacter metallireducens. Molecular Microbiology, 58*(5), 1238–1252. https://doi.org/10.1111/j.1365-2958.2005.04909.x

Wu, C.Y., Zhuang, L., Zhou, S.G., Yuan, Y., Yuan, T. & Li, F.B. (2013). Humic substance-mediated reduction of iron(III) oxides and degradation of 2,4-D by an alkaliphilic bacterium, *Corynebacterium humireducens* MFC-5. *Microbial Biotechnology, 6*(2), 141–149. https://doi.org/10.1111/1751-7915.12003

Yoshikawa, M., Zhang, M., Kawabe, Y. & Katayama, T. (2021). Effects of ferrous iron supplementation on reductive dechlorination of tetrachloroethene and on methanogenic microbial community. *FEMS Microbiology Ecology, 97*(5), fiab069. https://doi.org/10.1093/femsec/fiab069

Xu, M., Guo, J., Cen, Y., Zhong, X., Cao, W. & Sun, G. (2005). *Shewanella decolorat*ionis sp. nov., a dye-decolorizing bacterium isolated from activated sludge of a wastewater treatment plant. *International Journal of Systematic and Evolutionary Microbiology, 55*(Pt 1), 363–368. https://doi.org/10.1099/ijs.0.63157-0

Yu, C., Xie, S., Song, Z., Xia, S. & Åström, M.E. (2021). Biogeochemical cycling of iron (hydr-)oxides and its impact on organic carbon turnover in coastal wetlands: A global synthesis and perspective. *Earth-Science Reviews, 218*, 103658. https://doi.org/10.1016/j.earscirev.2021.103658

Zacharof, M.P. & Lovitt, R.W. (2014). Recovery of volatile fatty acids (VFA) from complex waste effluents using membranes. *Water Science and Technology, 69*(3), 495–503. https://doi.org/10.2166/wst.2013.717

Zhang, X. (2012). Factors influencing iron reduction–induced phosphorus precipitation. *Environmental Engineering Science, 29*(6), 511–519. http://doi.org/10.1089/ees.2011.0114

Zhang, F., Battaglia-Brunet, F., Hellal, J., Joulian, C., Gautret, P. & Motelica-Heino, M. (2020). Impact of Fe(III) (oxyhydr)oxides mineralogy on iron solubilization and associated microbial communities. *Frontiers in Microbiology, 11*, 571244. https://doi.org/10.3389/fmicb.2020.571244

Zhang, H., Liu, F., Zheng, S., Chen, L., Zhang, X. & Gong, J. (2020). The differentiation of iron-reducing bacterial community and iron-reduction activity between riverine and marine sediments in the Yellow River estuary. *Marine Life Science & Technology, 2*, 87–96. https://doi.org/10.1007/s42995-019-00001

Chapter 7

Biohydrogen Production from Cellulose-Containing Wastes

Nataliia Golub
National Technical University of Ukraine
"Igor Sikorsky Kyiv Polytechnic Institute"

CONTENTS

7.1 INTRODUCTION

7.1.1 Hydrogen as a Future Energy Source

Hydrogen, H_2, has great potential to be used as renewable and clean energy alternative to fossil fuels such as natural gas, coal and oil, which are nonrenewable and resources of which are depleted. The main chemical element in fossil fuels is carbon. The use of biofuels reduces biosphere pollution due to reduction of the consumption of fossil fuels, and thus reducing greenhouse gas emissions and their impact on the environment leading to global climate changes (Hanaki & Portugal-Pereira, 2018).

DOI: 10.1201/9781003329671-7

Biological hydrogen can be produced from numerous kinds of biomass such as crop residues, food processing waste, organic industrial wastes and animal manures (Yue et al., 2021). The use of biofuels reduces biosphere pollution as the combustion of biomass produces significantly less nitrogen and sulfur oxides (0.2% opposite 2%–3% for coal), ash (3%–5% opposite 10%–15% for coal), and the amount of carbon dioxide is equal to the amount of carbon absorbed during growth, which returns to the carbon cycle and provides a continuous process of energy production. With the application of chemical or biochemical processes, biomass can be transformed into solid (charcoal), liquid (methanol, ethanol, biodiesel) and gaseous (methane, hydrogen) fuels. Renewable energy sources, namely, biogas, hydrogen, and alcohols obtained through enzymatic processes, are gaining more and more importance as a fuel.

Agricultural crops cannot be the universal solution to be used as renewable energy raw materials, as the rapid development of biofuel-oriented crop production may become a real threat to traditional (food-oriented) agriculture. Therefore, the development of technologies that use various origins wastes and new types of biofuels is a promising research area, and an increase in the hydrogen fuel role in the energy market by 2030 is predicted (Ajanovic & Haas, 2018; Nnabuife et al., 2022). By the middle of the XXI century, the share of hydrogen as an energy source will reach about 10% of the world's final energy consumption. Annual demand for hydrogen will reach just over 200 million tons in 2030 and 530 million tons in 2050 (Nnabuife et al., 2022). The choice of H_2 as an energy source is mainly caused by such advantages as: (a) environmental safety, as the product of hydrogen oxidation is water; (b) high gravimetric energy content of about 143 MJ/kg, meanwhile, possessing energy density of hydrocarbon fuels is 33 MJ/kg; (c) high thermal conductivity and low viscosity, which is necessary for pipeline transportation; (d) its use is characterized by zero CO_2 emissions that caused climate change minimization; (e) the possibility of using both as a one-component energy carrier and in the form of additives to other fuels; (f) unlimited reserves of raw materials, as in addition to common method of hydrogen obtaining by electrolysis it can also be produced from raw materials; (e) unlimited supply of raw materials, since in addition to producing hydrogen by electrolysis, it can also be produced from organic raw materials, which are wastes of various origins (Tashie-Lewis & Nnabuife, 2021). Therefore, hydrogen is an environmentally friendly energy source, obtained by processes that do not indirectly harm the environment. So in the case of electricity-based methods, the energy spent on its production should also be obtained through the use of environmentally friendly and renewable energy sources—such as solar, wind and hydropower. As per today, all hydrogen production technologies include methane steam conversion (50% of world production), coal gasification (20%), water electrolysis (4%) and oil refining processes (26%). All of the said processes except water electrolysis are accompanied by significant CO_2 emissions and require fossil fuels (Claassen & de Vrije, 2006). Therefore, research on hydrogen bioproduction using anaerobic fermentation of cellulose and starch-containing raw materials is being intensively conducted in recent years (Ding et al., 2008; Gomez-Flores et al., 2017; Gupta et al., 2014; Zagrodnik & Seifert, 2020). The application of biochemical processes for hydrogen production is based on the possibility of decentralized energy production at the various origin waste formation sites. Biotechnological processes are conducted in the liquid medium at low temperatures and atmospheric pressure.

7.1.2 Microbial Production of Hydrogen

Depending on the biochemical characteristics of microorganisms, hydrogen can be produced during anaerobic (strict anaerobes), as well as during aerobic (facultative anaerobes) conditions. In the process of enzymatic hydrogen production, the substrate can be composed of various raw materials, such as industrial and domestic wastewater containing organic components, waste from cereals and other crops, household waste, wood processing waste, manure and bird droppings. The transformation of organic waste into biohydrogen in the dark fermentation process is carried out by a wide range of microorganisms. Biochemical reactions include hydrolytic split of polymeric compounds to monomers and their subsequent conversion into low-molecular-weight organic acids and alcohols with simultaneous hydrogen biosynthesis. The fermentation of raw materials produces both H_2 and other products, and the maximum yield of hydrogen (4 moles per mole of glucose) in anaerobic fermentation (Nasirian et al., 2010):

$$C_6H_{12}O_6 + 2H_2O \rightarrow 2CH_3COOH + 2CO_2 + 4H_2, \Delta G = -184.2\ \text{kJ} \tag{7.1}$$

In the case of other fermentation process types, the yield of hydrogen is reduced; for example, during butyric acid fermentation, only 2 mol of H_2 is formed:

$$C_6H_{12}O_6 + 2H_2O \rightarrow C_3H_7COOH + 2CO_2 + 2H_2, \Delta G = -257.1\ \text{kJ} \tag{7.2}$$

Acetate and butyrate are the main metabolites, but butyrate is more dominant because of its lower Gibbs free energy value. The use of mixed microbial cultures provides a combination of hydrogen production processes:

$$4C_6H_{12}O_6 + 2H_2O \rightarrow 2CH_3COOH + 3C_3H_7COOH + 8CO_2 + 10H_2 \tag{7.3}$$

The decrease in hydrogen yield from theoretical yield during the fermentation process is due to the formation of enzymes aimed at the synthesis of cell biomass and ATP production. Also, in many organisms the release of hydrogen into the medium is decreased by the function of the proton cycle in the cell specifically due to the presence of hydrogenases that consume part of the produced hydrogen (Frey, 2002). Problems that arise in the anaerobic fermentation of cellulose-containing waste to produce hydrogen relate to (a) the possibility of process deviation and the transition to methanogenesis; (b) consumption of produced hydrogen by microorganisms; (c) inhibition of the process by metabolites; (d) low hydrogen content in biogas; (e) low process rate; and (f) lack of trace elements needed for microbial growth.

7.2 PREPARATION OF CELLULOSE-CONTAINING RAW MATERIALS AND INOCULUM FOR THE ANAEROBIC FERMENTATION FOR HYDROGEN PRODUCTION

Specific microbial community is formed during fermentation of raw material depending on its type and process conditions. The bioconversion process is also influenced by the structure of the raw material, which may contain hardly converted components. To speed up the process of hydrogen production, pretreatment of feedstock is recommended. Agricultural, household and forest solid wastes contain as a major component biodegradable organic. The chemical composition of raw materials depends on the plant type, growing and storage conditions. So, straw contains from 22% to 49% of carbohydrates that are difficult to hydrolyze, from 16% to 26% of lignin, recalcitrant component of the plant cell wall, degradation of which by microorganisms is very low; wood contains up to 30% of lignin and up to 56% of cellulose and hemicellulose. To obtain hydrogen from such raw materials, it is required to increase the availability of nutrients to microorganisms by improving the cellulose and hemicellulose accessibility using physicochemical, chemical and biological pretreatment methods of feedstock (Kucharska et al., 2018; Kumar & Sharma, 2017; Mosier et al., 2005; Taherzabeh & Karimi, 2008). Pretreatment of lignocellulosic biomass includes mechanical grinding, treatment with steam, alkali or acid and other methods. The mechanism of action of these processes is different, but they all contribute to the degradation of cellulose and hemicellulose, and increase carbohydrate availability that is more susceptible to enzymatic hydrolysis and microbial conversion. The application of different pretreatment methods depends on the chemical composition of raw materials and the formation of by-products during pretreatment that can act as inhibitors for a further enzymatic process (Klinke et al., 2004; Parawira & Tekere, 2011; Taherzabeh & Karimi, 2008; Yang et al., 2018).

Mechanical grinding is used to increase the surface area on which cellulolytic enzymes act. Therefore, this method must be used for solid waste. Energy consumption is from 1 to 30 kWh per ton of raw material with a 3–6 mm size of the grains received (Parveen et al., 2009). Heat treatment leads to the formation of phenol and furan derivatives, which can inhibit hydrogen-producing microorganisms (Klinke et al., 2004). Acid or alkaline treatment can change the pH of the medium, which will adversely affect the growth of microorganisms. Other compounds may also be formed that inhibit the development of microorganisms and slow down the fermentation process (Parawira & Tekere, 2011; Yang et al., 2018). Steam treatment of lignocellulose-containing raw materials can be used as an alternative to chemical hydrolysis. The advantages of steam application include (a) limited chemical reagents use and as a result no equipment corrosion; (b) a low environmental impact; and (c) more complete sugar recovery compared to other pretreatment methods (Baruah et al., 2018; Pielhop et al., 2016). Steam pretreatment at low temperatures (130°C–150°C) allows activation of the polymer components for subsequent hydrolysis and decompose polysaccharides into sugars. Dissolution of hemicellulose is observed at 150°C (Kucharska et al., 2018). Decomposition of hemicellulose under steam treatment also increases the specific surface area of material particles. It was shown that pretreatment of crushed different raw materials with steam for 1 h without excessive pressure led to an increase in production of hydrogen by anaerobic fermentation (Table 7.1). Pretreatment of agricultural raw material with steam increased its decomposition rate and consequently the hydrogen yield by 6–10 times in comparison with untreated waste.

TABLE 7.1 CONTENT OF HYDROGEN, %, IN BIOGAS PRODUCED BY ANAEROBIC FERMENTATION OF RAW MATERIALS WITHOUT AND WITH STEAM PRETREATMENT (OUR DATA)

	Content of H_2 in Biogas, %	
Raw Material	**Without Pretreatment**	**With Pretreatment**
Wheat straw	1.4±0.1	16.4±1.5
Barley straw	6.1±0.5	54.3±4.3
Corn (stems, leaves)	4.8±0.4	41.2±3.2
Sunflower (stems, heads)	1.3±0.1	11.1±1.0
Rapeseed straw	0.3±0	1.7±0.1
Pine sawdust	0±0	0.7±0.0
Birch sawdust	0.1±0	1.1±0.1
Filter paper	3.2±0.2	7.1±0.5

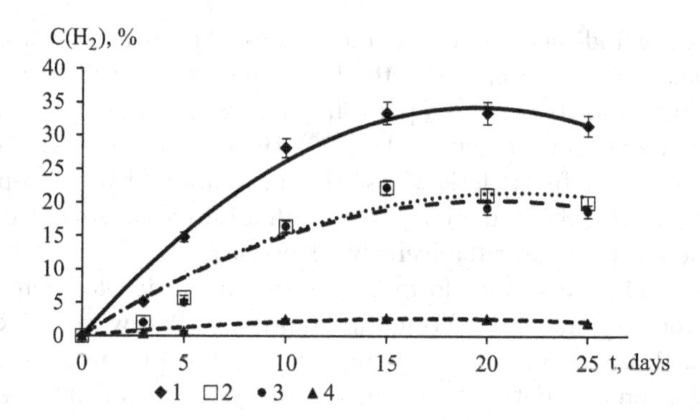

Figure 7.1 Change in the hydrogen content (C (H_2)) in biogas during the conversion of wheat straw depending on the methods of its pretreatment: 1, alkali treatment; 2, acid treatment; 3, steam treatment; 4, without treatment.

It is known that alkali pretreatment is effective for raw materials with 20% of lignin content or above causing an increase in the rate of enzymatic hydrolysis (Zhang, 2021). Figure 7.1 shows the hydrogen content in biogas using different pretreatment methods of wheat straw. For wheat straw, the highest yield of hydrogen occurs after alkali pretreatment.

Hydrogen production depends on the chemical composition of raw materials. The highest hydrogen yield was observed for barley straw, which contains the lowest amount of lignin, 11%, and the highest amount of easily hydrolyzed polysaccharides, 44%. Increasing the amount of lignin and reducing the amount of easily hydrolyzed polysaccharides led to a decrease in the rate of decomposition of cellulose-containing raw materials and hydrogen production, as it is observed in corn straw, wheat straw and sunflower residues. As can be seen from Table 7.1, the content of hydrogen in produced biogas is higher when barley straw and corn residues were used compared with pure cellulose (filter paper) application. This can be explained by the fact that barley straw contains significant amounts of hemicelluloses, polysaccharides that are not as complicated as cellulose and could be easier to be enzymatically hydrolyzed; it increases the amount of nutrients available for microbial hydrogen production. Also, barley straw contains twice higher protein than other raw materials, which also stimulates biohydrogen production. The microbiological transformation of raw materials into bioenergy is also affected by the presence of minerals in the feedstock (Su et al., 2018). Thus, when used waste does not contain enough trace elements for the growth and metabolism of anaerobic microorganisms, it is necessary to add organic waste rich in minerals, such as manure or poultry droppings. Therefore, the efficiency of microbial production of hydrogen is influenced by the composition of raw materials, namely by the contents of lignin, hydrolyzable polysaccharides and mineral salts. The duration of the batch fermentation process also affects the hydrogen yield. As the culture ages, the metabolism changes with the transition to the formation of products such as ethanol and butanol, and accordingly, reduces the yield of hydrogen. Table 7.2 shows the yields of hydrogen from fermentation of agricultural waste pretreated using different methods.

Analyzing the data provided in Table 7.2, it can be stated that the yield of hydrogen depends on the type of raw material, pretreatment methods and inoculum. Hydrogen production is affected by the concentration of ammonium, phosphate and minerals, and at optimal content of these ions in the medium, the hydrogen yield increases up to three times (Lay et al., 2005). To ensure the balance of nutrients in the fermentation medium needed for the development and activity of microorganisms involved in the production of hydrogen, it is advisable to use a mixture of various wastes (Ding et al., 2008). Thus, the use of a mixture of cellulose-containing wastes and wastes containing residues of fats and proteins leads to increased hydrogen yield (Boni et al., 2013). Addition of lime mud from paper-making process, 3 g, as a source of different metal cations, mainly Ca, Fe and Mn, to food waste, 200 g, and then fermented to produce H_2 resulted in accelerating the fermentation process and obtaining the highest hydrogen production of 137.6 mL H_2/g of volatile solids (VS) (Zhang et al., 2013). Increased concentration of Fe^{2+} ions caused improvement in hydrogen yield because iron is a much-required component of hydrogenase and ferredoxin protein, which are important for hydrogen bioproduction. Thus, increase in Fe^{2+} concentration from 0 to 350 mg/L resulted in an increase in hydrogen yield in batch tests by mixed cultures at 35°C from 196.6 mL/g glucose to 311.2 mL/g glucose (Wang & Wan, 2008). Concentration of 300 mg/L of Mg^{2+} ions, which are also an active site component of several enzymes, caused an increase in substrate utilization of up to 76% and hydrogen yield of up to 3.5 mol H_2/mol glucose (Srikanth & Mohan, 2012).

There are recommendations on what pretreatment method of renewable raw material for biohydrogen production, depending on its chemical composition, could be used. For raw materials containing up to 20% lignin, more than 30% of polysaccharides and other substances that are easily hydrolyzed, and have also sufficient content of minerals, 5%–8% of dry weight with all needed trace elements for microorganisms, as well as sufficient nitrogen and phosphorus contents, application of steam is considered to be the best. Materials with such composition include barley and corn straw. This pretreatment does not require additional costs for chemicals, and at the same time does not introduce inhibitory components and no changes of pH are needed, and at the end, maximum hydrogen yields are obtained. According to our studies, when using barley straw as a substrate, the maximum yield of hydrogen reaches 48 g H_2/kg of dry biomass on the 10th day of batch conversion by the mixed microbial culture,

TABLE 7.2 HYDROGEN PRODUCTION FROM AGRICULTURAL WASTES

Raw Materials	Hydrogen Yield	Pretreatment of Raw Materials	Temperature Regime	Process Mode, Inoculum	Reference
Corn stalks	150 mL H_2/g of substrate, 7.6 mL H_2/h	0.2% HCl, boiling for 30 min	mesophilic 35°C	batch, a mixture of microorganisms	Rodríguez-Valderrama et al. (2020)
Cornstalk	126 mL H_2/g of substrate	0.6% HCl, 90°C, 2 h	mesophilic 36°C	batch, river anaerobic sludge	Wang et al. (2010)
Corn straw	68 mL H_2/g substrate	1.5 MPa, 10 min	mesophilic 35°C	batch, *Clostridium butyricum*	Li & Chen (2007)
Corn stover	2.8–3.0 mol H_2/mol substrate	1.2% sulfuric acid, 2 h	mesophilic 36°C	batch, anaerobic sludge	Datar et al. (2007)
Food residues and manure	16.5 mL H_2/g VS	without pretreatment	thermophilic 55°C	batch, microflora from a landfill	Karlsson et al. (2008)
Wheat straw	41.5 mL H_2/g dry biomass	without pretreatment	thermophilic 70°C	batch, bacterial strain DSM8903	Ivanova et al. (2009)
Air-dried sugarcane bagasse	17.7 mL H_2/g dry biomass	without pretreatment	thermophilic 70°C	batch, bacterial strain DSM8903	Ivanova et al. (2009)
Maize leaves	14.9 mL H_2/g dry biomass	without pretreatment	thermophilic 70°C	batch, bacterial strain DSM8903	Ivanova et al. (2009)
Maize leaves	37.6 mL H_2/g dry biomass	enzymatic with cellulase	thermophilic 70°C	batch, bacterial strain DSM8903	Ivanova et al. (2009)

(Continued)

TABLE 7.2 (*Continued*) HYDROGEN PRODUCTION FROM AGRICULTURAL WASTES

Raw Materials	Hydrogen Yield	Pretreatment of Raw Materials	Temperature Regime	Process Mode, Inoculum	Reference
Rice bran	31–61 mL H_2/g VS	without pretreatment	mesophilic 35°C	continuous, an anaerobic microbial consortium	Noike & Mizuno (2000)
Wheat bran	10–43 mL H_2/g VS				
Bean curd processing waste	14–21 mL mL H_2/g VS				
Cow waste slurry	392 mL H_2/L slurry	without pretreatment	thermophilic 60°C	batch, cow waste slurry	Yokoyama et al. (2007)
	248 mL H_2/L slurry		thermophilic 75°C		
Dairy manure	31.5 mL H_2/g TVS	0.1%–0.2% HCl, boiling 30 min	mesophilic 36°C	batch, preincubated dairy manure	Xing et al. (2010)
Cabbage	26.3–61.7 mL H_2/g VS	without pretreatment	mesophilic 37°C	batch, anaerobic digested sludge boiled for 15 min	Okamoto et al. (2000)
Carrot	44.9–70.7 mL H_2/g VS				
Rice	19.3–96.0 mL H_2/g VS				
Food waste	62.6 mL H_2/g VS	pH 12.5, 60 rpm, 35°C, 24 h under anaerobic conditions	mesophilic 35°C	continuous, HRTs 30 h; heat-treated sludge from anaerobic digester of WWTP	Kim & Shin (2008)
Air-dried switch grass	100 mL H_2/g TVSFe	1% H_2SO_4 for 12 h, steam treatment	mesophilic 37°C	continuous, HRT 10 h, 1.75 g/L linoleic acid, anaerobic culture	Veeravalli et al. (2014)

(*Continued*)

TABLE 7.2 (*Continued*) HYDROGEN PRODUCTION FROM AGRICULTURAL WASTES

Raw Materials	Hydrogen Yield	Pretreatment of Raw Materials	Temperature Regime	Process Mode, Inoculum	Reference
High-solids food wastes	77 mL H_2/g TVS	without pretreatment	mesophilic 37°C	batch, *Clostridium*-rich grass compost treated at 80°C for 3 h	Lay et al. (2005)
Food waste, and slaughterhouse waste	145 mL H_2/g VS	without pretreatment	mesophilic 36°C	batch, activated aerobic sludge treated at 100°C for 30 min	Boni et al. (2013)
Food waste added with lime sludge	137.6 mL H_2/g VS	without pretreatment	thermophilic 55°C	batch, sewage sludge from WWTP, 80°C for 20 min	Zhang et al. (2013)
Vegetable waste	86 mL H_2/g VS	without pretreatment	mesophilic 28°C	batch, mixture *Buttiauxella* sp. *Rahnella* sp. *Raoultella* sp.	Marone et al. (2012)
Kitchen waste	72 mL H_2/g VS	without pretreatment	mesophilic	continuous, biogas plant slurry heated at 100°C, 30 min	Jayalakshmi et al. (2009)

Strain DSM8903—the extreme thermophilic bacterium Caldicellulosiruptor saccharolyticus; VS—volatile solid; TVS—total volatile solid; WWTP—municipal wastewater treatment plant; HRT—hydraulic retention time.

and the concentration of hydrogen in biogas contains 54%. In the case when raw materials contain more than 20% lignin, and less than 30% polysaccharides, it is necessary to use alkali pretreatment. The duration of pretreatment and the concentration of alkali must be increased with the increase in lignin content. Alkali treatment causes the destruction of the lignin–cellulose framework of raw materials and allows the release and precipitation of lignin, which facilitates the access of microorganisms to cellulose fibers and other nutrients. Such raw materials include wheat, sunflower and rape canola straw. The concentration of hydrogen in the gas formed during the processing of wheat straw reaches 35%.

Biological pretreatment has both advantages and disadvantages over physicochemical methods. The advantages include a greater focus on environmental protection and energy conservation, fewer compounds are formed that inhibit subsequent enzymatic processes, and a wider range of substances can be obtained (Chen et al., 2010, Zheng et al., 2009). The biodecomposition of biomass could be carried out by white-rot fungi such as *Pleurotus ostreatus, P. pulmonarius, P. sapidus, P. eryngii, Phanerochaete chrysosporium, Phlebia tremellosa, Ceriporia lacerate, Stereum hirsutum* and *Polyporus brumalis*; brown-rot fungi such as *Serpula lacrymans, Postia placenta, Coniophora puteana* and *Gloeophyllum trabeum*; and soft-rot fungus *Trichoderma reesei* (Teleomorph *Hypocrea jecorina*), which are the most efficient producers of cellulases and hemicellulases; and also filamentous fungi of the genus *Aspergillus*; and bacteria belonging to the genera *Streptomyces, Thermomonospora, Sphingomonas, Bacillus, Novosphingobium, Cupriavidus, Clostridium, Erwinia, Microbispora* and *Comamonas* (Chen et al., 2010, Kurakake et al., 2007; Zhuo et al., 2018). The main obstacle for large-scale application of biological pretreatment is the low destruction rate. Thus, it can be used only after physicochemical pretreatment.

In an industrial scale, anaerobic bioconversion of organic waste into hydrogen can be shifted toward methane production. Various methods are used to selectively inhibit the growth of methanogenic microorganisms based on their physiological characteristics: inability to form spores, toxic effects of oxygen, narrower pH range and the presence of specific inhibitors (2-bromethane sulfonic acid, iodopropane and acetylene) (Iyer et al., 2004; Khanna, 2013; Kim et al., 2004; Mu et al., 2007; Pendyala et al., 2012; Siriporn et al., 2012; Zhu & Béland, 2006). Most often, when hydrogen is obtained from waste, fermentation is carried out at pH 5.5, which leads to complete inhibition of methanogenesis. However, a decrease in pH leads to a decrease in the number of microbial species involved in the production of hydrogen, which leads to a decrease in the rate of its formation. To ensure high yield of hydrogen production, pretreatment of inoculum should be done. There is no consensus on which method of pretreatment of the inoculum to eliminate hydrogen-consuming bacteria is the best. Thermal pretreatment of the inoculum helps to inhibit the hydrogen consumers and to harvest spore-forming anaerobic bacteria, for example, species belonging to the genera *Clostridium* and *Bacillus*, which are used as seeds (Stabnikova et al., 2010). Thus, the pretreatment of inoculum (a granular sludge from the methanogenic anaerobic reactor) at 90°C for 10 min used for hydrogen production on sugarcane vinasse gave maximum hydrogen yield, 4.75 mmol H_2/g COD, the highest number of copies per mL of the gene Fe-hydrogenase, the enzyme responsible for hydrogen production catalyzing the reversible oxidation of molecular hydrogen and enrichment of microbial consortium with bacteria of the genera *Clostridium, Bacillus* and *Enterobacte*r (Magrini et al., 2021). In the case of acid or alkali treatment of the anaerobic sludge used as an inoculum for hydrogen fermentation from food waste, H_2 yield increased by 70% (Luo et al., 2022). Efficiency of different types of inoculum (anaerobic sludge from methanogenic reactor) pretreatment, to suppress methanogenesis and to enrich inoculum with H_2-producing bacteria, was studied (Mu et al., 2007). It was shown that heat pretreatment at 102°C for 90 min resulted in highest H_2 yield of 2.00 H_2 mol/mol glucose than acid treatment at pH 3.0–4.0 for 24 h (1.30 H_2 mol/mol glucose) or alkali treatment at pH 12.0 for 24 h (0.48 H_2 mol/mol glucose). Butyric acid fermentation occurred with heat and alkali-treated sludge, and mixed-type fermentation was observed when sludge treated with acid was used as seeds for hydrogen production (Mu et al., 2007). However, it was shown that although the heat and acid treatment methods completely repressed methanogens, hydrogen production was partially suppressed also (Zhu & Béland, 2006). Treatment of inoculum with specific inhibitors such as 2-bromoethane sulfonic acid and iodopropane resulted in selective inhibition of methanogens, meanwhile, no significant effects on hydrogen production were observed (Zhu & Béland, 2006). The similar effects were observed when anaerobic sludge was treated by flushing with air for 30 min (Zhu & Béland, 2006). A comparative study of pretreatment with heat, acid, alkali, linoleic acid or 2-bromoethanesulfonic acid of anaerobic mixed culture to be used as an inoculum for hydrogen production under mesophilic conditions (37°C) at an initial pH of 6.0 showed that the maximum yield of H_2, 1.69 mol H_2/mol glucose, was obtained for inoculum treated with linoleic acid, which correlated with the maximum specific activity of hydrogenase (Pendyala et al., 2012).

Reduction of the substrate retention time in the continuous reactor also leads to an increase in hydrogen production. Thus, when the hydraulic retention time (HRT) in the bioreactor operated at pH 5.5 to inhibit consumption of H_2 by methanogens was reduced from 30 h to 10 h, the production rate and yield of hydrogen increased from 80 mL/h and 0.91 mol H_2/mol glucose up to 436 mL/h and 1.61 mol H_2/mol glucose (Iyer et al., 2004). This changed the composition of the microbial population in the bioreactor: at HRT 30 h representatives of *Bacillaceae, Clostridiaceae* and E*nterobacteriaceae* were detected; meanwhile, at HRT 10 h, only *Clostridiaceae* was found.

Based on the data above, it can be stated that to create an effective industrial technology for hydrogen production from cellulose-containing raw materials, it is necessary to carry out pretreatment of both raw materials and inoculum. Pretreatment of inoculum is necessary to neutralize hydrogen consumers; meanwhile, raw material pretreatment ensures the destruction of the lignocellulose framework and receives low-molecular-weight sugars to improve enzyme access to substrates and their further processing into hydrogen.

7.3 EFFECTS OF OPERATING PARAMETERS ON HYDROGEN PRODUCTION

Microbial hydrogen formation depends on many factors such as the partial pressure of hydrogen, the qualitative composition of nutrients, the ratio of elements C: N: P, acidity and the presence of trace elements. During the process of organic raw material utilization by the microbial community, in addition to CO_2 and H_2, organic acids, alcohols, as well as energy-rich substances are formed, which provide resources required for cell growth and activity (Figure 7.2).

The enzyme system of microorganisms is inhibited by the end products of biosynthesis. In addition to organic acids formed during hydrogen production, dissolved CO_2 also increases the acidity of the medium, which will shift the metabolic pathways toward the synthesis of products such as ethanol, acetone, butanol or alanine, which reduce acidity. Changing metabolic pathways from the production of acids to neutral substances causes a reduction of hydrogen yield. The amounts of CO_2 and organic acids formed during fermentation depend on both the content of raw materials and the concentration of microorganisms in the bioreactor (Moussa et al., 2022).

Corn stems and leaves were shredded, pretreated with steam for 1 h and were added to the medium for hydrogen production in different quantities from 0 to 100 g/L. The river sludge taken on the depth of 1 m was used as an inoculum; the pH of the medium without corn waste was 7.5. Mesophilic batch process was performed at 35°C under stirring of 100 rpm for 10 days. To reduce the activity of methanogenic microorganisms, the reactor was started in the presence of oxygen. When the content of dry raw material was lower than 60 g/L, the pH value did not fall below 6, but at the content of raw materials above 80 g/L, the pH of the medium reached 5.5 on the 7th day of the process (Golub et al., 2017). The change in the pH values of the medium during the conversion at different contents of dry feedstock in the reactor is shown in Figure 7.3.

Carbohydrates are converted during anaerobic fermentation not only into hydrogen but to other products such as organic acid and alcohol and their composition was changed depending on the content of corn waste in the reactor (Table 7.3).

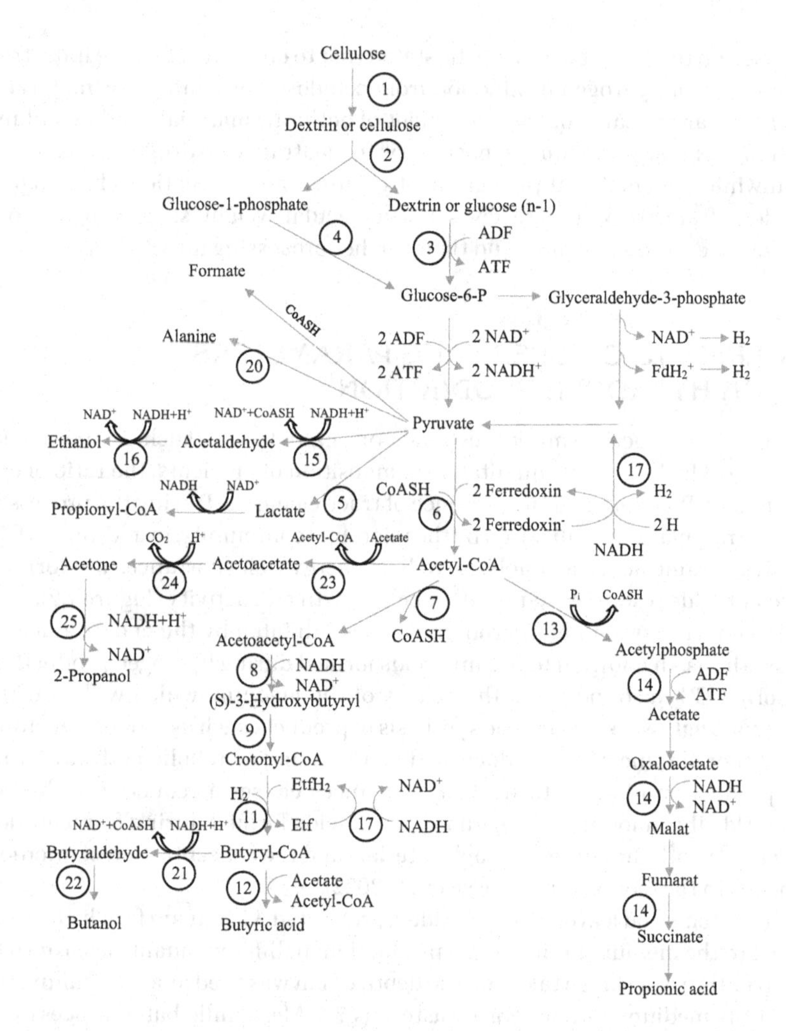

Figure 7.2 Directions of glucose catabolism and the main end products formed in enzymatic conversion processes:1—endo-1,4-β-glucanase and exo-1,4-β-glucanase (exo-cellobiohydrolase); 2—exo-1,4-β-glucosidase and cellobiase (β-glucosidase); 3—glucokinase, 4—phosphoglucomutase, 5—D- or L-lactate dehydrogenase; 6—pyruvate synthase; 7—acetyl-CoA-acetyltransferase; 8—3-hydroxybutyryl-CoA dehydrogenase; 9—crotonyl-CoA hydrate; 10—hydroxybutyryl-CoA dehydrogenase; 11—diaphorase; 12—butyrate:CoA-transferase, 13—phosphate-acetyltransferase; 14—acetatkinase; 15—acetaldehydrogenase; 16—alcohol dehydrogenase; 17—NADH:ferredoxin oxidoreductase and hydrogenase; 18—malate dehydrogenase; 19—fumarate reductase; 20—glutamate dehydrogenase and alanine aminotransferase; 21—butyraldehyde dehydrogenase; 22—1 -butanol dehydrogenase; 23—acetoacetate: CoA-transferase; 24—acetoacetate decarboxylase; 25—2-propanol dehydrogenase.

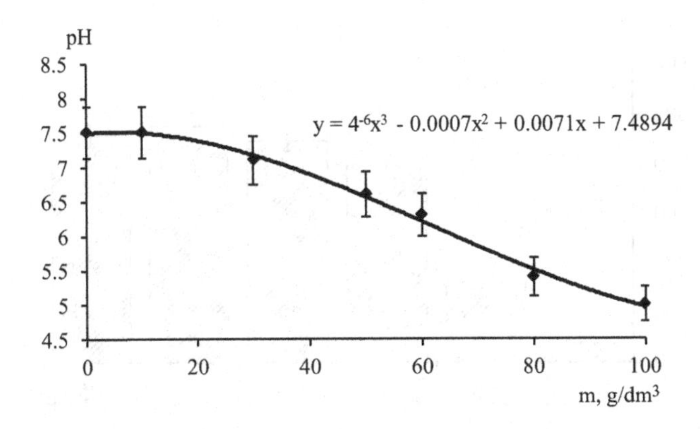

Figure 7.3 The effect of the dry corn content on the pH value of the fermented medium.

TABLE 7.3 CONCENTRATION OF PRODUCTS IN THE FERMENTED MEDIUM DEPENDING ON CORN WASTE CONTENT (OUR DATA)

	Corn Waste Content, g/L					
Products, %	**10**	**30**	**50**	**60**	**80**	**100**
Acetic acid	41±2	45±2	44±2	40±2	35±2	29±1
Butyric acid	48±2	43±3	43±2	43±2	37±2	36±2
Propionic acid	2±0	3±0	3±1	5±1	8±1	9±1
Ethanol	9±1	9±1	10±1	12±1	20±1	26±1

Values are shown as mean±standard deviation.

The soluble metabolites in the reactor were mainly butyric acid (36%–48%), acetic acid (29%–45%) and ethanol (9%–26%). Similar data (butyric acid (45%–50%), acetic acid (20%–30%) and propionic acid (10%–20%)) of the total volatile fatty acid were also obtained by Jayalakshmi and coauthors (2009) who studied hydrogen production from kitchen waste. The increase in the content of corn waste resulted in a decrease in the pH; meanwhile, the highest yield of hydrogen in the fermented medium with 40 g/L of corn waste was observed at the value of pH 7.0. It is known that pH value affects the action of hydrogenases, the main enzyme under the action in which the production of hydrogen takes place. Hydrogenases are stable within the pH range from 6 to 9. When the pH decreases, the rate of macromolecular substances decomposition increases, which leads to increased formation of low-molecular-weight fatty acids. They

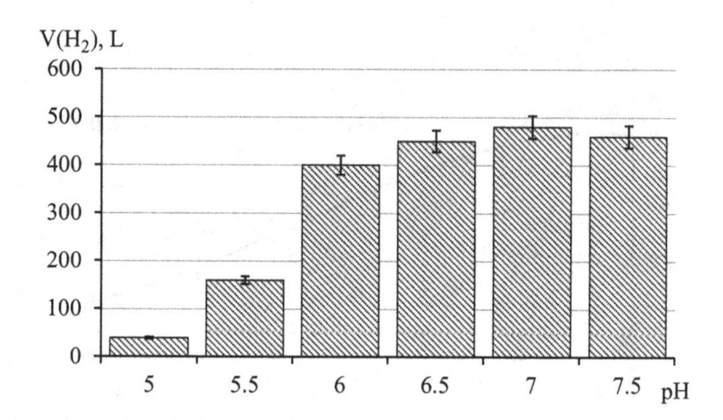

Figure 7.4 The effect of medium pH on the yield of hydrogen, V (H_2), at the content of raw materials 40 g/L.

penetrate the cells and change the pH of the intracellular medium, which leads to the inhibition of hydrogenase and fall in gene activation responsible for its biosynthesis. Changes in the pH lead also to the inhibition of some bacterial species and to lower diversity of the microbial community originating a decrease in the rate of substrate decomposition. The yield of hydrogen depends on the pH of the process performance with the content of corn waste at 40 g/L (Figure 7.4).

Yield is affected by both the concentration of raw materials in the reactor and the value of the pH medium. At the same time, the pH of the medium strongly depends on the concentration of raw materials and the accumulation of different by-products during the fermentation process. So, to have a high hydrogen formation rate process pH should be maintained at its optimal value for H_2 production.

The formation of hydrogen from cellulose-containing raw materials and food waste in fermentation processes is limited by the hydrolytic activity of microorganisms involved in its degradation (Guo et al., 2010). The process of waste fermentation can be carried out in one reactor, where the destruction of raw materials and energy production occurs simultaneously, or in two reactors in which the processes of hydrolysis and biosynthesis of energy carriers are separated. The advantages of the two-stage process include increased process stability, the ability to control and influence the metabolism of microorganisms, smaller reactor volumes and high resistance to toxic compounds effects and high loads. Anaerobic digestion of food waste in two-stage solid–liquid system was successfully used for conversion of solid food waste into leachate containing organic acids in the acidogenic reactor and methane production

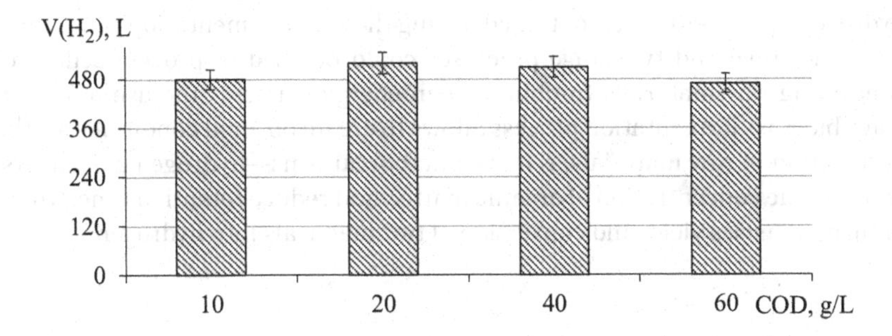

Figure 7.5 The yield of hydrogen, V (H_2), depending on the concentration of COD in the fermented medium.

in the second stage (Stabnikova et al., 2008; Luo & Wong, 2019; Paritosh et al., 2021). The separation of solid-phase biodegradation processes and the formation of hydrogen from low-molecular-weight substances allow to increase the yield from 0.5–1.0 to 2.8 mol H_2/mol glucose (Hafez et al., 2009).

Using a two-stage process of hydrogen production, the enzymatic decomposition of macromolecular compounds in the first reactor is carried out at low pH values, 5.5±0.5. This increases the rate of raw materials decomposition and inhibits the activity of methanogens, consumers of hydrogen, increasing consequently the hydrogen yield. Under such conditions, acid-forming bacteria, most of which belong to the genera *Bacillus*, *Micrococcus*, *Pseudomonas*, and *Clostridium*, are dominant. They break down complex organic compounds such as fiber, proteins, and fats into simpler components, and in the fermented medium there are primary fermentation products, namely, volatile fatty acids, lower alcohols, hydrogen, carbon dioxide, acetic acid and formic acid. The second stage of fermentation is the process of biohydrogen production carried out using the leachate obtained in the first stage. The yield of hydrogen at the second stage of mesophilic anaerobic fermentation of corn waste, depending on the concentration of COD, is shown in Figure 7.5.

In the first days of the process, there is mainly an accumulation of acetic (40%±2%) and butyric (20%±2%) acids. The duration of the process in the second reactor with an organic matter content of COD, 20 g/L should not exceed 3–4 days. When the concentration of organic matter decreases, the retention time of the substrate in the reactor is reduced. Increase in organic matter content by COD for more than 50 g/L led to a decrease in hydrogen yield and lowers the pH of the solution. Under such conditions, the process continues with the formation of ethanol (15%–20%). With the application of separate processes of raw materials conversion, the rate of hydrogen formation increases by five times in relation to the one-stage process. This increases the amount of hydrogen up to 8–10 times, obtained during the same fermentation time. Thus, both one-stage and two-stage processes could be used to process cellulose-containing raw materials to obtain clean energy carriers. The use of a two-stage biomass fermentation process allows to create optimal conditions for the destruction of raw materials, remove microorganisms—hydrogen consumers, increase the concentration of organic matter, and reduce retention time, which in turn increases yield, and conversion of raw materials into hydrogen.

7.4 EFFECT OF HYDROGEN PARTIAL PRESSURE ON CELL METABOLISM

Hydrogen partial pressure (HPP) caused by dissolved hydrogen affects the metabolic pathways during the fermentation of raw materials (Beckers et al., 2012; Nasirian et al., 2010). When the gas produced is removed from the reactor, the yield of hydrogen increases, since microbial hydrogen consumption decreases and cell metabolic pathways change (Figure 7.6).

Increasing the partial pressure of hydrogen promotes the formation of compounds such as ethanol and butyric acid. Mass spectrometric analysis of the solution formed without removing the gas from the reactor zone confirmed the content of ethanol 15% and butyric acid 20%, while acetic acid was not detected. At the same time, hydrogen that was formed during the fermentation of raw materials can be used as an energy source by many microorganisms that restore sulfates, nitrates, nitrites, fumarate, carbon dioxide and other compounds with the following reactions:

$$N_2 + 3H_2 \rightarrow 2NH_3; \tag{7.4}$$

$$2NO_3^- + 6H_2 \rightarrow N_2 + 6H_2O; \tag{7.5}$$

$$\text{Fumarate} \rightarrow \text{fumarate reductase}\left(FADH_2\right) \rightarrow \text{succinate}; \tag{7.6}$$

$$CO_2 + 2H_2 \rightarrow \text{The substance of the cell}\left(CH_2O\right) + H_2O \tag{7.7}$$

Also, the removal of the formed biogas promotes the removal of CO_2, which reduces the acidity of the solution and increases the yield of hydrogen. The formation of acetic acid with hydrogen and CO_2 increases the concentration

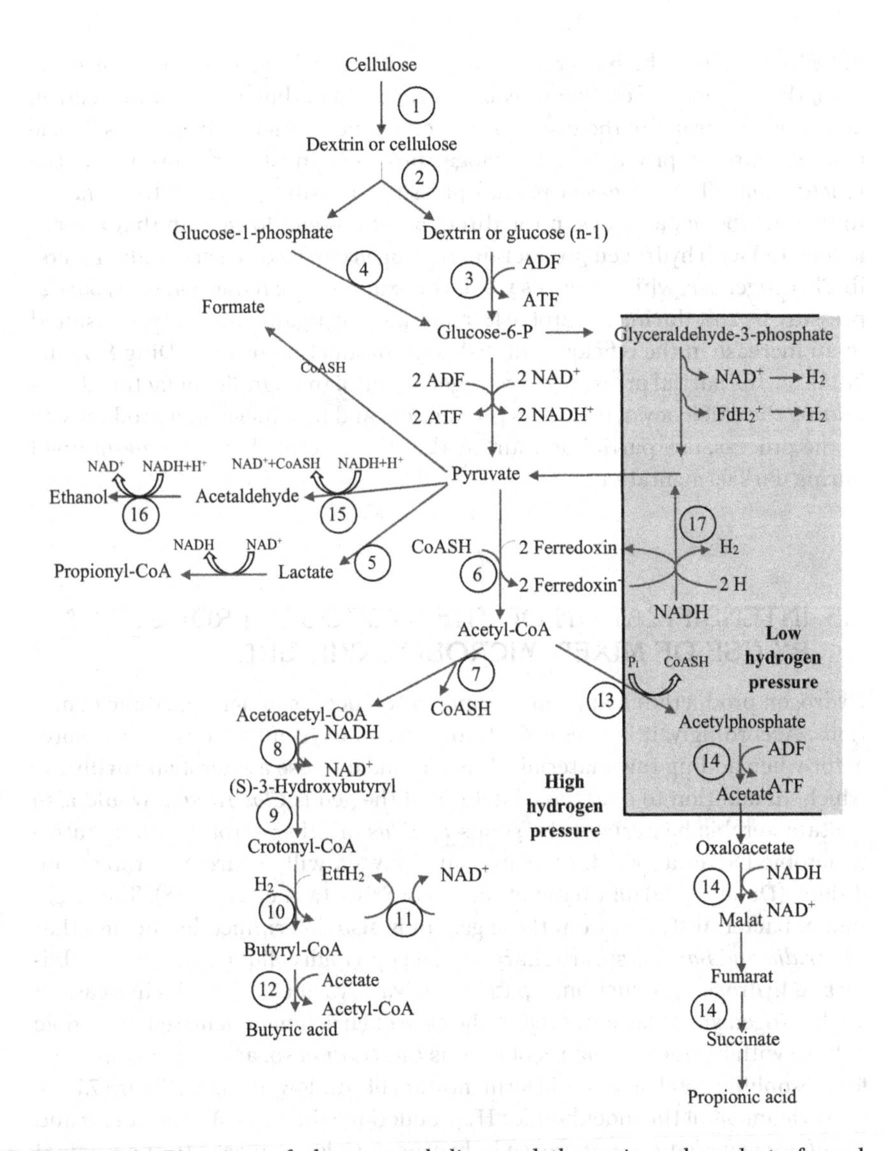

Figure 7.6 Directions of glucose catabolism and the main end products formed in the conversion process : 1 – endo-1,4-β-glucanase and exo-1,4-β-glucanase (exo-cellobiohydrolase); 2 – exo-1,4-β-glucosidase and cellobiase (β-glucosidase); 3 – glucokinase, 4 – phosphoglucomutase; 5 – D- or L-lactate dehydrogenase; 6 – pyruvate synthase; 7 – acetyl-CoA-acetyltransferase; 8 – 3-hydroxybutyryl-CoA dehydrogenase; 9 – crotonyl-CoA hydrate; 10 – diaphorase; 11 – NADH:feredoxin oxidoreductase and hydrogenase; 12 – butyrate:CoA-transferase; 13 – phosphate-acetyltransferase; 14 – acetatkinase; 15 – acetaldehydrogenase; 16 – alcohol dehydrogenase; 17 – NADH:ferredoxin oxidoreductase and hydrogenase.

of hydrogen ions, which decreases the activity of hydrogenases at pH below 6. Also, the presence of carbon dioxide serving as a carbon source and electron acceptor can promote the use of hydrogen in energy metabolism. Thus, in the case of hydrogen production, the biogas produced must be removed from the reactor zone. The increased partial pressure of hydrogen leads to a change in the metabolic pathways in the direction of product formation that are not associated with hydrogen production. Hydrogen can also be used under reversible hydrogenase, which reduces yield. For example, the reduction of H_2 partial pressure by 20% during anaerobic fermentation of organic wastewater resulted in an increase in the efficiency of hydrogen production by 54% (Ding & Zhao, 2018). So, the partial pressure of hydrogen is one of the significant factors determining the efficiency of hydrogen production, and to achieve high productivity in the process, low partial pressure within the reactor should be maintained during dark fermentation.

7.5 INTENSIFICATION OF THE HYDROGEN PRODUCTION BY USE OF MIXED MICROBIAL CULTURE

Hydrogen production using dark fermentation occurs under anaerobic conditions. Accordingly, it is necessary to dispose of oxygen that enters the bioreactor when adding raw materials. It is advisable to use a microbial coculture, which, in addition to anaerobic bacteria of the genus *Clostridium*, would also contain aerobic bacteria of the genus *Bacillus* or other aerobic or facultative anaerobic bacteria, which by consuming oxygen will ensure anaerobic conditions (Du et al., 2020; Chang et al., 2008; Srivastava et al., 2018). The selection of bacterial strains from these genera is also determined by the fact that *Clostridia* and *Bacillus* sp. are characterized by a high cellulose-degrading ability and hydrogen production capability (Srivastava et al., 2018). An increase in the hydrogen yield was observed in the case of enrichment of mixed anaerobic culture with bacterial biomass of strains *Clostridium* sp. and *Bacillus* sp. used for mesophilic batch anaerobic fermentation of sunflower waste (Figure 7.7).

Enrichment of the inoculum for H_2 production with cells of *Clostridium* and *Bacillus* genera led to an increase in hydrogen yield compared to the original anaerobic microbial culture. The highest content of hydrogen in biogas and its yield during fermentation was observed for the inoculum enriched with cells of *Clostridium* sp. and *Bacillus* sp. taken in the ratio of 1:2.5 (Drapoi & Golub, 2021). In the presence of *Bacillus* cells, the yield of hydrogen increased due to the depletion of oxygen in the environment and at the ratio of added bacterial cells of *Clostridium* to *Bacillus* 1:2.5; the hydrogen content in biogas on the 2nd day of fermentation reached 87%.

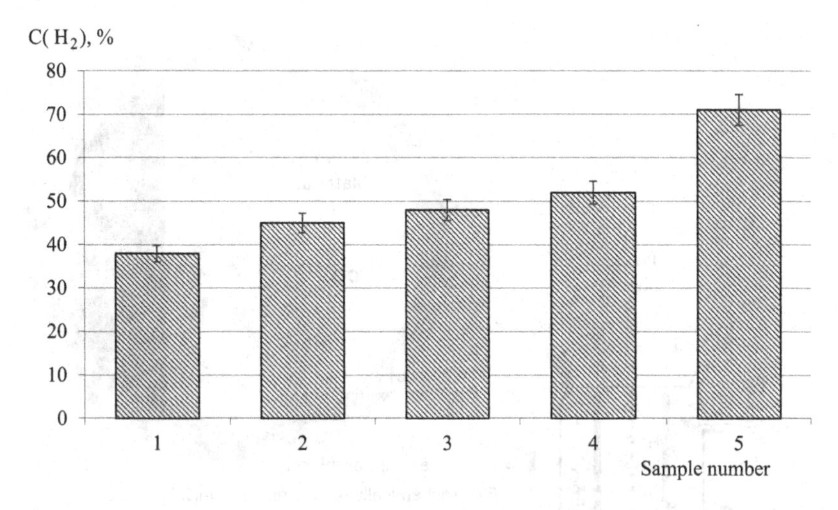

Figure 7.7 Change in hydrogen content, C(H_2), in biogas at different ratios of *Clostridium* and *Bacillus* added to the mixed anaerobic culture: 1, no addition (control); 2, enriched with *Clostridium*; 3, enriched with *Bacillus*; 4, enriched of *Clostridium* and *Bacillus* in the ratio of 1:1; 5, enriched with *Clostridium* and *Bacillus* in the ratio of 2.5:1; 6, enriched with *Clostridium* and *Bacillus* in the ratio of 1:2.5.

7.6 HYDROGEN PRODUCTION IN A MICROBIAL FUEL CELL

To increase hydrogen yield in the processing of cellulose-containing food and other organic wastes, successive fermentation processes can be used in anaerobic conditions in the dark, in anaerobic conditions in the light (using purple bacteria) or in a microbial fuel cell (MFC). The major substrates that have been used to be treated in MFC include different wastewaters and lignocellulosic materials (Pant et al., 2009; Rahmani et al., 2022). Organic acids obtained during fermentation are sources of nutrients for exogenous microorganisms in MFC, where subsequent disposal of raw materials occurs. The biofilm formed at the MFC anode contains various types of microorganisms that are involved in the transformation of the substrate. Electrons produced in glycolysis, pyruvate oxidation and the Krebs cycle are involved in the electron transport chain of the membrane, where oxygen oxidation occurs under aerobic conditions, and their transfer to the MFC anode is possible under anaerobic environments. Protons produced in the respiratory chain, under the action of the potential difference through the proton permeable membrane, enter the cathode, where they participate in the electrochemical process. Electrons move to the cathode through an external electrical circuit (Figure 7.8).

At the anode due to the metabolism of immobilized exoelectrogenic microorganisms under anaerobic conditions, electrons are transferred to the

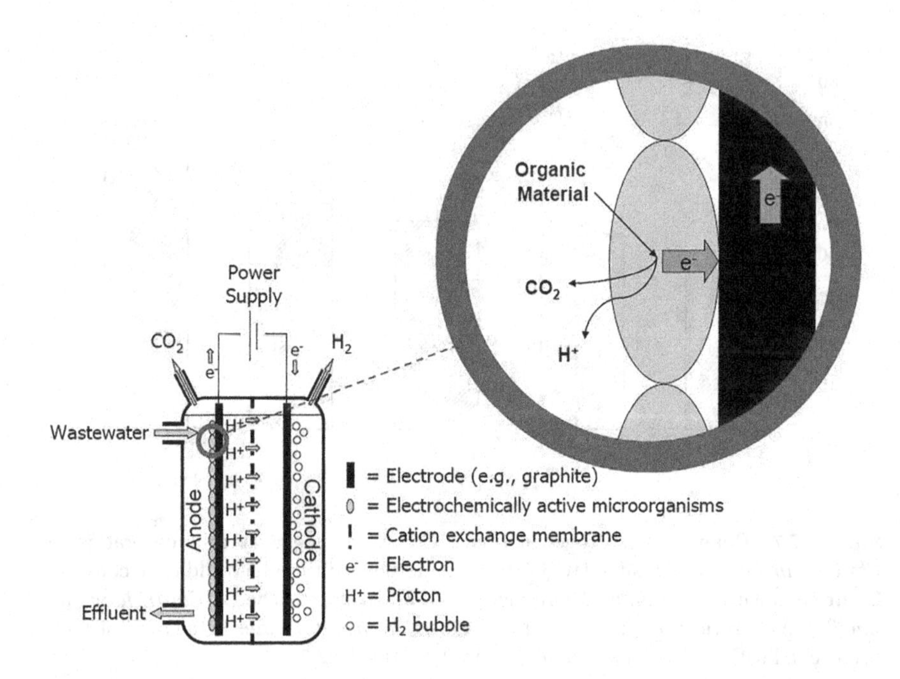

Figure 7.8 Scheme of a microbial fuel cell with hydrogen production.

conductive electrode, resulting in a current in the MFC, and at the same time, there is further utilization of raw materials. That is, the use of sequential processes increases the degree of conversion of cellulose-containing or other raw materials. By using additional voltage in the MFC, hydrogen can be obtained at the cathode. Theoretically, in MFC hydrogen can be obtained by applying an additional voltage of 0.12 V, which is equivalent to energy of 0.26 kWh/m^3 H_2. The amount of electricity required to produce hydrogen by electrolysis of aqueous solutions exceeds the energy obtained by burning hydrogen (3 kWh/m^3 H_2 or 33.5 kWh/kg H_2). Hydrogen with the purity of 99.9% can be produced in MFC. That is, under anaerobic conditions in the MFC it is possible to get hydrogen at a much lower energy cost. Table 7.4 shows the yield of hydrogen in the MFC using different types of organic waste, different anode materials and MFC designs.

Combination of MFC with anaerobic fermentation could significantly increase hydrogen productivity. Thus, Huang and coauthors (2020) showed that in the processing of food waste to obtain hydrogen by sequential use of the anaerobic treatment of substrate in the fermenter followed by the MFC process, hydrogen yield increased by 4–8 times from 49 to 511 mLH_2/g VS and reaching production of 7–9 mole of H_2 per mole of glucose.

TABLE 7.4 THE HYDROGEN PRODUCTION IN THE MICROBIAL FUEL CELL (MFC) FROM DIFFERENT WASTES

Waste Materials	Hydrogen Yield	Anode	MFC	Inoculate	Reference
Pig manure	168 mL H_2/L·d	Carbon felt	Dual chambers	Mixed culture	Shen et al. (2018)
Rice mill wastewater	3.6 mL/L·h	Carbon cloth	Dual chamber	*Rhodobacter* sp.	Keruthiga et al. (2021)
Glycerol solution	144 mL H_2/m^2·d	Graphite rods	Dual chamber	Mixed culture	Mahmoud et al. (2022)
Cane sugar effluent	1.6 mmol H_2/L·d	Graphite rectangle	Dual chamber	Mixed culture	Jayabalan et al. (2019)
Cassava starch processing wastewater	197 mL H_2/L·d	Graphite fiber felt	Single chamber	Mixed culture	Peerawat et al. (2017)
Winery wastewater	190 mL H_2/L·d	Graphite fiber brush	Single chamber	Mixed culture	Cusick et al. (2011)
Domestic wastewater	600 mL H_2/d	Carbon felt	Dual chamber	Mixed culture	Heidrich et al. (2014)
Wastewater with pig manure	0.9–1.0 L H_2/L·d	Graphite fiber brush	Single chamber	Mixed culture	Wagner et al. (2009)
Bovine serum albumin	420 mL H_2/L·d	Graphite fiber brush	Single chamber	Mixed culture	Lu et al. (2010)
Bass agave hydrolyzate	8.5 mL H_2/L·d	Carbon felt	Dual chamber	Mixed culture	García-Amador et al. (2019)
Hydrolyzate of wheat straw	0.61 m^3 H_2/m^3·d	Graphite rods	Dual chamber	Mixed culture enriched with bacteria of *Bacteroidetes phylum* and *Geobacter sulfurreducens*	Thygesen et al. (2011)

7.7 CONCLUSIONS

To intensify the process of hydrogen production by microorganisms, the following techniques are used: pretreatment of raw materials, regulation of the inoculum composition, change of the physicochemical parameters of the process, and combination of various processes. Pretreatment of raw materials (grinding, treatment with steam, alkali, and acid, among others) facilitates the access of microorganisms to nutrients in the substrate. Application of such techniques allows to increase the hydrogen output by up to 10 times. Enrichment of the microbial association with species that create the necessary conditions for hydrogen producers and increasing their content in the inoculum makes it possible to increase the rate of hydrogen formation by three times. Variation of the physicochemical parameters of the process depending on the type of raw material makes it possible to eliminate hydrogen consumers, changes in the metabolism of microorganisms and prevents formation of metabolites that inhibit the processes of waste utilization and hydrogen production, as well as to increase the rate of raw materials utilization. The combination of different processes (anaerobic fermentation and bioelectrochemical hydrogen production) allows to increase the utilization degree of organic raw materials and to increase the yield of hydrogen from a unit of biomass. These factors determine the potential for hydrogen produced by anaerobic fermentation from organic waste as an alternative energy carrier.

REFERENCES

Ajanovic, A. & Haas, R. (2018). Economic prospects and policy framework for hydrogen as fuel in the transport sector. *Energy Policy*, 123, 280–288. https://doi.org/10.1016/j.enpol.2018.08.063.

Baruah, J., Nath, B.K., Sharma, R., Kumar, S., Deka, R.C., Baruah, D.C. & Kalita, E. (2018). Recent trends in the pretreatment of lignocellulosic biomass for value-added products. *Frontiers in Energy Research. Frontiers Media S.A.* https://doi.org/10.3389/fenrg.2018.00141.

Beckers, L., Hiligsman, S., Masset, J., Hamilton, C. & Thonart, P. (2012). Effects of hydrogen partial pressure on fermentative biohydrogen production by a chemotropic *Clostridium* bacterium in a new horizontal rotating cylinder reactor. *Energy Procedia*, 29, 34–41. https://doi.org/10.1016/j.egypro.2012.09.006.

Boni, M.R., Sbaffoni, S. & Tuccinardi, L. (2013). The influence of slaughterhouse waste on fermentative H_2 production from food waste: Preliminary results. *Waste Management*, 33(6), 1362–1371. https://doi.org/10.1016/j.wasman.2013.02.024.

Chang, J., Chou, C.H., Ho, C., Chen, W., Lay, J. & Huang, C. (2008). Syntrophic co-culture of aerobic *Bacillus* and anaerobic *Clostridium* for bio-fuels and bio-hydrogen production. *International Journal of Hydrogen Energy*, 33, 5137–5146. https://doi.org/10.1016/j.ijhydene.2008.05.021.

Chen, S., Zhang, X., Singh, D., Yu, H. & Yang, X. (2010). Biological pretreatment of lignocellulosics: potential, progress and challenges. *Summary Biofuels*, 1(1), 177–199. https://doi.org/10.4155/bfs.09.13.

Claassen, P.A.M. & de Vrije, T. (2006). Non-thermal production of pure hydrogen from biomass: HYVOLUTION. *International Journal of Hydrogen Energy*, 31(11), 1416–1423. https://doi.org/10.1016/j.ijhydene.2006.06.006.

Cusick, R.D., Bryan, B., Parker, D.S., Merrill, M.D., Mehanna, M., Kiely, D., Liu, G. & Logan, B.E. (2011). Performance of a pilot-scale continuous flow microbial electrolysis cell fed winery wastewater. *Applied Microbiology and Biotechnology*, 89(6), 2053–2063. https://doi.org/10.1007/s00253-011-3130-9.

Datar, R., Huang, J., Manèss, P.C., Mohagheghi, A., Czemik, S. & Chornet, E. (2007). Hydrogen production from the fermentation of corn stover biomass pretreated with a steam-explosion process. *International Journal of Hydrogen Energy*, 32(8), 932–939. https://doi.org/10.1016/j.ijhydene.2006.09.027.

Ding, H.B., Liu, X.Y., Stabnikova, O. & Wang, J.Y. (2008). Effect of protein on biohydrogen production from starch of food waste. *Water Science and Technology*, 57(7), 1031–1036. https://doi.org/10.2166/wst.2008.080.

Drapoi, D. & Golub, N. (2021). Hydrogen production from cellulosic materials by natural microbial association from soil enriched by *Clostridium* and microorganisms. *The Scientific Heritage*, 4(60), 3–8. https://doi.org/10.24412/9215-0365-2021-60-1-3-8.

Du, Y., Zou, W., Zhang, K., Ye, G. & Yang, J. (2020). Advances and applications of Clostridium ao-culture systems in biotechnology. *Frontiers in Microbiology*, 16(11), 560223. https://doi.org/10.3389/fmicb.2020.560223.

García-Amador, R., Hernández, S., Ortiz, I. & Cercado B. (2019). Use of hydrolysate from agave bagasse for bio-hydrogen production in microbial electrolysis cells. *Revista Mexicana De Ingeniería Química*, 18(3), 865–874. https://doi.org/10.1016/j.biortech.2018.12.101.

Golub, N., Denisyuk, P. & Drapoi, D. (2017). Determination of optimal conditions for the release of hydrogen during fermentation of vegetable raw materials. *International Scientific Journal Alternative Energy and Ecology*, 216–218(04–06), 80–89. https://doi.org/10.15518/isjaee.2017.04-06.080-088.

Gomez-Flores, M., Nakhla, G. & Hafez, H. (2017). Hydrogen production and microbial kinetics of *Clostridium termitidis* in mono-culture and co-culture with *Clostridium beijerinckii* on cellulose. *AMB Express*, 7(1), 84. https://doi.org/10.1186/s13568-016-0256-2.

Guo, X.M., Trably, E., Latrille, E., Carrère, H. & Steye, J.P. (2010). Hydrogen production from agricultural waste by dark fermentation: A review. *International Journal of Hydrogen Energy*, 35(19), 10660–10673. https://doi.org/10.1016/j.ijhydene.2010.03.008.

Gupta, M., Velayutham, P., Elbeshbishy, E., Hafez, H., Khafipour, E., Derakhshani, H., El Naggar, M.H., Levin, D.B. & Nakhla, G. (2014). Co-fermentation of glucose, starch, and cellulose for mesophilic biohydrogen production. *International Journal of Hydrogen Energy*, 39(36), 20958–20967. https://doi.org/10.1016/j.ijhydene.2014.10.079.

Frey, M. (2002). Hydrogenases: Hydrogen-activating enzymes. *ChemBioChem*, 3, 153–160.

Hafez, H., Baghchehsaraee, B., Nakhla, G., Karamanev, D., Margaritis, A. & El Naggar, H. (2009). Comparative assessment of decoupling of biomass and hydraulic retention times in hydrogen production bioreactors. *International Journal of Hydrogen Energy*, 34, 7603–7611. https://doi.org/10.1016/j.ijhydene.2009.07.060.

Hanaki, K. & Portugal-Pereira, J. (2018). The effect of biofuel production on greenhouse gas emission reductions. In Takeuchi, K., Shiroyama, H., Saito, O., Matsuura, M. (Eds.), *Biofuels and Sustainability*. Science for Sustainable Societies. Springer, Tokyo. https://doi.org/10.1007/978-4-431-54895-9_6.

Heidrich, E.S., Edwards, S.R., Sarah, J.D., Cotterill, E. & Curtis, T.P. (2014). Performance of a pilot scale microbial electrolysis cell fed on domestic wastewater at ambient temperatures for a 12 month period. *Bioresource Technology*, 173, 87–95. https://doi.org/10.1016/j.biortech.2014.09.083.

Huang, J. Feng, H., Huang, L., Uing, X., Ying, X., Shen, D., Chen, T., Shen, X., Zhou, Y. & Xu, Y. (2020). Continuous hydrogen production from food waste by anaerobic digestion (AD) coupled single-chamber microbial electrolysis cell (MEC) under negative pressure. *Waste Management*, 103(15), 61–66. https://doi.org/10.1016/j.wasman.2019.12.015.

Ivanova, G., Rakhely, J. & Kovacs, K.L. (2009). Thermophilic biohydrogen production from energy plants by *Caldicellulosiruptor saccharolyticus* and comparison with related studies. *International Journal of Hydrogen Energy*, 34(9), 3659–3670. https://doi.org/10.1016/j.ijhydene.2009.02.082.

Iyer, P., Bruns, M.A., Zhang, H., Ginke, S.V. & Logan, B.E. (2004). H_2-producing bacterial communities from a heat-treated soil inoculum. *Applied Microbiology and Biotechnology*, 66(2), 166–173. https://doi.org/10.1007/s00253-004-1666-7.

Jayalakshmi, S., Joseph, K. & Sukumaran, V. (2009). Biohydrogen generation from kitchen waste in an inclined plug flow reactor. *International Journal of Hydrogen Energy*, 34(21), 8854–8858. https://doi.org/10.1016/j.ijhydene.2009.08.048.

Jayabalan, T., Matheswaran, M. & Mohammed, S.N. (2019). Biohydrogen production from sugar industry effluents using nickel based electrode materials in microbial electrolysis cell. *International Journal of Hydrogen Energy*, 44(32), 17381–17388. https://doi.org/10.1016/j.ijhydene.2018.09.219.

Karlsson, A., Vallin, L. & Eglertsson, J. (2008). Effects of temperature, hydraulic retention time and hydrogen extraction rate on hydrogen production from the fermentation of food industry residues and manure. *International Journal of Hydrogen Energy*, 33(3), 953–962. https://doi.org/10.1016/j.ijhydene.2007.10.055.

Keruthiga, K., Mohamed, S.N., Gandhi, N.N. & Muthukumar, K. (2021). Sugar industry waste-derived anode for enhanced biohydrogen production from rice mill wastewater using artificial photo-assisted microbial electrolysis cell. *International Journal of Hydrogen Energy*, 46(39), 20425–20434. https://doi.org/10.1016/j.ijhydene.2021.03.181.

Khanna, N. (2013). Biohydrogen production by dark fermentation. *Energy and Environment*, 2(4), 401–421. https://doi.org/10.1002/wene.15.

Kim, I.S., Hwang, M.H., Jang, N.J., Hyun, S.H. & Lee, S.T. (2004). Effect of low pH on the activity of hydrogen utilizing methanogen in bio-hydrogen process. *International Journal of Hydrogen Energy*, 29(11), 1133–1140. https://doi.org/10.1016/j.ijhydene.2003.08.017

Kim, S.H. & Shin, H.S. (2008). Effects of base-pretreatment on continuous enriched culture for hydrogen production from food waste. *International Journal of Hydrogen Energy*, 33(19), 5266–5274. https://doi.org/10.1016/j.ijhydene.2008.05.010.

Klinke, H.B., Thomsen, A.B. & Ahring, B.K. (2004). Inhibition of ethanol-producing yeast and bacteria by degradation products produced during pretreatment of biomass. *Applied Microbiology and Biotechnology*, 66(1), 10–26. https://doi.org/10.1007/s00253-004-1642-2.

Kucharska, K., Rybarczyk, P., Hołowacz, I., Łukajtis, R., Glinka, M. & Kamiński, M. (2018). Pretreatment of lignocellulosic materials as substrates for fermentation processes. *Molecules*, 23(11), 2937. https://doi.org/10.3390/molecules23112937.

Kumar, A.K. & Sharma, S. (2017). Recent updates on different methods of pretreatment of lignocellulosic feedstocks: A review. *Bioresources and Bioprocessing*, 4, 7. https://doi.org/10.1186/s40643-017-0137-9.

Kurakake, M., Ide, N. & Komaki, T. (2007). Biological pretreatment with two bacterial strains for enzymatic hydrolysis of office paper. *Current Microbiology*, 54(6), 424–428. https://doi.org/10.1007/s00284-006-0568-6.

Lay, J.J., Fan, K.S., Hwang, J.I., Chang, J.I. & Hsu, P.C. (2005). Factors affecting hydrogen production from food wastes by *Clostridium richcomposts*. *Environmental Engineering*, 131(4), 595–602. https://doi.org/10.1061/%28ASCE%290733-9372%282005%29131%3A4%28595%29.

Li, D. & Chen, H. (2007). Biological hydrogen production from steam-exploded straw by simultaneous saccharification and fermentation. *International Journal of Hydrogen Energy*, 32(12), 1742–1748. https://doi.org/10.1016/j.ijhydene.2006.12.011.

Lu, L., Xing, D., Xie, T., Ren, N. & Logan, B.E. (2010). Hydrogen production from proteins via electrohydrogenesis in microbial electrolysis cells. *Biosensors and Bioelectronics*, 25(12), 2690–2695. https://doi.org/10.1016/j.bios.2010.05.00.

Luo, L., Sriram, S., Johnravindar, D., Martin, T.L.P., Wong, J.W.C. & Pradhan, N. (2022). Effect of inoculum pretreatment on the microbial and metabolic dynamics of food waste dark fermentation. *Bioresource Technology*, 358, 127404. https://doi.org/10.1016/j.biortech.2022.127404.

Luo, L. & Wong, J.W.C. (2019). Enhanced food waste degradation in integrated two-phase anaerobic digestion: Effect of leachate recirculation ratio. *Bioresource Technology*, 291, 121813. https://doi.org/10.1016/j.biortech.2019.121813.

Magrini, F.E., de Almeida, G.M., da Maia Soares, D., Fuentes, L., Ecthebehere, C., Beal, L.L., da Silveira, M.M. & Paesi, S. (2021). Effect of different heat treatments of inoculum on the production of hydrogen and volatile fatty acids by dark fermentation of sugarcane vinasse. *Biomass Conversion and Biorefinery*, 11, 2443–2456. https://doi.org/10.1007/s13399-020-00687-0.

Marone, A., Massini, G., Patriarca, C., Signorini, A., Varrone, C. & Izzo, G. (2012). Hydrogen production from vegetable waste by bioaugmentation of indigenous fermentative communities. *International Journal of Hydrogen Energy*, 37(7), 5612–5622. https://doi.org/10.1016/j.ijhydene.2011.12.159.

Mahmoud, R.H., Gomaa, O.M. & Hassan, R.Y.A. (2022). Bio-electrochemical frameworks governing microbial fuel cell performance: technical bottlenecks and proposed solutions. *Royal Society of Chemistry Advances*, 10, 5749–5764. https://doi.org/10.1039/d1ra08487a.

Mosier, N., Wyman, C., Dale, B., Elander, R., Lee, Y.Y., Holtzapple, M. & Ladisch, M. (2005). Features of promising technologies for pretreatment of lignocellulosic biomass. *Bioresource Technology*, 96(6), 673–686. https://doi.org/10.1016/j.biortech.2004.06.025.

Moussa, R.N., Moussa, N. & Dionisi, D. (2022). Hydrogen production from biomass and organic waste using dark fermentation: An analysis of literature data on the effect of operating parameters on process performance. *Processes*, 10, 156. https://doi.org/10.3390/pr10010156.

Mu, Y., Yu, H.Q. & Wang, G. (2007). Evaluation of three methods for enriching H_2-producing cultures from anaerobic sludge. *Enzyme and Microbial Technology*, 40(4, 5), 947–953. https://doi.org/10.1016/j.enzmictec.2006.07.033.

Nasirian, N., Almassi M., Minaei, S. & Widmann, R. (2010). Continuous fermentative hydrogen production under various process conditions. *Journal of Food Agriculture and Environment*, 88(3&4), 968–972. https://www.academia.edu/66296116/Continuous_fermentative_hydrogen_production_ under_various_ process_conditions.

Nnabuife, S.G., Ugbeh-Johnson, J., Okeke, N.E. & Ogbonnaya, C. (2022). Present and projected developments in hydrogen production: A technological review. *Carbon Capture Science & Technology*, 3, 100042. https://doi.org/10.1016/j.ccst.2022.100042.

Noike, T. & Mizuno, O. (2000). Hydrogen fermentation of organic municipal wastes. *Water Science and Technology*, 42(12), 155–162. https://doi.org/10.2166/wst.2000.0261.

Okamoto, M., Miyahara, T., Mizuno, O. & Noike, T. (2000). Biological hydrogen potential of materials characteristic of the organic fraction of municipal solid wastes. *Water Science and Technology*, 41(3), 25–32. https://doi.org/10.2166/wst.2000.0052.

Pant, D., Van Bogaert, G., Diels, L. & Vanbroekhoven, K. (2009). A review of the substrates used in microbial fuel cells (MFCs) for sustainable energy production. *Bioresource Technology*, 101, 1533–1543. https://doi.org/10.1016/j.biortech.2009.10.017.

Parawira, W. & Tekere, M. (2011). Biotechnological strategies to overcome inhibitors in lignocellulose hydrolysates for ethanol production. *Critical Reviews in Biotechnology*, 31(1), 20–31. https://doi.org/10.3109/07388551003757816.

Parveen, K., Barrett, D.M., Delwiche, M.J. & Stroeve, P. (2009). Methods for pretreatment of lignocellulosic biomass for efficient hydrolysis and biofuel production. *Industrial & Engineering Chemistry Research*, 48(8), 3713–3729. https://doi.org/10.1021/IE801542G.

Peerawat, K., Kongjan, P., Utarapichat, B. & Reungsang. A. (2017). Continuous hydrogen production from cassava starch processing wastewater by two-stage thermophilic dark fermentation and microbial electrolysis. *International Journal of Hydrogen Energy*, 42(45), 27584–27592. https://doi.org/10.1016/j.ijhydene.2017.06.145.

Pendyala, B., Chaganti, S.R., Lalman, J.A., Shanmugam, S.R., Heath, D.D. & Lau, P.C.K. (2012). Pretreating mixed anaerobic communities from different sources: Correlating the hydrogen yield with hydrogenase activity and microbial diversity. *International Journal of Hydrogen Energy*, 37(17), 12175–12186. https://doi.org/10.1016/j.ijhydene.2012.05.105.

Pielhop, T., Amgarten, J., Rohr, P.R. & Studer, M.H. (2016). Steam explosion pretreatment of softwood: the effect of the explosive decompression on enzymatic digestibility. *Biotechnology for Biofuels*, 9, 152. https://doi.org/10.1186/s13068-016-0567-1.

Paritosh, K., Kumar, V., Pareek, N., Sahoo, D., Fernandez, Y.B., Coulon, F., Radu, T., Kesharwani, N. & Vivekanand, V. (2021). Solid state anaerobic digestion of water poor feedstock for methane yield: an overview of process characteristics and challenges. *Waste Disposal and Sustainable Energy*, 3, 227–245. https://doi.org/10.1007/s42768-021-00076-x.

Rahmani, A.R., Navidjouy, N., Rahimnejad. M., Alizadeh, S., Samarghandi, M.R. & Nematollahi, D. (2022). Effect of different concentrations of substrate in microbial fuel cells toward bioenergy recovery and simultaneous wastewater treatment. *Environmental Technology*, 43(1), 1–9. https://doi.org/10.1080/09593330.2020.1772374.

Rodríguez-Valderrama, S., Escamilla-Alvarado, C., Magnin, J.P., Rivas-Garcia, P., Valdez-Vazquez, I., & Rios-Leal, E. (2020). Batch biohydrogen production from dilute acid hydrolyzates of fruits-and-vegetables wastes and corn stover as co-substrates. *Biomass and Bioenergy*, 140, 105666. https://doi.org/10.1016/j.biombioe.2020.105666.

Shen, R., Jiang, Y., Ge, Z., Lu, J., Zhang, Y., Liu, Z. & Ren, Z.J. (2018). Microbial electrolysis treatment of post-hydrothermal liquefaction wastewater with hydrogen generation. *Applied Energy*, 212, 509–515. https://doi.org/10.1016/j.apenergy.2017.12.065.

Siriporn, Y., Sompong, O.T. & Poonsuk, P. (2012). Effect of initial pH, nutrients and temperature on hydrogen production from palm oil mill effluent using thermotolerant consortia and corresponding microbial communities. *International Journal of Hydrogen Energy*, 37(18), 13806–13814. https://doi.org/10.1016/j.ijhydene.2012.03.151.

Srikanth, S. & Mohan, S. (2012). Regulatory function of divalent cations in controlling the acidogenic biohydrogen production process. *RSC Advances*, 2(16), 6576–6589. https://doi.org/10.1039/C2RA20383A.

Srivastava, N., Srivastava, M., Gupta, V.K., Ramteke, P.W. & Mishra, P.K. (2018). A novel strategy to enhance biohydrogen production using graphene oxide treated thermostable crude cellulase and sugarcane bagasse hydrolyzate under co-culture system. *Bioresource Technology*, 270, 337–345. https://doi.org/10.1016/j.biortech.2018.09.038.

Stabnikova, O., Liu, X.Y. & Wang, J.Y. (2008). Anaerobic digestion of food waste in a hybrid anaerobic solid-liquid system with leachate recirculation in an acidogenic reactor. *Biochemical Engineering Journal*, 41(2), 198–201. https://doi.org/10.1016/j.bej.2008.05.008.

Stabnikova, O., Wang, J.Y. & Ivanov, V. (2010). Value-added biotechnological products from organic wastes. In L.K. Wang, V. Ivanov, J.H. Tay, Y.T. Hung (Eds.), *Handbook of Environmental Engineering*. Vol. 10. Environmental Biotechnology (pp. 343–394). Humana Press, Totowa, NJ. https://doi.org/10.1007/978-1-60327-140-0_8.

Su, X., Zhao, W. & Xia, D. (2018). The diversity of hydrogen-producing bacteria and methanogens within an *in situ* coal seam. *Biotechnology for Biofuels*, 11, 245. https://doi.org/10.1186/s13068-018-1237-2.

Taherzabeh, M.J. & Karimi, K. (2008). Pretreatment of lignocellulosic wastes to improve ethanol and biogas production: A review. *International Journal of Molecular Sciences*, 9, 1621–1651. https://doi.org/10.3390/ijms9091621.

Tashie-Lewis, B.C. & Nnabuife, S.G. (2021). Hydrogen production, distribution, storage and power conversion in a hydrogen economy - A technology review. *Chemical Engineering Journal Advances*, 8, 100172. https://doi.org/10.1016/j.ceja.2021.100172.

Thygesen, A., Marzorati, M., Boon, N., Thomsen, A.B. & Verstraete, W. (2011). Upgrading of straw hydrolysate for production of hydrogen and phenols in a microbial electrolysis cell (MEC). *Applied Microbiology and Biotechnology*, 89(3), 855–865. https://doi.org/10.1007/s00253-010-3068-3.

Veeravalli, S., Chaganti, S., Lalman, J. & Heath, D. (2014). Optimizing hydrogen production from a switch grass steam exploded liquor using a mixed anaerobic culture in an upflow anaerobic sludge blanket reactor. *International Journal of Hydrogen Energy*, 39(7), 3160–3175. https://doi.org/10.1016/j.ijhydene.2013.12.057.

Wagner, R.C., Regan, J.M., Oh, S.E., Zuo, Y. & Logan, B.E. (2009). Hydrogen and methane production from swine wastewater using microbial electrolysis cells. *Water Research*, 43(5), 1480–1488. https://doi.org/10.1016/j.watres.2008.12.037.

Wang, J. & Wan, W. (2008). Effect of Fe^{2+} concentration on fermentative hydrogen production by mixed cultures. *International Journal of Hydrogen Energy*, 33(4), 1215–1220. https://doi.org/10.1016/j.ijhydene.2007.12.044.

Wang, Y., Wang, H., Feng, X., Wang, X. & Huang, J. (2010). Biohydrogen production from cornstalk wastes by anaerobic fermentation with activated sludge. *International Journal of Hydrogen Energy*, 35(7), 3092–3099. https://doi.org/10.1016/j.ijhydene.2009.07.024.

Xing, Y., Li, Z., Fan, Y. & Hou, H. (2010). Biohydrogen production from dairy manures with acidification pretreatment by anaerobic fermentation. *Environmental Science and Pollution Research*, 17(2), 392–399. https://doi.org/10.1007/s11356-009-0187-4.

Yang, Y., Hu, M., Tang, Y., Geng, B., Qui, M., He, Q., Chen, S., Wang, X. & Yang, S. (2018). Progress and perspective on lignocellulosic hydrolysate inhibitor tolerance improvement in *Zymomonas mobilis*. *Bioresources and Bioprocessing*, 5(1), 1–12. https://doi.org/10.1186/s40643-018-0193-9.

Yokoyama, H., Waki, M., Moriya, N., Yasuda, T., Tanaka, Y. & Haga, K. (2007). Effect of fermentation temperature on hydrogen production from cow waste slurry by using anaerobic microflora within the slurry. *Applied Microbiology and Biotechnology*, 74(2), 474–483. https://doi.org/10.1007/s00253-006-0647-4.

Yue, M., Lambert, H., Pahon, E., Roche, R., Jemei, S. & Hissel, D. (2021). Hydrogen energy systems: A critical review of technologies, applications, trends and challenges. *Renewable and Sustainable Energy Reviews*, 146(C). https://doi.org/10.1016/j.rser.2021.11118.

Zagrodnik, R. & Seifert, K. (2020). Direct fermentative hydrogen production from cellulose and starch with mesophilic bacterial consortia. *Polish Journal of Microbiology*, 69(1), 109–120. https://doi.org/10.33073/pjm-2020-015.

Zhang, J., Wang, Q. & Jiang, J. (2013). Lime mud from paper-making process addition to food waste synergistically enhances hydrogen fermentation performance. *International Journal of Hydrogen Energy*, 38(6), 2738–2745. https://doi.org/10.1016/j.ijhydene.2012.12.048.

Zhang, Z. (2021). Waste pretreatment technologies for hydrogen production. In Q. Zhang, C. He, J. Ren, M.E. Goodsite (Eds.), *Waste to Renewable Biohydrogen*. Volume 1: Advances in Theory and Experiments (pp. 109–122). Academic Press, Science. https://doi.org/10.1016/B978-0-12-821659-0.00004-6.

Zheng, Y., Pan, Z. & Zhang, R. (2009). Overview of biomass pretreatment for cellulosic ethanol production. *International Journal of Agricultural and Biological Engineering*, 2(3), 51–58. https://doi.org/10.3965/j.issn.1934-6344.2009.03.051-068.

Zhu, H. & Béland, M. (2006). Evaluation of alternative methods of preparing hydrogen producing seeds from digested wastewater sludge. *International Journal of Hydrogen Energy*, 31(14), 1980–1988. https://doi.org/10.1016/j.ijhydene.2006.01.019.

Zhuo, S., Yan, X., Liu, D., Si, M., Zhang, K., Liu, M., Peng, B. & Shi, Y. (2018). Use of bacteria for improving the lignocellulose biorefinery process: importance of pre-erosion. *Biotechnology for Biofuels*, 11, 146. https://doi.org/10.1186/s13068-018-1146-4.

Chapter 8

Bioconversion of Poultry Waste into Clean Energy

Sergey Zhadan
LLC "H2Holland Ukraine"

Yevhenii Shapovalov
Department of Educational Knowledge
System Creation National Center

Anatoliy Salyuk
Department of Food Chemistry

Stanislav Usenko
Department of Environmental Safety and Occupational Health

CONTENTS

DOI: 10.1201/9781003329671-8

8.1 INTRODUCTION

8.1.1 Characteristics of Poultry Manure as a Substrate for Anaerobic Digestion

Poultry manure is considered a suitable substrate for biogas generation using anaerobic digestion (Bi et al., 2020; Ivanov et al., 2019; Jurgutis et al., 2020; Sürmeli et al., 2019). It is characterized by a greater degree of biodegradation than other animal waste (Hill, 1983). Biogas yield from poultry manure varies from 0.219 to 0.281 m^3/kg of total solids (TS), with an average methane content of about 60% (Mazur et al., 2014). However, biogas plants that use poultry manure as a monosubstrate are not widespread in Europe due to the peculiarities of its chemical composition. At the same time, the vast scale of poultry waste has a great potential for utilization with further energy production (Mazur et al., 2014).

Studies on anaerobic processing of poultry waste showed that high nitrogen content often causes problems associated with the toxicity of ammonium for anaerobic microorganisms (Abouelenien et al., 2013; Bi et al., 2021; Bujoczek et al., 2000; Coban et al., 2014; Fakkaew & Polprasert, 2021; Karaalp et al., 2013; Zhadan et al., 2021). The toxic effect of ammonium is associated with the production of undissociated ammonia. It diffuses into cell membranes and ionizes to form NH_4^+ ammonium ions leading to the pH imbalance inside and outside the bacterial cell. This negatively affects both the transport of substances and the activity of enzymes (Niu et al., 2013). It was shown that at a concentration of ammonium 4275 mg NH_4^+-N/L in poultry slurry, the gas yield was only 10% of the maximum value (Webb & Hawkes, 1985). During anaerobic digestion of farm animal dung, 50%–75% of all nitrogen is converted to ammonium (Kirchmann & Witter, 1992). Meanwhile, poultry manure has a higher nitrogen content in comparison with those from other farm animals.

It is known that the anaerobic digestion of wastes with high content of sulfur is accompanied by the formation of toxic hydrogen sulfide, H_2S, which can diffuse into the cell membranes of microorganisms and inhibits methane production (Chen et al., 2008; Koster et al., 1986; Stabnikov & Ivanov, 2006). The inhibitory effect of H_2S may be associated with the formation of sulfide and disulfide bonds between polypeptide chains and impaired sulfur assimilation (Vogels, 1988). The content of sulfur in poultry manure is around 0.5 g/kg, which is higher than in the manure from other farm animals (Schulte & Kelling, 1992), and the concentration of hydrogen sulfide may exceed 250 mg S/L during its anaerobic digestion (Sürmeli et al., 2019). Meanwhile, it is shown that free hydrogen sulfide at a concentration of 250 mg S/L at the pH ranging from 6.4 to 7.2 caused 50% inhibition of methane production (Koster et al., 1986). Thus, high levels of nitrogen and sulfur can make it difficult to use chicken manure

TABLE 8.1 THE CHEMICAL COMPOSITION OF CHICKEN MANURE

	Mean Value		Range
Content	g/kg of Wet Mass	% of Total Solids	g/kg of Wet Mass
Water	657	–	369–770
C	289	84.26	224–328
Total N	46	13.41	18.2–72.0
Organic N	38	11.08	–
NH_4^+	14.4	4.20	0.21–29.9
NO_3-N	0.4	0.12	0.03–1.5
Total P	20.7	6.03	13.5–34.0
K	20.9	6.09	12.5–32.5
Cl	24.5	7.14	6–60
Ca	38.9	11.34	36.2–59.6
Mg	4.7	1.37	1.8–6.6
Na	4.2	1.22	2.0–7.4
Mn	0.3	0.09	0.26–0.38
Fe	0.32	0.009	0.08–0.56
Cu	0.53	0.02	0.04–0.07
Zn	0.35	0.10	0.29–0.39
As	0.03	0.01	–

Source: Adapted from Edwards and Daniel (1992).

for biogas production. Moreover, a synergistic negative effect of sulfide and ammonia on the anaerobic digestion of chicken manure was shown (Sürmeli et al., 2019). The chemical composition of chicken manure is shown in Table 8.1.

Although anaerobic digestion is an effective method of processing livestock waste, a few studies have been conducted on the anaerobic treatment of poultry manure (Abouelenien et al., 2009; Ojolo et al., 2007; Sürmeli et al., 2019). In the present research on the anaerobic fermentation of chicken manure as monosubstrate to obtain biogas, we focused on two important characteristics of this process, namely, the moisture content of the substrate and the temperature of its fermentation. Anaerobic digestion, depending on the substrate moisture content, can be called dry or wet (Sinkora & Havlicek, 2011). If the substrate moisture content is more than 85%, it is considered wet digestion, and the substrate with a solid content higher than 15% is treated in dry mode (Uddin & Wright, 2022). The wet digester is the most common type of digester, and in most studies on the anaerobic treatment of poultry waste, the substrate has been in a liquid state (Singh et al., 2010). The moisture content of native poultry manure is approximately 75% (Karaalp et al., 2013; Kukic et al., 2010; Nakashimada et al., 2013). To ensure a stable process and to prevent its inhibition by ammonium nitrogen and sulfide, dilution of treating waste by water to the moisture content of 94% or more is usually used (Bujoczek et al., 2000; Singh et al., 2010). However, an increase in the substrate moisture leads to an increase in the size of the biogas plant, construction costs, and operating costs, such as energy consumption for heating of raw materials and temperature maintenance in the digester (Karaalp et al., 2013; Sinkora & Havlicek, 2011). The effluent from the biogas reactor is directly proportional to the amount of water necessary for the preliminary dilution of poultry manure (Shapovalov et al., 2020). Typically, the digestate is supposed to be used as an organic mineral fertilizer in the field. However, fertilizer is complicated to store, transport and apply. Therefore, in modern poultry farms with high productive capacity, it is challenging to implement effluent generated throughout the year and must be stored when it cannot be deposited in the fields (Nie et al., 2015); meanwhile, placing in a lagoon may affect the environment (Nie et al., 2015; Wu et al., 2016). Compared to conventional chemical fertilizers, transportation of the digestate is not cost-effective due to its low efficiency and high water content. At the same time, there may be a problem with the distribution of a significant amount of liquid fertilizers among many small farms (Wu et al., 2016). Thus, the moisture content of the initial substrate for anaerobic fermentation affects not only the process parameters but also has significant ecological importance.

Anaerobic digestion can be performed at three temperature modes: psychrophilic, mesophilic and thermophilic (Angelidaki & Sanders, 2004; Chen et al., 2008; Jaimes-Estévez et al., 2021). Most studies on the anaerobic digestion of poultry manure have been carried out in the mesophilic mode (Bujoczek et al., 2000, Edwards & Daniel, 1992; Jurgutis et al., 2020). However, no comparative study of poultry waste anaerobic digestion in mesophilic and thermophilic conditions was done.

8.2 MESOPHILIC ANAEROBIC DIGESTION OF POULTRY MANURE

8.2.1 Batch Process

Anaerobic digestion of poultry manure was performed in laboratory bioreactors with a volume of 50 mL. All experiments were performed in triplicate and data were expressed as the mean ± standard deviation. The moisture content of the poultry manure slurry used for anaerobic digestion varied from 72% to

99%. Each batch reactor contained 20 g of poultry manure slurry (further substrate) with a 10% of mass fraction of an inoculum (anaerobic activated sludge). The anaerobic digestion was performed in the mesophilic mode at 35°C and in the thermophilic mode at 50°C for 50 days. The amount of generated biogas was measured daily. In the mesophilic mode, the increase in poultry manure moisture content from 72% to 94% resulted in gradual increase in the gas yield and methane production per unit of volatile solids (VS); meanwhile, an increase in the substrate moisture from 94% to 99% led to decrease of the gas yield (Figure 8.1a).

In the mesophilic mode, a decrease in the substrate moisture from 99% to 96% led to increased biogas production per unit mass. Starting at poultry manure moisture content of 94%, the gas yield decreased. Methane production decreased steadily with the decrease in substrate moisture. The biogas yield varied from 66.2 mL/g VS to 301.8 mL/g VS, and methane yield—from 11.9 mL/g VS to 150.0 mL/g VS. In the mesophilic mode, the maximum biogas yield per unit mass was 301.8 mL/g VS at the substrate moisture content of 96%, and methane yield—150.0 mL/g VS at the substrate moisture content of 99%. Mazur et al. (2014) carried out anaerobic digestion at the substrate's moisture content level varying from 90% to 98% and also received the maximum biogas yield at 96% and maximum methane yield—at 98%. Similar results were obtained in the work of Bujoczek et al. (2000), which describes the fermentation of fresh manure at a moisture content of 78.3%, 90% and 95%. With a decrease in moisture content, the gas yield decreased and was maximum at a moisture content of 95% (Bujoczek et al., 2000). The decrease in biogas and methane yield, associated with the decrease in substrate moisture content in the mesophilic mode, may be caused by the increase in the concentration of inhibitors, the formation of local zones of their accumulation, the more significant osmotic pressure of the medium and lower content of bioavailable moisture. At the same time, the initial increase in biogas yield can be explained by the saturation of the microbial consortium with the substrate.

In the mesophilic mode, a decrease in the substrate moisture content from 99% to 88% led to increased biogas and methane per unit volume production. Starting at the manure moisture content of 86%, biogas and methane yield decreased. Biogas production ranged from 1.8 mL/mL to 23.9 mL/mL, and methane production—from 1.0 mL/mL to 10.8 mL/mL. The maximum biogas yield was 23.9 mL/mL, and methane yield was 10.8 mL/mL at a substrate moisture content of 88% (Figure 8.3). The maximum rate of methanogenesis in the mesophilic conditions increased exponentially with an increase in the moisture content of the substrate, and the maximum rate of methane production was 22.1 mL CH_4/g VS·day. The obtained values are higher than those obtained by Bujoczek et al. (2000), who studied the anaerobic digestion of poultry manure in the mesophilic mode. The maximum rates of methane production at

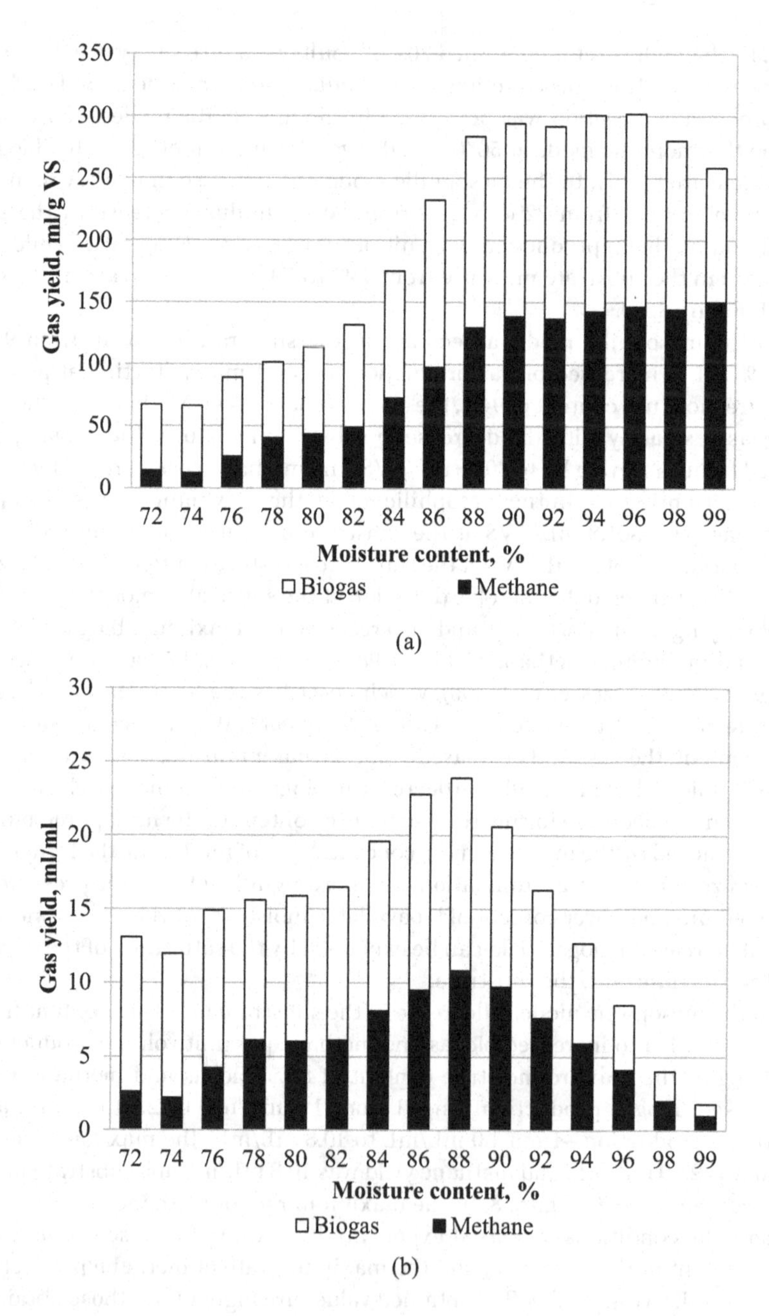

Figure 8.1 Gas and methane yields from VS (a) and reactor unit volume (b) in the mesophilic fermentation at different substrate moisture contents.

substrate moisture content 84, 90 and 94% were 6.1, 10.2 and 17.0 mL CH_4/g VS day, respectively. Bujoczek et al. (2000) found that the maximum rate of methanogenesis was 1.8 mL CH_4/g VS·day at a substrate moisture content of 84%; 5.9 mL CH_4/g VS·day at a substrate moisture content of 90% and 9.2 mL CH_4/g VS·day at a substrate moisture content of 95%. The volume ratio of produced methane to biogas increased with an increase in the moisture content of the substrate. Methane content in the produced gas varied from 17.7% to 58.0%. The characteristics of poultry manure before and after mesophilic anaerobic digestion, which were obtained in the present study, are shown in Table 8.2.

The content of volatile solids and total solids destruction increased with the increase in moisture content in the substrate and varied from 25.09% to 96.43% and from 17.56% to 67.50%, respectively. The initial pH of the substrate decreased from 6.4 to 5.8 with an increase in moisture content from 72% to 99%, which corresponded to a decrease in the content of ammonium nitrogen. After methane fermentation, the pH value increased with substrate moisture decreasing from pH 8.5 at 99% of moisture content to 9.2 at 72%, 74% and 82% of moisture content. At the beginning of the process, the electrical conductivity of the substrate increased with a decrease in moisture content from 1372 μS/cm to 8914 μS/cm. As a result of anaerobic digestion, the electrical conductivity increased due to the mineralization of organic substances and varied from 3880 μS/cm to 30940 μS/cm at 78% of moisture content. The initial concentration of ammonium nitrogen increased with a decrease in the substrate moisture content and varied from 136 mg/L to 3808 mg/L. Ammonium nitrogen content after anaerobic digestion increased and ranged from 430 mg/L at moisture content of 99% to 9310 mg/L at moisture content of 76%. The initial content of free ammonia in the substrate ranged from 0.10 mg/L to 11.28 mg/L. After anaerobic digestion, its concentration increased over the entire range of substrate moisture values due to an increase in the concentration of ammonium nitrogen and the pH. The content of free ammonia varied from 113.2 mg/L at moisture content of 99% to 5742.8 mg/L at moisture content of 74%. The amount of volatile fatty acids (VFA) after anaerobic digestion at a substrate moisture content of 96%–99% was lower than the initial concentration. The VFA concentration varied from 60 mg/L at moisture content of 99% to 12633 mg/L at moisture content of 74%.

8.2.2 Continuous Process

For continuous anaerobic digestion in a continuous stirred tank reactor (CSTR), poultry manure must have the required consistency and substrate should be present in liquid form as it is needed for pumping. For this reason, the moisture content of the initial substrate should be higher than 85% (Kerekrety et al., 1991). Using the results of batch mesophilic methane fermentation of poultry manure slurry with different moisture content from 86% to 99%, mathematical

TABLE 8.2 CHARACTERISTICS OF POULTRY MANURE BEFORE AND AFTER BATCH MESOPHILIC ANAEROBIC DIGESTION

Moisture Content,	Total Nitrogen, mg/L		Ammonium Nitrogen, mg/L		Free Ammonia, mg/L		VFA, mg/L		Free VFA, mg/L		pH		Conductivity, μS/cm	
%	Initial	Treated	Initial	Treated	Initial	Treated	Initial	Treated	Initial	Treated	Initial	Treated	Initial	Treated
99	460	437	136	430	0.10	113.19	146	60	12.00	0.01	5.8	8.5	1372	3880
98	920	874	272	872	0.21	270.53	292	184	22.52	0.03	5.8	8.6	2252	5088
96	1840	1748	544	1752	0.50	838.71	585	340	38.84	0.02	5.9	8.9	3721	10345
94	2760	2622	816	2425	0.80	1160.88	877	1062	54.59	0.08	5.9	8.9	5186	13065
92	3680	3496	1088	2810	1.15	1490.62	1170	3077	68.25	0.18	6.0	9.0	5989	16968
90	4600	4370	1360	4015	1.64	2060.65	1462	7116	74.84	0.44	6.0	9.0	6761	17962
88	5520	5244	1632	3799	2.27	1949.79	1754	6640	78.72	0.41	6.1	9.0	7538	23352
86	6440	6118	1904	4204	2.83	2491.90	2047	8104	86.00	0.36	6.1	9.1	7860	23884
84	7360	6992	2176	4956	3.39	3046.82	2339	9504	94.02	0.39	6.1	9.2	8231	25200
82	8280	7866	2448	4895	4.69	2819.36	2632	9715	86.65	0.47	6.2	9.1	8214	27160
80	9200	9016	2720	5960	5.45	3331.73	2924	10738	92.06	0.55	6.3	9.1	8330	29260
78	10120	9918	2992	7026	6.73	4086.19	3216	11582	90.55	0.54	6.3	9.1	8524	30940
76	11040	10819	3264	9310	7.34	5570.08	3509	12206	98.81	0.54	6.3	9.1	8557	28798
74	11960	11721	3536	8950	9.78	5742.81	3801	12570	87.46	0.46	6.4	9.2	8713	27300
72	12880	12622	3808	8589	11.28	5511.17	4094	12633	88.05	0.46	6.4	9.2	8914	30660

modeling of anaerobic digestion was performed and the maximum growth rate, μ_{max}, and the half-saturation constant K_S, a concentration of substrate at which the growth rate is equal ½ of the maximum growth rate for methanogens, were determined according to (Kucheruk, 2016) by equation (8.1):

$$(S_0 - S_\tau)/(X_j \cdot \tau_j) = [K_S \cdot Y_{X/S} \cdot \ln(S_0 - S_\tau)/\mu_{max} \cdot \tau_j] + \mu_{max} \quad (8.1)$$

where S_0 is an initial mass of VS of substrates, g VS; S_τ is a mass of VS at the time τ_j, g VS; X_j is cell mass of methanogens at time τ_j, g; K_S is the half-saturation constant, g VS/L; $Y_{X/S}$ is the yield of cell mass of methanogens, 0.0419 g/g VS (Kucheruk, 2016); and μ_{max} is a maximum growth rate of methanogens, d^{-1}.

The substrate half-saturation constant K_S increases with the decrease in moisture content from 99% to 86% and ranges from 2.973 g VS/L to 44.343 g VS/L. On the contrary, the maximum growth rate of methanogens μ_{max} decreases with a decrease in moisture content from 99% to 86% and is in the range from 0.146 d^{-1} to 0.275 d^{-1} (Table 8.3).

The obtained kinetic parameters of methanogenesis in mesophilic conditions were used to simulate methane production in a continuous mode. To estimate the specific rate of methane release, P_{CH4}, L CH_4/L·d, the following equation (8.2) was used (Kucheruk, 2016):

$$P_{CH_4} = (D \cdot Y_{P/S}/0.7 \cdot \rho_{CH_4}) \cdot [S_0 - D \cdot K_S/(\mu_{max} - D)] \quad (8.2)$$

where D is the hydraulic load of the reactor, d^{-1}; $Y_{P/S}$ is methane yield from acetate, g CH_4/g Ac; ρ_{CH4} is a specific gravity of methane, 0.717 g/L; S_0 is the

TABLE 8.3 KINETIC PARAMETERS OF METHANOGENESIS FOR MESOPHILIC AND THERMOPHILIC METHANE FERMENTATION

The Moisture Content of the Substrate, %	Mesophilic Fermentation		Thermophilic Fermentation	
	K_S, g VS/L	μ_{max}, d^{-1}	K_S, g VS/L	μ_{max}, d^{-1}
99	2.937	0.275	3.588	0.459
98	6.886	0.245	6.706	0.706
96	13.195	0.238	13.706	0.811
94	19.815	0.222	20.319	0.746
92	26.853	0.189	26.047	0.557
90	32.972	0.186	32.319	0.467
88	39.809	0.186	38.890	0.376
86	44.343	0.146	45.612	0.323

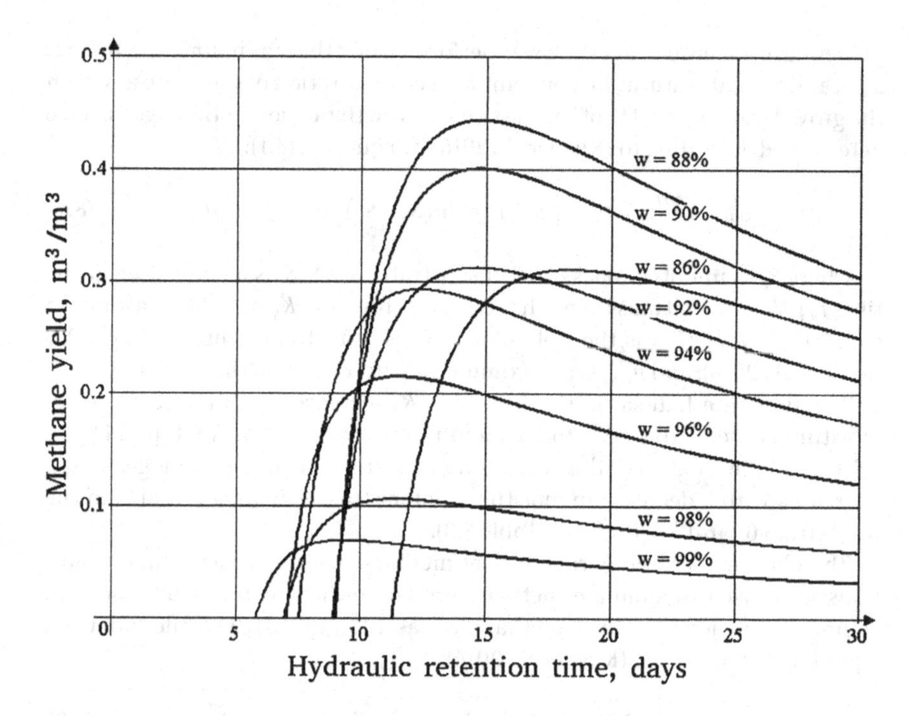

Figure 8.2 Methane yield per unit volume of the continuous anaerobic digester operated in mesophilic mode at a temperature of 35°C depending on the substrate moisture content and hydraulic retention time.

initial equivalent concentration of acetate in the input substrate, g-equal Ac/L; K_S is a half-saturation constant, g-equal Ac/L; and μ_{max} is a maximum growth rate of methanogens, d^{-1}.

To define the data on optimal hydraulic retention time, the hydraulic load of the reactor, *D*, value was reciprocal. The rate of methane release in the mesophilic methane fermentation increases with a decrease in the substrate moisture content from 99% to 88% and decreases with a subsequent decrease in moisture content to 86% (Figure 8.2).

The maximum methane yield per unit volume of the reactor is observed at a substrate moisture content of 88% and a hydraulic retention time (HRT) of 14.75 days. In mesophilic methane fermentation, the regularities of the change in the rate of methane release per unit volume of the reactor depend on the hydraulic retention time and substrate moisture. The values of HRT at different substrate moisture contents in the mesophilic methane fermentation corresponding to a specific operating mode of the biogas plant have been determined (Table 8.4).

TABLE 8.4 OPERATIONAL MODES OF A BIOGAS PLANT UNDER MESOPHILIC CONDITIONS

	Hydraulic Retention Time, Days			
The Moisture Content of the Substrate, %	Methane Production Equal to Zero	Maximum methane Production	Maximum Value of $[P_{CH4} \cdot Q_{CH4}]$	95% CH_4 Yield Potential
99	3.77	6.23	8.13	36.76
98	2.39	3.91	5.08	22.70
96	2.10	3.44	4.48	17.94
94	2.27	3.72	4.83	23.04
92	2.99	4.88	6.32	28.25
90	3.55	5.79	7.51	34.57
88	4.42	7.21	9.34	39.02
86	5.15	8.41	10.09	46.78

It is possible to use the proposed results to define parameters of moisture content of substrate and hydraulic retention time that correspond to different operation modes of anaerobic digester: to reach the methane yield rate that drops to zero due to the critical washout of the population of methanogens; the methane yield rate is the highest (under these conditions, the payback period of a biogas plant is the shortest); the value of criterion complex $P_{CH4} \cdot Q_{CH4}$ is the highest (this mode allows simultaneous maximization of biogas yield from the unit of apparatus volume, P_{CH4}, and the unit of substrate VS mass, Q_{CH4}); the degree of methane yield is 95%, and it makes no sense to increase the duration of the process further (this mode may be helpful in the production of fertilizers due to higher content of mineral substances).

8.3 THERMOPHILIC ANAEROBIC DIGESTION OF POULTRY MANURE

8.3.1 Batch Process

In the thermophilic mode, biogas and methane production increased per unit mass with a decrease in the substrate moisture content from 99% to 92%. With the decrease in substrate moisture content from 90% to 78%, the gas and methane yield decreased. At an increase in substrate moisture content from 72% to 78%, biogas and methane yield were relatively at the same level. Biogas production varied from 12.1 mL/g VS to 382.3 mL/g VS, and methane production—from 2.0 mL/g VS to 207.9 mL/g VS. In the thermophilic mode, the maximum biogas and methane yield per unit mass were 382.3 mL/g VS and 207.9 mL/g VS, respectively, with a substrate moisture content of 92% (Figure 8.3a).

The initial increase in biogas and methane yield with a decrease in substrate moisture content in the thermophilic mode is possibly related to the increased concentrations of nutrients needed for the methanogenic bacteria. A further decrease is possibly related to an increase in the concentrations of inhibitors, the formation of local zones of their accumulation, an increase in the osmotic pressure of the medium and a decrease in the content of biologically available moisture. In the thermophilic anaerobic digestion, the biogas yield per unit volume increased with a decrease in the substrate moisture content from 99% to 84% and was relatively at the same level at a substrate moisture content of 72%–78%. Biogas production decreased when the moisture content decreased from 82% to 78%. In general, biogas production varied from 2.1 (at the moisture content level of 76% and 99%) to 25.8 mL/mL (at the moisture content level of 86%), and methane production from 0.4 (at the moisture content level of 72% and 76%) to 13.0 mL/mL (at the moisture content level of 88%) (Figure 8.3b). The maximum rate of methanogenesis under thermophilic conditions increased with an increase in substrate moisture content. The growth was exponential. The maximum rate of methanogenesis was 37 mL CH_4/g VS·day at the moisture content of 98%. The volume ratio of produced methane to biogas increased with an increase in substrate moisture content. The methane content in the produced gas varied from 16.8% to 62.9%. The increase in methane content in biogas from the beginning of the experiment was faster at higher moisture content in the substrate. The characteristics of poultry manure before and after thermophilic anaerobic digestion are shown in Table 8.5.

The biogas yield was proportional to the VS reduction in poultry manure. Reduction of VS decreased with a decrease in substrate moisture content and varied from 1.83% to 96.55%. The destruction of TS varied from 1.28% to 67.59%. The initial pH of the substrate decreased with an increase in moisture content from 6.45 to 5.79, which corresponded to an increase in the content of ammonium nitrogen. After anaerobic digestion, the pH value increased with a decrease in the substrate moisture content from 99% to 92% and decreased with a subsequent decrease of moisture content from 90% to 72%. The pH ranged from 5.3 at the moisture content of 72%, 74% and 76% to 9.4 at the moisture content of 92%. The initial electrical conductivity of the substrate increased with a decrease in moisture content from 1200 μS/cm to 8964 μS/cm. As a result of anaerobic digestion, it increased due to the mineralization of organic substances and varied from 3964 μS/cm at the moisture content of 99% to 19780 μS/cm at the moisture content of 72%. The initial concentration of ammonium nitrogen increased with a decrease in substrate moisture content from 143 mg/L to 3990 mg/L.

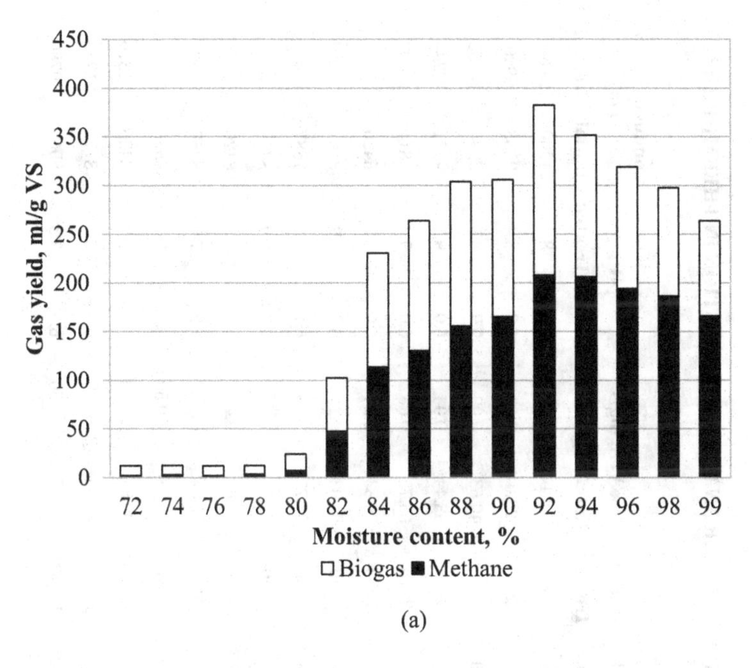

(a)

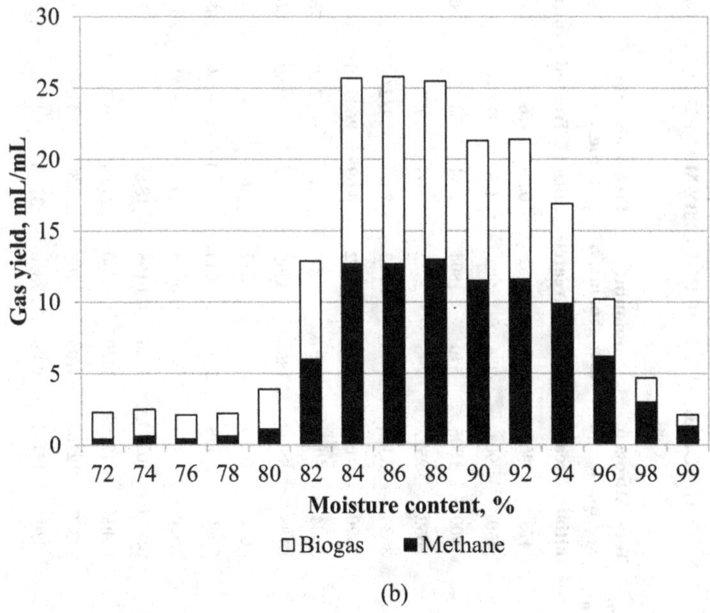

(b)

Figure 8.3 Gas and methane yields from VS (a) and reactor unit volume (b) in the thermophilic fermentation at different substrate moisture contents.

TABLE 8.5 CHARACTERISTICS OF POULTRY MANURE BEFORE AND AFTER ANAEROBIC DIGESTION IN THERMOPHILIC MODE

Moisture Content, %	Total Nitrogen, mg/L		Ammonium Nitrogen, mg/L		Free Ammonia, mg/L		VFA, mg/L		Free VFA, mg/L		pH		Conductivity, μS/cm	
	Initial	Treated	Initial	Treated	Initial	Treated	Initial	Treated	Initial	Treated	Initial	Treated	Initial	Treated
99	475	451	143	451	0.25	164.97	145	60	13.1	0.02	5.8	8.3	1200	3964
98	950	903	285	902	0.6	379.44	290	210	22.6	0.05	5.9	8.4	2533	6990
96	1900	1805	570	1804	1.34	830.50	580	420	40.63	0.09	5.9	8.5	3992	11455
94	2652	2522	796	2522	2.05	1978.40	831	885	53.44	0.04	6.0	9.1	4871	13244
92	3536	3363	1061	3222	3.06	2832.00	1140	3210	65.78	0.08	6.0	9.4	5811	15807
90	4420	4199	1326	3082	4.09	2608.40	1449	7080	78.33	0.23	6.0	9.3	6420	15160
88	5304	5037	1591	4483	5.14	3617.30	1757	6840	90.93	0.29	6.1	9.2	8107	20380
86	6188	5879	1856	3362	6.89	1229.80	2066	8100	93.75	2.49	6.1	8.3	7887	19840
84	7072	6720	2122	3642	8.63	2554.70	2374	7140	98.64	0.54	6.2	8.9	8457	28180
82	7956	7558	2387	4644	11.14	2404.80	2683	14990	97.62	2.47	6.2	8.6	8120	21480
80	9500	9358	2850	5648	14.25	155.20	2898	12230	98.64	76.16	6.2	7.0	8257	25372
78	10450	10293	3135	6168	18.82	8.90	3188	9281	90.78	1010.50	6.3	5.7	8533	21376
76	11400	11229	3420	6760	21.01	4.70	3478	12700	96.84	2582.70	6.3	5.3	8436	21370
74	12350	12165	3705	7269	27.33	4.00	3767	16315	87.65	3967.90	6.4	5.3	8771	21280
72	13300	13101	3990	7816	32.24	4.70	4057	13582	86.27	3078.50	6.5	5.3	8964	19780

The content of ammonium nitrogen as a result of anaerobic digestion increased and ranged from 451 mg/L at the moisture content of 99% to 7816 mg/L at the moisture content of 72%. The initial free ammonia content in the substrate increased with a decrease in moisture content and ranged from 0.25 mg/L to 32.24 mg/L. After anaerobic digestion, its concentration at a substrate moisture content from 72% to 78% was lower than the initial, which is associated with a decreased pH. Free ammonia concentration varied from 4.00 mg/L at the moisture content of 74% to 3617.27 mg/L at the moisture content of 88%. With a decrease in the substrate moisture content, the initial concentration of VFA increased and ranged from 145 mg/L (at the moisture content of 99%) to 4075 mg/L (at the moisture content of 72%). After anaerobic digestion, the concentration of VFA at a substrate moisture content of 96%–99% was lower than the initial, and at the moisture content in the 72%–94% range, the VFA content was higher. The VFA concentration varied from 60 mg/L at the moisture content of 99% to 16315 mg/L at the moisture content of 74%. After anaerobic digestion, the concentration of free VFA in substrates with moisture content in the range from 82% to 99% was lower than initial and with moisture content in the range from 72% to 80%—it was higher. The content of free VFA ranged from 0.02 mg/L at the moisture content of 99% to 3967.93 mg/L at the moisture content of 74%.

8.3.2 Continuous Process

Based on the fact that for anaerobic digestion in continuous mode poultry manure must have the required consistency, mathematical modeling of anaerobic digestion was carried out for the moisture content of the substrate in the range from 86% to 99%. Using the results of anaerobic digestion of poultry manure slurry with different moisture content in the batch mode, the kinetic parameters, namely, the maximum rate k_{max} and the half-saturation constant K_S of methanogens, were determined. The substrate half-saturation constant K_S increased with a decrease in moisture content from 99% to 86% and ranged from 3.588 g VS/L to 45.612 g VS/L. The maximum growth rate of methanogens μ_{max} increased with a decrease in moisture content from 99% to 96%, and it decreased with a subsequent decrease in moisture content to 86%. The μ_{max} value ranged from 0.323 d^{-1} to 0.811 d^{-1} (Table 8.3).

The values of the hydraulic retention time at different substrate moisture contents in the thermophilic anaerobic digestion, corresponding to a specific operating mode of the biogas plant, have been determined (Table 8.6).

The obtained kinetic parameters of acetoclastic methanogenesis in thermophilic conditions were used to model the production of methane in a continuous operation. The rate of methane yield in the thermophilic process with a decrease in substrate moisture content from 99% to 92% increases, and it decreases with a subsequent decrease in moisture content to 86% (Figure 8.4).

TABLE 8.6 OPERATIONAL MODES OF A BIOGAS PLANT UNDER THERMOPHILIC CONDITIONS

The Moisture Content of the Substrate, %	Hydraulic Retention Time, Days			
	Methane Production Equal to Zero	Maximum Methane Production	Maximum Value of $[P_{CH4} \cdot Q_{CH4}]$	95% CH_4 Yield Potential
99	5.82	9.38	12.06	55.09
98	6.95	11.41	14.86	69.33
96	7.03	11.49	14.91	67.22
94	7.54	12.33	15.99	71.21
92	8.92	14.60	18.97	81.82
90	9.00	14.70	19.07	82.93
88	9.02	14.75	19.14	91.93
86	11.28	18.34	23.73	98.17

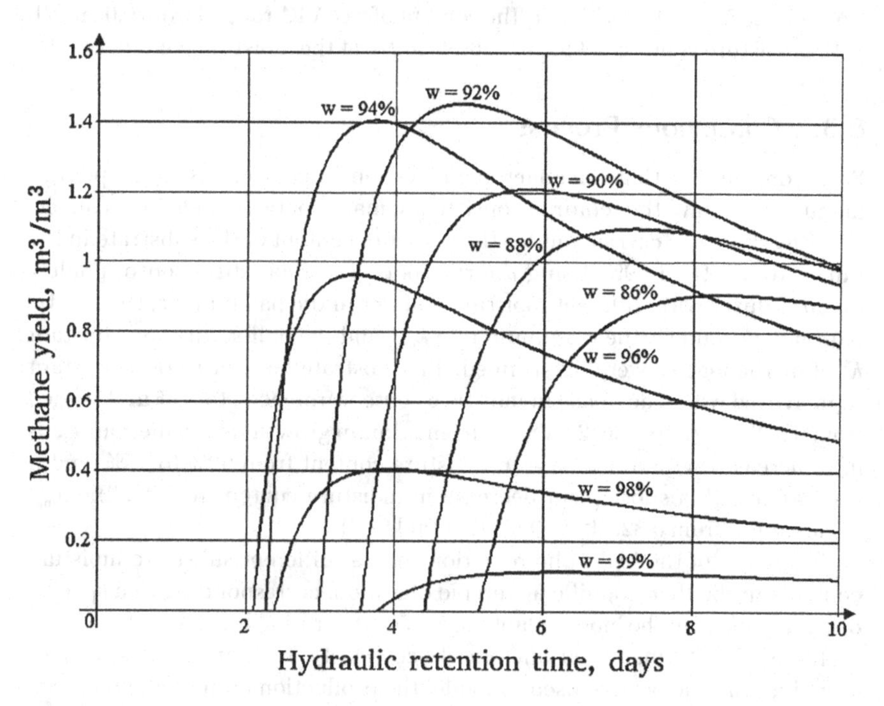

Figure 8.4 Methane yield per unit volume of the continuous anaerobic digester operated in thermophilic mode at a temperature of 35°C depending on the substrate moisture and hydraulic retention time.

The maximum methane yield per unit volume of the apparatus is observed at a substrate moisture content of 92% and a hydraulic retention time of 4.88 days. The regularities of the change in the rate of methane release from a unit volume of the apparatus, depending on the hydraulic retention time of the reactor and the moisture content of the substrate, in the thermophilic mode are following the results of studies by Huang and Shih (1981), Steinsberger and Shih (1984), and Williams and Shih (1989).

8.4 COMPARISON OF ANAEROBIC MESOPHILIC AND THERMOPHILIC DIGESTION OF POULTRY MANURE

8.4.1 Batch Process

In general, in the range of moisture content from 72% to 82%, the production of biogas and methane per unit of VS and unit of reactor volume in the mesophilic mode was higher than in the thermophilic mode. At higher moisture content (from 84 to 99%), on the contrary, it was lower in the mesophilic mode. In the latter mode, a sharp decrease in biogas and methane yield was observed between the values of substrate moisture content from 88% to 82%; and the same trend was shown in the thermophilic mode, from 84% to 82%. The maximum rate of methane production under both mesophilic and thermophilic conditions increased with the increase in substrate moisture content. The thermophilic mode was characterized by a higher maximum rate of methane production at a substrate moisture content above 90%. While under thermophilic conditions, the maximum rate of methanogenesis was relatively at the same level in the range of moisture content values from 72% to 78%, this was not observed under mesophilic conditions (Figure 8.5).

In the mesophilic mode, there was a sharp decrease in the methane content in the produced gas when substrates with moisture content lower than 76% were used. In the thermophilic mode, the sharp decrease was when a substrate with a moisture content of 80% was used (Figure 8.6).

The increase in methane content in biogas during the first few days of the experiment in mesophilic and thermophilic conditions was more significant at higher substrate moisture content values. The increase in the content of methane in biogas obtained during mesophilic fermentation was slower than during that at thermophilic conditions. The electrical conductivity after treatment increased due to the mineralization of organic matter in both temperature modes. The values obtained in the mesophilic mode were similar to those obtained in the thermophilic mode at a substrate moisture content from 84% to 99%. At a moisture content of 72%–82%, the electrical conductivity

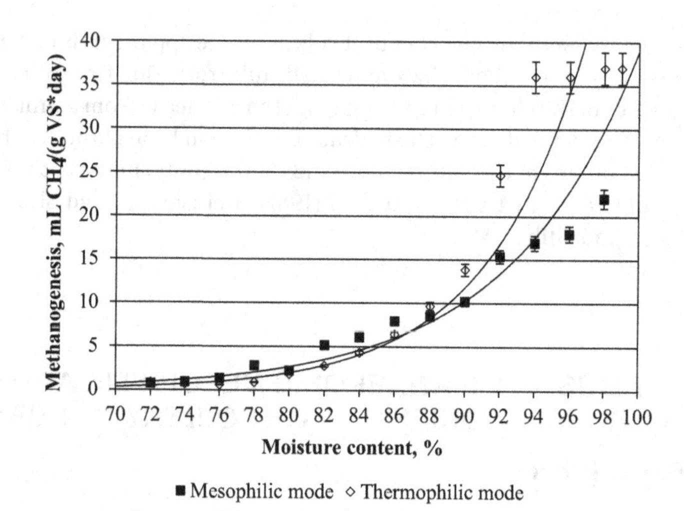

Figure 8.5 The maximum rate of methane production at different substrate moisture in mesophilic and thermophilic anaerobic fermentations.

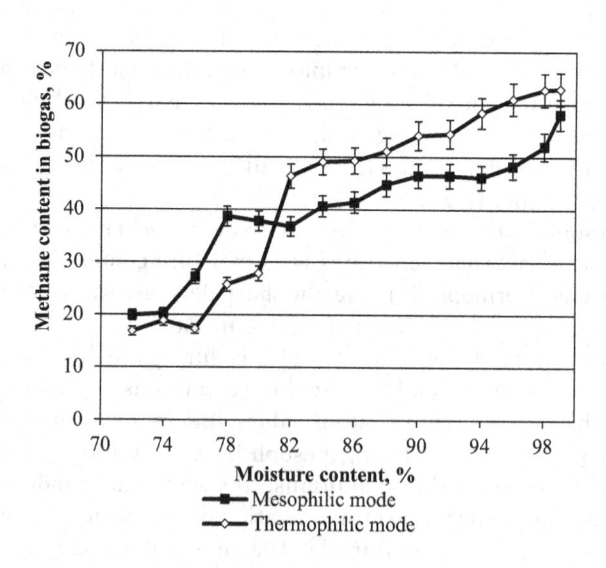

Figure 8.6 Methane content in the biogas at different substrate moisture in mesophilic and thermophilic anaerobic fermentations.

in the mesophilic mode was significantly higher than in the thermophilic mode. This may be caused by the greater degree of destruction of organic substances. The content of ammonium nitrogen, obtained as a result of anaerobic digestion, increased in both temperature modes. The degree of ammonification decreased with an increase in moisture content. Ammonium nitrogen concentration at low substrate moisture content was higher in the mesophilic mode; this may be due to the greater degree of destruction of organic substances.

In the mesophilic and thermophilic modes after anaerobic digestion, the amount of VFA at a substrate moisture content from 96% to 99% was lower, and from 72% to 94% was higher than the initial values. At low substrate moisture content, the amount of VFA in the mesophilic mode was lower than in the thermophilic mode.

Anaerobic digestion of poultry manure at a moisture content from 72% to 80% was more efficient in the mesophilic mode than in thermophilic conditions since it was characterized by a higher biogas and methane yield, both per unit mass and unit volume, including a higher concentration of methane in the produced gas and a higher degree of destruction of the substrate VS. In mesophilic conditions, the processed manure had an alkaline pH value and high electrical conductivity, which indicated a high degree of mineralization. The waste processed in the mesophilic mode, in contrast to the thermophilic one, did not have an unpleasant odor.

8.4.2 Continuous Process

The half-saturation constant increased with a decrease in substrate moisture content for both temperature modes; under mesophilic and thermophilic conditions, its values did not differ significantly. The maximum rate of methane production was 1.7–3.4 times higher in thermophilic anaerobic fermentation than in the mesophilic one (Figure 8.7).

The maximum methane yield per unit volume of the reactor in the mesophilic mode is observed at a substrate moisture content of 88%, in a hydraulic retention time of 14.75 days; the maximum yield was shown in the thermophilic mode at a substrate moisture content of 92% and hydraulic retention time of 4.88 days. The regularities of anaerobic digestion of poultry manure in continuous mode, obtained as a result of modeling, in both the mesophilic and thermophilic conditions, are in good agreement with the experimental data published by other authors (Huang & Shih, 1981; Miller & Parkin, 1983; Steinsberger & Shih, 1984; Webb & Hawkes, 1985; Williams & Shih, 1989), which indicate both the reliability of our results and the adequacy of the used model.

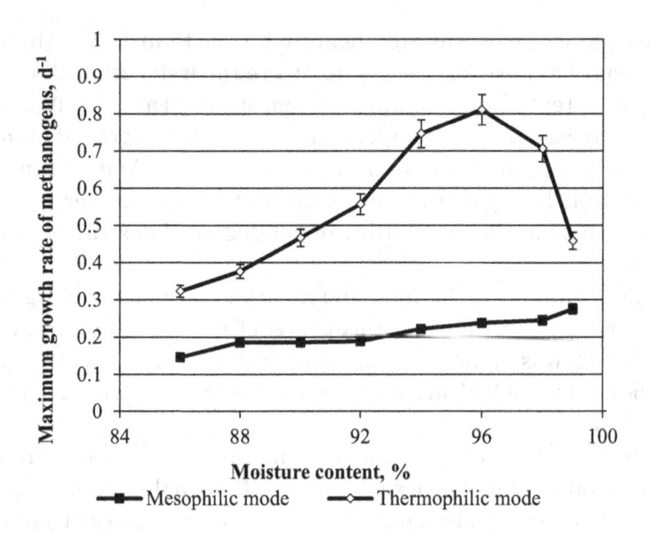

Figure 8.7 Maximum growth rate of methanogens in mesophilic and thermophilic anaerobic fermentations.

8.5 CONCLUSIONS

The patterns of the anaerobic digestion of poultry manure in a periodic mode in the mesophilic and thermophilic conditions have been determined, and their comparison has been carried out. In the range of substrate moisture content from 72% to 82%, the production of biogas and methane per unit of VS and unit volume in the mesophilic mode is higher than in the thermophilic mode; and at higher moisture content from 84% to 99% is, on the contrary, lower. As expected, it was shown that mesophilic conditions are more suitable for anaerobic digestion of poultry manure at low substrate moisture which results in higher biogas and methane production.

The kinetic parameters of the methanogenesis of anaerobic digestion of poultry manure in the batch mode were determined and this basic information was used to model methane production in the continuous mode. The maximum methane yield per unit volume of the bioreactor operated in the mesophilic mode was at a substrate moisture of 88% and a hydraulic retention time of 14.75 days. Meanwhile, the maximum methane yield per unit volume of the bioreactor operated in the thermophilic mode was at a substrate moisture of 92% and a hydraulic retention time of 4.88 days.

In the future, regularities should be obtained for the anaerobic digestion of poultry manure with different moisture content and in different temperature conditions when extracting ammonium nitrogen from the substrate.

REFERENCES

Abouelenien, F., Elsaidy, N. & Nakashimada, Y. (2013). Simultaneous ammonia removal and methane production from poultry manure under dry thermophilic condition. *Journal of American Science*, 9, 90–94.

Abouelenien, F., Nakashimada, Y. & Nishio, N. (2009). Dry mesophilic fermentation of poultry manure for production of methane by repeated batch culture. *Journal of Bioscience and Bioengineering*, 107, 293–295. https://doi.org/10.1016/j.jbiosc.2008.10.009.

Angelidaki, I. & Sanders, W. (2004). Assessment of the anaerobic biodegradability of macropollutants. *Re/Views in Environmental Science & Bio/Technology*, 3, 117–129.

Bi, S., Qiao, W., Xiong, L., Mahdy, A., Wandera, S.M., Yin, D. & Dong, R. (2020). Improved high solid anaerobic digestion of poultry manure by moderate in situ ammonia stripping and its relation to metabolic pathway. *Renewable Energy*, 146, 2380–2389.

Bi, S., Westerholm, M., Hu, W., Mahdy, A. & Dong, R. (2021). The metabolic performance and microbial communities of anaerobic digestion of poultry manure under stressed ammonia condition: A case study of a 10-year successful biogas plant. *Renewable Energy*, 167, 644–651. https://doi.org/10.1016/j.renene.2020.11.133.

Bujoczek, G., Oleszkiewicz, J., Sparling, R. & Cenkowski, S. (2000). High solid anaerobic digestion of poultry manure. *Journal of Agricultural Engineering Research*, 76, 51–60. https://doi.org/10.1006/jaer.2000.0529.

Chen, Y., Cheng, J.J. & Creamer, K.S. (2008). Inhibition of anaerobic digestion process: A review. *Bioresource Technology*, 99(10), 4044–4064. https://doi.org/10.1016/j.biortech.2007.01.057.

Coban, V., Sarac, H., Kaya, D., Eyidogan, M. & Cagman, S. (2014). Effects of various nitrogen rates to biogas production yield in anaerobic digestion of only poultry manure. *Life Science Journal*, 11(2), 186–190. http://www.lifesciencesite.com.

Edwards, D.R. & Daniel, T.C. (1992). Environmental impacts of on-farm poultry waste disposal—A review. *Bioresource Technology*, 41, 9–33. https://doi.org/10.1016/0960-8524(92)90094-E.

Fakkaew, K. & Polprasert, C. (2021). Air stripping pre-treatment process to enhance biogas production in anaerobic digestion of poultry manure wastewater. *Bioresource Technology Reports*, 14, 100647. https://doi.org/10.1016/j.biteb.2021.100647.

Hill, D.T. (1983). Simplified Monod kinetics of methane fermentation of animal wastes. *Agricultural wastes*, 5, 1–16. https://doi.org/10.1016/0141-4607(83)90009-4.

Huang, J.J. & Shih, J.C. (1981). The potential of biological methane generation from poultry manure. *Biotechnology and Bioengineering*, 23(10), 2307–2314. https://doi.org/10.1002/bit.260231013.

Ivanov, V., Stabnikov, V., Stabnikova, O., Salyuk, A., Shapovalov, E., Ahmed, Z. & Tay, J.H. (2019). Iron-containing clay and hematite iron ore in slurry-phase anaerobic digestion of chicken manure. *AIMS Materials Science*, 6(5), 821–832. https://doi.org/10.3934/matersci.2019.5.821.

Jaimes-Estevez, J., Zafra, G., Marti-Herrero, J., Pelaz, G., Moran, A., Puentes, A., Gomez, C., Castro, L.P. & Hernandez, H.E. (2021). Psychrophilic full scale tubular digester operating over eight years: complete performance evaluation and microbiological population. *Energies*, 14(1), 151. https://doi.org/10.3390/en14010151.

Jurgutis, L., Slepetiene, A., Volungevicius, J. & Amaleviciute-Volunge, K. (2020). Biogas production from poultry manure at different organic loading rates in a mesophilic full scale anaerobic digestion plant. *Biomass and Bioenergy*, 141(12), 105693. https://doi.org/10.1016/j.biombioe.2020.105693.

Karaalp, D., Caliskan, G. & Azbar, N. (2013). Performance evaluation of a biogas reactor processing poultry manure with high solids content. *Journal of Selcuk University Natural and Applied Science*, 768–773.

Kerekrety, J., Petrovicova, B., Boda, K. & Adamec, O. (1991). Anaerobic treatment of excrements from large-scale animal farms. *Environmental Biotechnology*, 42, 277–286. https://doi.org/10.1016/S0166-1116(08)70333-9.

Kirchmann, H. & Witter, E. (1992). Composition of fresh, aerobic and anaerobic farm animal dungs. *Bioresource Technology*, 40(2), 137–142. https://doi.org/10.1016/0960-8524(92)90199-8.

Koster, I.W., Rinzema, A., De Vegt, A.L. & Lettinga, G. (1986). Sulfide inhibition of the methanogenic activity of granular sludge at various pH-levels. *Water Research*, 20(12), 156–1567. https://doi.org/10.1016/0043-1354(86)90121-1.

Kucheruk, P.P. (2016). Investigation of kinetic parameters during batch anaerobic digestion of a mixture of manure effluents and corn silage. *Renewable Energy*, 1, 73–78.

Kukic, S., Bracun, B., Kralik, D., Burns, R.T., Rupcic, S. & Jovicic, D. (2010). Comparison between biogas production from manure of laying hens and broilers. *Poljoprivreda*, 16, 67–72.

Mazur, R., Mazurkiewicz, J., Lewicki, A. & Kujawiak, S. (2014). Poultry manure as a substrate for methane fermentation: Problems and solutions. *Biogas World: Internationale Fachmesse für Biogastechnologien und dezentrale Energieversorgung*, 01–03 April 2014, Berlin.

Miller, S.W. & Parkin, G.F. (1983). Response of methane fermentation to continuous addition of selected industrial toxicants. In *Proc. 37th Purdue Industrial Waste Conf.* (Bell, J. M., Ed), Ann Arbor Science Publishers, Ann Arbor, pp. 729–743.

Nakashimada, Y., Abouelenien, F. & Nishio, N. (2013). Novel uric acid degrading bacteria isolated from dry methanogenic sludge of poultry manure. *HU-AIST Biomass Seminar*. https://unit.aist.go.jp/chugoku/even/BiomassOpenSeminar/No16/report1.pdf.

Nie, H., Jacobi, H.F., Strach, K., Xu, C., Zhou, H. & Liebetrau, J. (2015). Mono-fermentation of poultry manure: Ammonia inhibition and recirculation of the digestate. *Bioresource Technology*, 178, 238–246. https://doi.org/10.1016/j.biortech.2014.09.029.

Niu, Q., Qiao, W., Qiang, H., Hojo, T. & Li, Y.Y. (2013). Mesophilic methane fermentation of poultry manure at a wide range of Ammonia concentration: Stability, inhibition and recovery. *Bioresource Technology*, 137, 358–367. https://doi.org/10.1016/j.biortech.2013.03.080.

Ojolo, S.J., Oke, S.A., Animasahun, K. & Adesuyi, B.K. (2007). Utilization of poultry, cow and kitchen wastes for biogas production: A comparative analysis. *Journal of Environmental Health Science & Engineering*, 4, 223–228.

Schulte, E. & Kelling, K. 1992. Soil and applied sulfur. Understanding plant nutrients: Soil and applied sulfur. University of Wisconsin-Extension Publication No. 2525. http://corn.agronomy.wisc.edu/Management/pdfs/a2525.pdf.

Shapovalov, Y., Zhadan, S., Bochmann, G., Salyuk, A. & Nykyforov, V. (2020). Dry anaerobic digestion of poultry manure: A review. *Applied Sciences*, 10, 7825. https://doi.org/10.3390/app10217825.

Singh, K., Lee, K., Worley, J., Risse, L. & Das, K. (2010). Anaerobic digestion of poultry litter: A review. *Applied Engineering in Agriculture*, 26(4), 677–688. https://doi.org/10.13031/2013.32061.

Sinkora, M. & Havlicek, M. (2011). Monitoring of dry anaerobic fermentation in experimental facility with use of biofilm reactor. *Acta Universitatis Agriculturae et Silviculturae Mendelianae Brunensis*, 59, 343–354.

Sürmeli, R.Ö., Bayrakdar, A., Molaey, R. & Çalli, B. (2019). Synergistic effect of sulfide and ammonia on anaerobic digestion of chicken manure. *Waste and Biomass Valorization*, 10, 609–615. https://doi.org/10.1007/s12649-017-0090-z.

Stabnikov, V.P. & Ivanov, V.N. (2006). The effect of various iron hydroxide concentrations on the anaerobic fermentation of sulfate-containing model wastewater. *Applied Biochemistry & Microbiology*, 42, 284–288. https://doi.org/10.1134/S0003683806030112.

Steinsberger, S.C. & Shih, J.C.H. (1984). The construction and operation of a low-cost poultry waste digester. *Biotechnology and Bioengineering*, 26, 537–543. https://doi.org/10.1002/bit.260260520.

Vogels, G.D., Keltjens, J.T., & van der Drift, C. (1988). Biochemistry of methane production. In A.J.B. Zehnder, (Ed.), Biology of anaerobic microorganisms (pp. 707–770). Wiley, New York.

Webb, A.R. & Hawkes F.R. (1985). The anaerobic digestion of poultry manure: Variation of gas yield with influent concentration and ammonium-nitrogen levels. *Agricultural Wastes*, 14(2), 135–136. https://doi.org/10.1016/S0141-4607(85)80025-1.

Williams, C. & Shih, J. (1989). Enumeration of some microbial groups in thermophilic poultry waste digesters and enrichment of a feather-degrading culture. *Journal of Applied Bacteriology*, 67, 25–35. https://doi.org/10.1111/j.1365-2672.1989.tb04951.x.

Wu, S., Ni, P., Li, J., Sun, H., Wang, Y., Luo, H., Dach, J. & Dong, R. (2016). Integrated approach to sustain biogas production in anaerobic digestion of poultry manure under recycled utilization of liquid digestate: Dynamics of ammonium accumulation and mitigation control. *Bioresource Technology*, 205, 75–81. https://doi.org/10.1016/j.biortech.2016.01.021.

Uddin, M.M. & Wright, M.M. (2022). Anaerobic digestion fundamentals, challenges, and technological advances. *Physical Sciences Reviews*. https://doi.org/10.1515/psr-2021-0068.

Zhadan, S., Shapovalov, Y., Tarasenko, R. & Salyuk, A. (2021). Development of an ammonia production method for carbon-free energy generation. *Eastern-European Journal of Enterprise Technologies*, 5, 113. https://doi.org/10.15587/1729-4061.2021.243068.

Chapter 9

Vinasse Utilization into Valuable Products

Tetiana Ivanova and Sergey Tsygankov
Institute of Food Biotechnology and Genomics

Larysa Titova, Larysa Dzyhun, and Inna Klechak
National Technical University of Ukraine

Nina Bisko
Kholodny Institute of Botany

CONTENTS

DOI: 10.1201/9781003329671-9

9.1 INTRODUCTION

The history of ethyl alcohol production goes back for millennia. The alcohol distilleries are growing extensively worldwide due to widespread industrial applications of ethanol such as in chemicals, pharmaceuticals, cosmetics, beverages, food, perfumery (Kharayat, 2012) and biofuel (Gebreeyessus, 2019). Bioethanol is a form of renewable energy that can be produced from agricultural feedstock by the sugar fermentation process. In recent years, bioethanol production has increased because the European Union has implemented a renewable fuels program, which obliges Member States to use renewable fuels as a share in transport fuels (Mikucka & Zielińska, 2020). In 2019, the global bioethanol output reached its historical maximum, 112 billion liters (Torroba, 2021). In the following year 2020, the production of bioethanol slightly decreased due to the pandemic, to 101 billion liters (Torroba, 2021). Corn and sugarcane are the most used feedstocks, accounting for 63% and 30% of all raw materials, respectively, used for bioethanol production in 2020 (Torroba, 2021). Sugar beet is grown in regions with temperate climates and is one of the main feedstocks used for ethanol production in Europe (Hoang & Nghiem, 2021). Different wastes are generated in the process of ethanol production depending on the sugar source type. Vinasse (stillage or bard) is generated as the lower fraction in distillation columns in large quantities of 10–15 L by each liter of ethanol produced (Ramos et al., 2022). It has been forecasted that about 158 billion liters of vinasse will need to be managed in 2023 (Carrilho et al., 2016). Maize, barley and wheat vinasses have a high proportion of insoluble solids and could be used as animal feed (España-Gamboa et al., 2011). Sugarcane, sugar beet, grape, agave or sweet sorghum vinasses are dark brown, low-pH liquids with high concentrations of soluble solids and organics (España-Gamboa et al., 2011). The technological methods used for the treatment of such distillery stillage are thermal (evaporation/combustion), physicochemical (coagulation/flocculation, electrocoagulation, adsorption, advanced oxidation and membrane process) and biological which can be divided into anaerobic or aerobic (Mikucka & Zielińska, 2020).

The common practice used for vinasses with high concentrations of soluble solids is to fertilize crops in the fields, but without stillage pretreatment the chemical oxygen demand (COD) of groundwater is significantly increased, and there is contamination of water resources with vinasse-derived ions, especially nitrate and potassium (Cerri et al., 2020). At the same time, the digestate after anaerobic fermentation of vinasses has a lower content of COD, phenolic substances and nitrates but can be a source of potassium, magnesium, phosphorus and the potential to be used in the fields as a fertilizer (Moraes et al., 2014). The drawback of biological vinasse wastewater treatment is ineffective removal of the dark color which reduces sunlight penetration, photosynthetic activity and dissolved oxygen concentration in water reservoirs causing hazardous conditions for aquatic life (Gebreeyessus, 2019). Vinasse combustion at low operating costs leads to the maximum removal of organic components with the formation of minimum residue while generating energy. The disadvantages of this method are high investment costs, the need for preliminary concentration with partial removal of moisture and the formation of sulfur and nitrogen emissions (Gebreeyessus, 2019; Mikucka & Zielińska, 2020). The composition and different approaches of vinasse wastewater treatment are further described in this chapter.

9.2 THE COMPOSITION OF VINASSES

As far as concentrated maize, barley and wheat vinasses in the form of distillers dried grains with soluble substances are successfully used for animal feed (España-Gamboa et al., 2011), the description of their chemical composition and properties will be not included in this chapter. On the contrary, vinasses which in most cases are not properly utilized, attract the attention; therefore, the composition and ways of their treatment will be further discussed. One of the first problems with sugarcane, sugar beet, grape, agave and sweet sorghum vinasses utilization is a dark brown color (España-Gamboa et al., 2011) causing hazardous conditions for aquatic life when disposed in large volumes unprocessed as a liquid fertilizer or dumped to the lagoons. The color in such vinasse is driven by the presence of melanoidins from Maillard's reaction between reducing sugars and amino groups, phenolics (tannic and humic acids) from the feedstock, furfurals from acid hydrolysis and caramels from overheated sugars (Pant & Adholeya, 2007). The high content of organic matter in the vinasse, including phytotoxic, antibacterial and recalcitrant compounds such as phenols and polyphenols, as well as a high COD concentration, can lead to eutrophication of groundwater when the vinasse enters the environment (España-Gamboa et al., 2011). The ratio of biological oxygen demand (BOD) to COD of vinasses is substantially variable and in some cases does not exceed 0.5. It means that a significant part of organic substances in such wastewaters is not biodegradable (Table 9.1).

Acidic nature of vinasses discarded onto soil results in the reduction of soil alkalinity (Pant & Adholeya, 2007). As shown in practice, fresh vinasse has low pH, which gradually increases due to natural biodegradation processes. According to different reports, pH values of vinasses from various feedstock are as follows: sugarcane vinasse, from 3.9 (Christofoletti et al., 2013) to 4.84 (Reis & Hu, 2017); sugar beet stillage, from 4.7 (Romaniuk et al., 2022) to 5.98 (own data); sweet sorghum stillage, from 4.5 (Christofoletti et al., 2013) to 5.1 (Reyes-Cabrera et al., 2017); grape (wine) vinasse, from 2.9 (Christofoletti et al., 2013) to 4.1 (Tena et al., 2021); and agave (tequila) vinasse, from 3.4 (España-Gamboa et al., 2011) to 4.5 (López-López at al., 2010).

TABLE 9.1 PHYSICOCHEMICAL COMPOSITION OF VINASSES FROM DIFFERENT FEEDSTOCK

Compounds	Sugarcane	Sugar Beet	Sweet Sorghum	Grape (Wine)	Agave (Tequila)
Moisture, %	89.64[a]	89.15[c]	–*	–	–
Organic matter, %	3.96[a]	5.5–7.82[c,d]	–	–	–
Protein, %	2.92[a]	3.72[d]	–	–	–
Carbohydrate, %	3.42[a]	–	–	–	–
Ash, %	3.61[a]	3.03[c]	–	–	–
C/N ratio	10[a]	5.4–5.9[c,e]	–	–	–
COD, g/L	30.4–95[b]	55.5–91.1[b]	79.9[b]	26–50.2[b]	28.1–66.3[b,g]
BOD, g/L	16.7–39.5[b]	27.5–44.9[b]	46[b]	14.54–16.3[b]	20.6[b]
Phenols, mg/L	230–390[a]	–	–	29–474[b]	44–81[b]
NH_4^+, mg/L	23.9[a]	–	–	–	–
Acetaldehyde, g/L	0.697[a]	–	–	–	–
Ethanol, g/L	3.83[a]	–	–	–	–
Glycerol, g/L	5.86[a]	–	–	–	–
Acetic acid, mg/L	1,560[a]	–	–	1,304.3[f]	1,900[g]
Propionic acid, mg/L	–	–	–	22.2[f]	500[g]
Butyric acid, mg/L	–	–	–	7.4[f]	300[g]
Formic acid, g/L	0.582[a]	–	–	–	1.0[g]
Lactic acid, g/L	7.74[a]	–	–	–	2.9[g]

* , not determined; COD, chemical oxygen demand; BOD, biological oxygen demand.

[a] Reis & Hu (2017).

[b] España-Gamboa et al. (2011).

[c] Own data.

[d] Romaniuk et al. (2022).

[e] Tejada et al. (2008).

[f] Tena et al. (2021).

[g] Toledo-Cervantes et al. (2020).

Unpleasant smell of the vinasse is mainly attributed to ammonia products of degradation (Table 9.1) as well as sulfur compounds such as skatole and indole (Pant & Adholeya, 2007). The odor of fresh vinasse may be increased by the high temperature of this effluent, for example, beet vinasse disposed at 70°C (Kulichkova et al., 2020), sugarcane vinasse at 75°C–100°C (Carrilho et al., 2016; Nakashima & Oliveira Junior, 2020) and tequila vinasse at 90°C (Carvajal-Zarrabal et al., 2012; Toledo-Cervantes et al., 2020). The elemental composition of vinasses is mainly formed at the stages of growing crops and in the industrial processes of bioethanol production. Apart from agricultural practices and soil composition, magnesium is added for clarification and calcium for carbonation in sugar production, and sulfuric acid is added to reduce pH in fermentation and yeast production (Parsaee et al., 2019). Vinasses can contain heavy metals, some of them, such as chromium, copper, nickel and zinc, appear during the processing of molasses (Parsaee et al., 2019), other toxic elements can accumulate in plants depending on the conditions of their cultivation (Table 9.2).

As can be seen from the information above, vinasses color, vapor, pH, organic and mineral composition may pose a threat to the environment, if disposed untreated on the industrial scale. The further information in this chapter will characterize some possible ways of the application and obtain valuable products from this effluent.

9.3 VINASSES FOR PLANT FERTILIZATION

The first use of vinasses for crop fertirrigation began in the 1950s and still stands as the predominant application of vinasses (Christofoletti et al., 2013; Reis & Hu, 2017). The concept behind fertirrigation consists of a combined effect of irrigation of the planted fields with the simultaneous fertilization of dissolved nutrients to the crops (Reis & Hu, 2017). Beneficial effects of vinasse fertirrigation for the growth parameters of different crops were observed at limited amounts applied, and the dosages should be determined based on the characteristics of each soil. The adverse effects of vinasse include a decrease in the germination rate of seeds, which depended on the crop and vinasse concentration. Besides decreasing the costs involved with chemical fertilizers, completely supplying phosphorus and being of low capital cost, direct application of vinasse in the soil can cause unpleasant odor, changes in soil quality due to unbalance of nutrient, organic overloading of soil and water bodies, leaching of metals present in the soil to the groundwater (Christofoletti et al., 2013; Fuess et al., 2017; Reis & Hu, 2017). Due to the transportation costs for faraway regions, fertirrigation is usually disposed in areas near the distillery, which may lead to the overuse of vinasse/area (Nakashima & Oliveira Junior, 2020). The use of concentrated vinasses in fertirrigation promotes similar effects in crop productivity when

TABLE 9.2 MINERAL ANALYSIS OF VINASSES FROM DIFFERENT FEEDSTOCKS

Compound (mg/kg)	Sugarcane	Sugar Beet	Sweet Sorghum	Grape (Wine)
K	2.06–78.53[a,b]	10.00–11.20[a,c]	70[e]	118–800[a]
SO_4^{2-}	710[a]	–*	–	120[a]
S	–	1.03–1.76[c,d]	3,400[e]	–
Na	810[b]	2.67–2.80[c,d]	–	–
P	9.12[b]	130[c]	360[e]	83[a]
Ca	719–720[a,b]	400–273[d,c]	30[e]	–
Mg	237–610[a,b]	50–137[d,c]	30[e]	–
Al	–	0.21–11.10[d,c]	138[e]	–
As	–	<0.1[c]	–	–
Ba	0.41[a]	–	–	–
Be	–	<0.5[c]	–	–
Bi	–	<0.5[c]	–	–
B	–	2.39–3.52[c,d]	6.5[e]	–
Cd	–	<0.02[d]; <0.1[c]	–	0.05–0.08[a]
Co	–	<0.5[c]; 1.53[d]	–	–
Cr	0.04[a]	<0.1[c]	–	–
Cu	<0.009[b]; 0.35[a]	0.40–0.41[c,d]	–	0.2–3.26[a]
Fe	11,840[b]	15.32–30.43[c,d]	–	–
Li	–	0.12[c]	–	–
Mn	167.0[b]	[c].75–10.50[c,d]	–	–
Mo	0.008[a]	<0.02[d]; 0.10[c]	–	–
Ni	0.03[a]	0.66[c]	–	–
Pb	–	<0.02[d]; <0.1[c]	–	0.55–1.34[a]
Sb	–	<0.5[c]	–	–
Sr	–	2.68–12.90[c,d]	–	–
Ti	–	<0.5[c]	–	–
Zn	<0.70[b]; 1.66[a]	3.59[d]	30[e]	–

* , not determined.
[a] Christofoletti et al. (2013).
[b] Crespi et al. (2011).
[c] Moraes et al. (2015).
[d] Own data.
[e] Reyes-Cabrera et al. (2017).

applied in the proportional levels of solids, and the condensate retrieved from the concentration can be repurposed to reduce water consumption of integrated industrial processes (Nakashima & Oliveira Junior, 2020). To produce optimized granulated organomineral fertilizer, Gurgel et al. (2015) supplemented concentrated vinasses with filter cake, boiler ash, soot from chimneys and mineral substances. Tested fertilizer was cheaper compared to mineral fertilizer and reduced losses caused by leaching in relation to application of only concentrated vinasses because of the improvement of soil physicochemical properties, cation exchange capacity and porosity (Gurgel et al. 2015). Alternatively, vinasses can be processed by biodigestion before using it as a fertilizer (Formann et al., 2020). The composting process, similar to anaerobic digestion, delivers an organic fertilizer with high humification index, which generates a more stable humus effect (Formann et al., 2020). Additionally, anaerobic digestion provides biogas (Formann et al., 2020) and higher energy efficiency compared to fertirrigation with vinasse (Nakashima & Oliveira Junior, 2020). Digestate after anaerobic fermentation of vinasses as a fertilizer has a lower organic and nitrate load but can be a potent source of potassium, magnesium and phosphorus (Moraes et al., 2014). As vinasse contains a high amount of potassium and sulfates (Table 9.2), therefore, potassium is widely extracted and used in the sulfated form for plant fertilization. However, potassium sulfate is a highly water-soluble salt, and this leads to eutrophication of surface waters (Arslanoğlu & Tümen, 2021). The production of potassium struvite ($KMgPO_4 \cdot 6H_2O$), slow-release fertilizer, was achieved through precipitation from concentrated vinasse. The procedure included mixing of phosphoric acid with $MgCl_2$ and concentrated to 63.5 °Bx sugar beet vinasse, adjustment of pH to 8–10, crystallization with the excess of NaOH solution, filtration, washing and drying. As the remaining residue from sugar beet vinasse still requires proper utilization, it was pyrolyzed under nitrogen with cellulosic agent grape marc for the increase in the potassium struvite yield and for obtaining activated carbon (Arslanoğlu & Tümen, 2021). For the improvement of K-struvite chemical precipitation from vinasse, electrodialysis with supernatant recycling was proposed (Silva et al., 2022). Natural polymers such as pectin and chitosan also have shown to provide a slow-release matrix for sugarcane vinasse, which hinders water evaporation rates from sandy soil under drought-prone conditions (Cerri et al., 2020). Generally, fertirrigation with vinasse is a simple and inexpensive practice, but it can lead to long-term effects on the physical properties of the soil, increasing the infiltration capacity and contaminating the groundwater (Christofoletti et al., 2013). Concentration of vinasse allows better transportability but has high energy demands, which needs to be in accordance with the plant's concept of mass flow and energy management (Formann et al., 2020). Biodigestion of vinasse and extraction or incorporation of nutrients to the slow-release matrix represent more advanced methods of vinasse utilization as a fertilizer.

9.4 PHYSICOCHEMICAL METHODS OF VINASSE TREATMENT

Physicochemical processes for the vinasse wastewater treatment separate the organic load and/or degrade the chemical pollutants by strong oxidizing agents while biological methods are implemented in milder conditions of temperature and pressure (Hoarau et al., 2018). Advanced oxidation processes, such as ultraviolet irradiation, the Fenton reaction and ozonation, involve the generation of high-reactivity radicals, which promote oxidation of wastewater pollutants and recalcitrant components (Reis et al., 2019). The Fenton reaction degrades organic compounds with hydroxyl radical (produced from the catalyzing reaction between Fe^{2+} or Fe^{3+} and hydrogen peroxide) to simpler intermediates under acidic conditions (Ghernaout et al., 2020; Hakika et al., 2019). It is considered as simple, efficient, eco-friendly, comparatively cheap and increasing the further biodegradability of wastewater. Thus, the Fenton reaction applied to sugarcane vinasse increased the BOD to COD ratio from 0.29 up to 0.48 and reduced the initial COD up to 48.10% (Hakika et al., 2019). At the same time, pH sensitivity, sludge formation, instability of the reagent, and undesirable reactions and loss of oxidant currently prevent Fenton process from its industrial implementation (Ghernaout et al., 2020).

Coagulation/flocculation treatment destabilizes the particles by adding coagulants, assists to their better aggregation by using flocculants and eliminates produced particles by filtration or sedimentation (Mahmoudabadi et al., 2022). A number of chemical agents such as aluminum sulfate, poly-aluminum chloride, iron sulfate, iron chloride and polyacrylamide have been used for the application of the method (Carvajal-Zarrabal et al., 2012; Syaichurrozi et al., 2020). It was suggested that some of the aforementioned chemical substances cause adverse reactions for human health (for example, polyacrylamide is a neurotoxin), which prompted the development of biodegradable flocculants (e.g., polysaccharide from *Proteus mirabilis* and *Bacillus mucilaginosus*, polyamide from *B. licheniformis* and *B. subtilis*, protein from *Bacillus* sp.). Thus, the combination of biodegradable poly-γ-glutamic acid with sodium hypochlorite and sand filtration in a coagulation/flocculation process of tequila vinasse resulted in the removal of color, 70% turbidity and 79.5% COD (Carvajal-Zarrabal et al., 2012).

Electrocoagulation uses metals as anode and cathode which are immersed in the vinasse, implementing the approaches of electrochemistry, coagulation and flotation in one method (Syaichurrozi et al., 2020). Aluminum or mild steel is commonly used as electrodes to produce floccules (España-Gamboa et al., 2011). At the same time, hydrogen gas is formed from the water at the cathode and forces the pollutants to float on the surface (Syaichurrozi et al., 2020). Electrocoagulation can be used for decolorization and further COD reduction of digestate from biogas reactors (España-Gamboa et al., 2011).

Adsorption on activated carbon is an expensive but efficient method of color and COD removal from vinasses due to extended surface area and a high degree of surface reactivity (España-Gamboa et al., 2011; Satyawali & Balakrishnan, 2007). Various agro-residues showed (Satyawali & Balakrishnan, 2007) lower color removal (up to 50% for phosphoric acid carbonized bagasse) compared to activated carbon (80%) for anaerobic-treated vinasse. Potassium ions were adsorbed from the vinasse using strong acid–cation exchange resin with a breakthrough capability of 56.79 mg K^+/mL and dissolved into sulfuric acid to regenerate the resin (Zhang et al., 2012).

The most efficient results were acquired utilizing the combination of treatments. Sequential anaerobic, aerobic and ozone treatment of sugarcane vinasse resulted in the removal of 95% COD, over 80% of Kjeldahl nitrogen, complete removal of phenolic compounds as well as the generation of fungal biomass and biogas (Reis et al., 2019). Almost complete (98.64%) COD removal from vinasse was achieved by the combination of advanced oxidation using porous α-Fe_2O_3 nanoparticles and coagulation/flocculation with polyacrylamide compared to electrocoagulation, combination of coagulation/flocculation and Fenton's oxidation, and oxidation/adsorption with ozone and granular activated carbon, 80%, 69.2% and 25% of COD removal according to Mahmoudabadi et al. (2022).

9.5 THERMAL METHODS FOR VINASSE TREATMENT

To make transportation more efficient, vinasse undergoes concentration by evaporation, direct or reverse osmosis. Selective membranes made of dense film are used for direct and reverse osmosis, and the difference in such processes is the driving force, osmotic or hydraulic pressure according to Peiter et al. (2019). Membrane filtration generally requires less energy than evaporation, but membranes can become fouled. Anaerobic membrane bioreactors represent an alternative that combines a biological approach and concentration by membrane filtration (Park et al., 2015). Evaporation of vinasse is used for concentration before burning in a boiler (Larsson & Tengberg, 2014). In falling film evaporators, the liquid stream is pumped to the surface of the evaporating column, forcing the liquid to flow down by gravity while simultaneously evaporating. The resulting condensate can be recirculated to the combustion tank to maintain a constant solids content or can be removed from the system. In the latter case, it is possible to collect condensate for diluting the concentrated product for further use, as well as for washing the evaporation equipment.

Biomass burning is considered CO_2 neutral, its use reduces overreliance on fossil fuels and creates a source of income for producers. On the other hand, burning biomass creates a number of technical obstacles that need to be solved, in particular, it can lead to significant fouling and slagging of the boiler

(Lachman et al., 2021). Currently, the thermal treatment of vinasse is widely used in the industry (Mikucka & Zielińska, 2020), at the same time, the stillage should be concentrated to a moisture content of less than 55%. The vinasse can be dried with hot air (180°C), and the caloric content of the obtained stillage is about 3,200 kCal/kg (Babu et al., 1995). The higher heating value of vinasse decreases significantly with the increase in the moisture content: 12.7–15.07 MJ/kg at 4% moisture and 6.4 MJ/kg at 68% moisture for sugarcane vinasse (Camargo et al., 2021). Concentrated vinasse is commonly mixed with agricultural wastes (about 20%) and burned in a boiler (Mikucka & Zielińska, 2020). Fluidized bed combustion has overcome limitations caused by stillage viscosity and high sulfate concentration (Satyawali & Balakrishnan, 2008). Incineration is also an effective method of vinasse disposal, at closed places, compared to distillation, as it produces potassium-rich ash that can be used for soil application. According to economic indicators, the technology is more suitable for large enterprises (Mane et al., 2006).

9.6 THE USE OF THE VINASSES FOR BIOGAS PRODUCTION

Anaerobic treatment of vinasses is preferred as compared to aerobic because it saves energy for aeration and cooling, produces low amounts of sludge, requires a lower quantity of nutrients and produces energy in the form of biogas (España-Gamboa et al., 2011) and/or biohydrogen (López-López at al., 2010; Toledo-Cervantes et al., 2020). Methane fermentation naturally occurs in the deep layers of disposed vinasse on the fields of filtration (lagoons). It releases ammonia and hydrogen sulfide as well as greenhouse gases (methane and carbon dioxide) into the atmosphere. As a substrate for biogas production, vinasses contain easily degradable proteins, amino acids, volatile fatty acids as well as macro- and micronutrients (Tables 9.1 and 9.2). Heavy metals can have an inhibitory effect on the activity of microorganisms, but low concentrations of some heavy metals are necessary as part of the active sites of enzymes (Fermoso et al., 2009). Thus, the essential trace elements involved in anaerobic reactions and transformations are: zinc (hydrogenase, formate dehydrogenase, superoxide dismutase), nickel (CO-dehydrogenase, acetyl-CoA synthase, methyl-CoM reductase (cofactor F_{430}), urease, hydrogenase), copper (superoxide dismutase, hydrogenase, nitrite reductase, acetyl-CoA synthase), selenium (hydrogenase, formate dehydrogenase, glycine reductase), cobalt (B_{12}-enzymes, CO-dehydrogenase, methyltransferase), molybdenum (formate dehydrogenase, nitrate reductase, nitrogenase), tungsten (formate dehydrogenase, formylmethanofuran-dehydrogenase, aldehyde-oxidoreductase) and manganese (stabilize methyltransferase, participate in redox reactions) (Fermoso et al., 2009).

Depending on the vinasse source, some trace elements can be found below the optimum values and supplementation with known concentrations result in stable fermentation. Addition of: cobalt, 5; iron, 2,000; copper, 40; manganese, 700; molybdenum, 7; wolfram, 1; nickel, 15; and zinc, 150 mg/kg of total solids (and urea) to sugarcane vinasse led to 79% higher methane production rate (Janke et al., 2016).

Vinasse is also a challenging substrate due to the high nitrogen and sulfur content, which is converted to inhibiting substances (NH_3/NH_4^+ and H_2S/HS^-) in anaerobic fermentation process and in the form of ammonia and hydrogen sulfide reduce the quality of biogas. High nitrogen content unbalances the C/N ratio causing a decrease in methane production (Maurya et al., 2019). The optimal C/N ratio for methane fermentation is considered at the range of 10–40 (Kulichkova et al., 2020). Thermodynamically, sulfate-reducing bacteria are more competitive acetate consumers than methanogens and their half-saturation constant, K_s (the concentration of substrate that supports a half-maximum growth rate) equals to 9.5 mg/L; meanwhile, this value is 32.8 mg/L for methanogens (Braun, 2007; Stabnikov & Ivanov, 2006). Thereby, the level of sulfides increases while methane production inhibits.

The addition of cosubstrates to optimize C/N ratio and microelement composition of vinasse was implemented (Kulichkova et al., 2020; Moraes et al., 2015; Romaniuk et al., 2022) as well as the dilution of vinasse to minimize the toxicity of inhibiting substances (Romaniuk et al., 2022). For the implementation of the addition of sweet sorghum bagasse to sugar beet vinasse on the industrial scale, an experimental facility project for biogas fermentation of complex substrates consisting of beet vinasse and lignocellulose biomass has been developed (Kulichkova et al., 2020).

Sweet sorghum bagasse and vinasse are produced by the company "Eco-Energy" as by-products from bioethanol production. Bagasse is gained after plant stem crushing to obtain sweet sorghum juice as a source of sugars for fermentation, which is expected to be performed under thermophilic conditions because the main substrate vinasse is obtained at a temperature of 70°C. Input components are 500 tons/day vinasse with about 11% dry matter, and 30 tons/day shredded sorghum bagasse with about 90% dry matter. The process of fermentation occurs in concrete continuously stirred tank reactors. The calculated output is 2,500 nm^3/day of biogas (for heat and electricity) and 510 t/day of digestate for crop fertilization. When applying untreated lignocellulosic feedstock as a cosubstrate, it requires small particle size and thorough mixing. Such additives tend to float on the surface and can be only partially degraded. Massive volumes of vinasse effluents result in high construction cost in biogas plants. Thus, the public enterprise Haysin Distillery in Ukraine produces about 800 m^3 of sugar beet vinasse in operation daily. Therefore, the goal of the other work was to investigate sugar beet vinasse compositional changes and methane-producing performance when it is concentrated four times on the dry matter basis (Ivanova et al., 2021).

Sugar beet vinasse was concentrated on a rotary evaporator (Unipan 350, Poland) at 65°C. The value of pH was measured on the portative pH/Eh meter (Milwaukee pH58, Hungary). The dry matter and volatile solids (VS) content of the samples (in %) were determined by differential weighing before and after drying at 105°C until the constant weight and by subsequent washing at 550°C until the constant weight, respectively, by using standard methods (APHA, 2005; Morozova et al., 2020). Carbon, nitrogen and sulfur contents were determined by reported methods (APHA, 2005; Volodko et al., 2020), and barium perchlorate titration according to Buděšinský & Krumlová (1967). Elemental analysis of vinasse was performed by emission spectroscopy (APHA, 2005) using inductively coupled plasma (ICP) Emission Spectrometer (Shimadzu ICPE-9000, Japan). Anaerobic cultivation was conducted in inoculated flasks on rotary shaker under mesophilic conditions. Biogas volume was determined by water displacement. Methane content in biogas was evaluated using multi-gas analyzer MGA-1.12 (Svitlo Sachtarya Ltd., Kharkiv, Ukraine). The data have been statistically analyzed by t test using Origin 2016, OriginLab Corporation, USA. Differences between means at 5% ($p<0.05$) level have been considered significant. The dry matter content of sugar beet vinasse increased from 10.85%±0.02% to 40.10%±0.05% after vacuum evaporation (Figure 9.1).

The process of sugar beet vinasse evaporation slightly increased the C/N ratio, from 5.4 to 6.0, which can cause a positive effect on the fermentation performance. Inorganic sulfur content (in the ash) after the concentration of vinasse was reduced in dry matter from 2.50% to 1.87%. Organic sulfur content

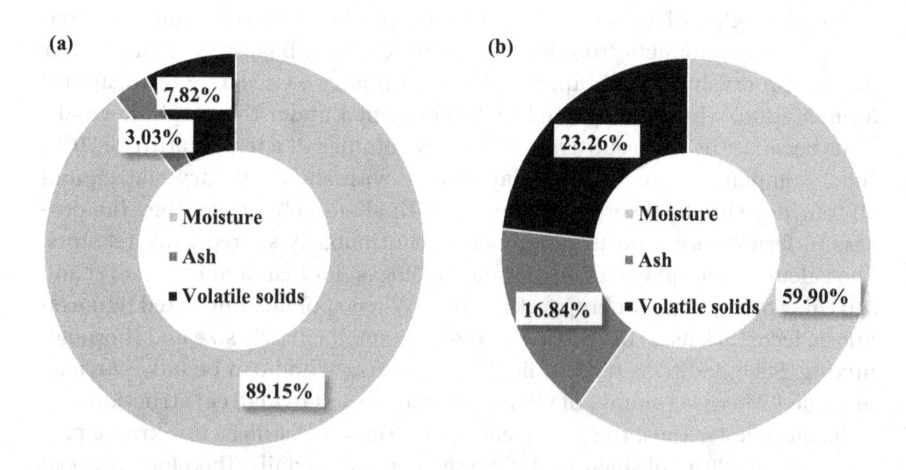

Figure 9.1 Composition of crude (a) and concentrated (b) beet vinasse.

was about 1% in dry matter of sugar beet vinasse before evaporation and it did not change significantly. Concentrated sugar beet vinasse contains 23.26% of VS which is about three times higher in comparison with the crude one (7.82%), and biogas production was three times higher in fermentation of concentrated sugar beet vinasse, 122.80±5.0 L/kg of concentrated vinasse and 40.9±2.1 L/kg of crude vinasse. Biogas from concentrated sugar beet vinasse contained 71.30% methane on average, which did not significantly differ from methane content in biogas from crude vinasse.

9.7 USE OF THE SUGAR BEET VINASSE FOR MEDICINAL MUSHROOM CULTIVATION

Sugar beet vinasse can be used for aerobic cultivation of bacteria, fungi and algae for color and organic load removal with the obtaining of biomass and some targeted products (España-Gamboa et al., 2011). The treatment of vinasse via submerged cultivation of the medicinal mycelial mushroom will be described in detail.

Mushrooms are fungi which form fruiting bodies that can be seen by the naked eye. Medicinal mushrooms have beneficial effects in human consumers as food, dietary supplements and medicine; also as biocontrol agents in plants, dietary pet and veterinary food supplements, cosmeceuticals and nutricosmetics (Huang et al., 2021; Ivanov et al., 2021; Valverde et al., 2015). Polysaccharides of medicinal mushrooms possess immunomodulating, anticancer (Ivanova et al., 2014; Singdevsachan et al., 2016), antibacterial, antiviral, antidiabetic and anti-asthmatic properties (Choudhary, 2020). Compared to endopolysaccharides, fungal exopolysaccharides require simpler methods for isolation and purification (Choudhary, 2020; Hamidi et al., 2022).

Medicinal properties of some fungal exopolysaccharides are well established. For example, β-glucan from cultured broth of *Schizophyllum commune* Fr. (schizophyllan) has been produced by several Japanese pharmaceutical companies as immunopotentiator and it is approved for clinical use in Japan (Zhang et al., 2013). Tramesan, recently extracted fucose-enriched branched glycan from the cultured liquid of *Trametes versicolor* (L.) Lloyd, acts as a pro-antioxidant and is suggested as a possible solution for defusing toxins released into foodstuff, necrotic lesions into plant crops or growth of cancer cells (Scarpari et al., 2017). Hypoglycemic (Sun et al., 2021) and antioxidant (Zhang et al., 2021) activities were reported for exopolysaccharides of *Cordyceps militaris* (L.) Fr. as well as anti-inflammation properties were also proved for exopolysaccharides synthesized by *Cordyceps sinensis* (Li et al., 2020). Nutrient medium composition significantly influences the production of biologically active substances by fungal cultures. Thus, it was shown that carbon and nitrogen sources significantly influence structural features and immunomodulatory activity of exopolysaccharides from *Ganoderma lucidum* (Curtis) P. Karst (Liu et al., 2022). The enhancement in production and antioxidant activity of exopolysaccharides synthesized by *Inonotus obliquus* (Ach. ex Pers.) Pilát culture was associated with the addition of lignocellulose substrate, corn stover, to the medium (Xiang et al., 2012).

The screening of 14 medicinal mushrooms (Rathore et al., 2017; Soković et al., 2018; Wasser & Weis, 1999) for their ability to produce biomass and exopolysaccharides during cultivation on beet vinasse, the wastewater from bioethanol production, was conducted. Mycelia of fungi from *Ascomycota* division (*C. militaris* IBK 1862, *O. sinensis* IBK 1928 and *Morchella esculenta* (L.) Pers. IBK 1843) as well as mycelia of fungi from *Basidiomycota* division (*Coprinus comatus* (O.F. Müll.) Pers. IBK 137, *Flammulina velutipes* (Curtis) Singer IBK 1878, *Fomes fomentarius* (L.) Fr. IBK 355, *Ganoderma applanatum* (Pers.) Pat. IBK 1701, *G. lucidum* IBK 1900, *I. obliquus* IBK 1877, *Lentinula edodes* (Berk.) Pegler IBK 502, *Pleurotus eryngii* (DC.) Quél. IBK 2015, *Pleurotus ostreatus* (Jacq.) P. Kumm. IBK 551, *S. commune* IBK 1768 and *T. versicolor* IBK 353) were provided by the Culture Collection of Mushrooms from the M.G. Kholodny Institute of Botany of the National Academy of Sciences of Ukraine (Bisko et al., 2016). All current fungal taxon names were checked in the online database MycoBank (http://www.MycoBank.org). The fungal cultures were maintained on wort agar slants at 4°C. Solid glucose–peptone (GP) medium containing, g/L: glucose, 25; peptone, 3; yeast extract, 2; KH_2PO_4, 1; K_2HPO_4, 1; $MgSO_4 \times 7\ H_2O$, 0.25; agar, 20, was used for fungi growth. The Petri dishes with GP medium were inoculated with selected fungi and incubated at 28°C±2°C. Three types of liquid media ((a) crude sugar beet vinasse; (b) sugar beet vinasse diluted 1:1 with distilled water and (c) sugar beet vinasse concentrated three times by evaporation) were autoclaved for 40 min at 1 atm. Three disks of agar medium with mycelia (8 mm diameter) from the edge of the Petri dish were used to inoculate 250-mL flasks containing 50 mL of liquid media. Mycelium in the exponential growth phase was used at a 10% level (v/v) for inoculation of medium. Submerged cultivation of *G. lucidum* IBK 1900 and *T. versicolor* IBK 353 was performed in a rotary shaker (120 rpm) on crude sugar beet vinasse at 28°C±2°C. The mycelia were filtered, washed with water and desiccated at 105°C to a constant weight. Amounts of synthesized biomass and exopolysaccharides were determined gravimetrically (Babitskaya et al., 2000). The data have been statistically analyzed by *t* test using Origin 2016, OriginLab Corporation, USA.

9.7.1 Biomass Accumulation

The results of fungi screening for biomass accumulation on the waste of bioethanol production revealed that on the diluted sugar beet vinasse fungal

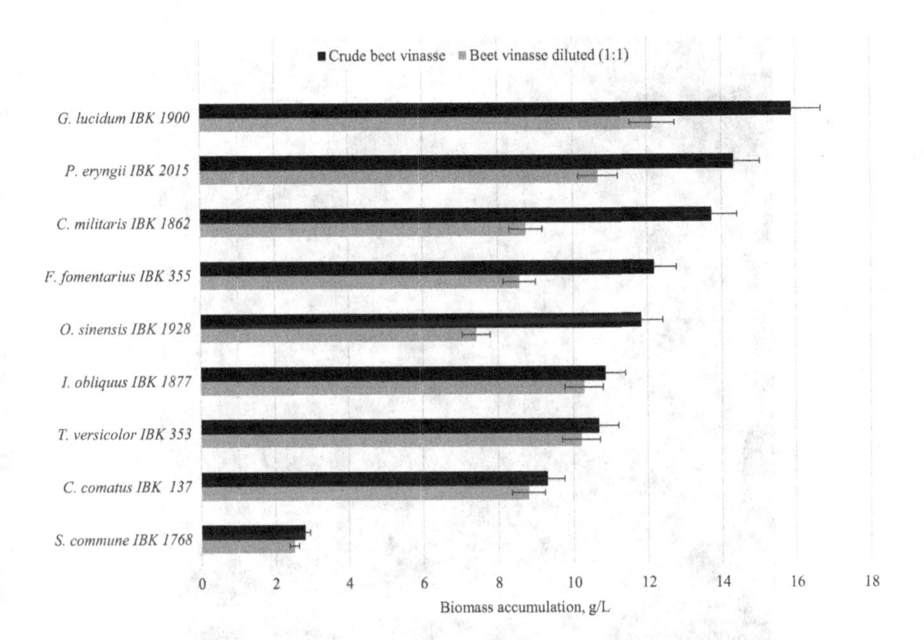

Figure 9.2 Accumulation of fungal biomass on the crude and diluted beet vinasse (values are significantly different, $p<0.05$).

mycelium accumulated 5%–37% less biomass than on crude sugar beet vinasse (Figure 9.2).

At the same time, three-fold concentrated sugar beet vinasse did not support the growth of the tested mycelia. Maximal biomass accumulation on crude sugar beet vinasse was achieved for *G. lucidum* IBK 1900 (15.89±0.79 g/L) and *P. eryngii* IBK 2015 (14.33±0.76 g/L). Though five mushroom species did not show any growth on sugar beet vinasse, the other nine species accumulated on average three to six times higher biomass on crude sugar beet vinasse than on liquid media with dried sorghum bagasse (50 g/L of water) and dried corn stillage (50 g/L of water). It was estimated that a better distribution of nutrients in liquid sugar beet vinasse medium resulted in higher fungal biomass accumulation compared to liquid media with solid particles of sorghum bagasse or dried corn stillage.

In submerged cultivation on crude sugar beet vinasse, the mycelium of *G. lucidum* IBK 1900 and *T. versicolor* IBK 353 grows in the form of smooth agglomerated sphere-like particles with a diameter of 0.3–3.0 mm during the stationary phase (Figure 9.3).

Inside the agglomerates, the colonies had a jelly-like consistency. In the dynamic of cultivation, the maximal biomass for *T. versicolor* IBK 353 was

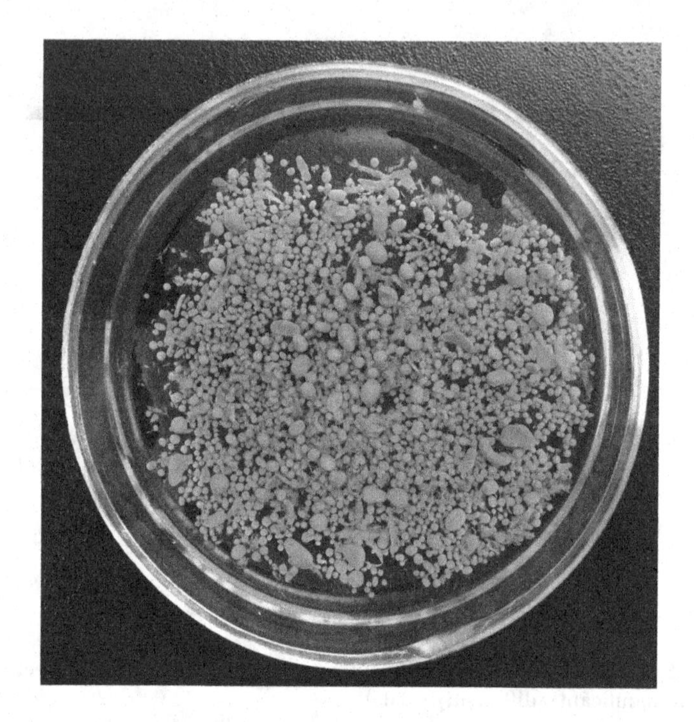

Figure 9.3 The appearance of *Trametes versicolor* IBK 353 mycelial agglomerates resuspended in the distilled water, the tenth day of submerged cultivation on crude beet vinasse.

achieved on the tenth day of submerged cultivation, 10.64±0.4 g/L, while *G. lucidum* IBK 1900 accumulated maximal biomass on the seventh day of cultivation, 10.42±0.5 g/L. Maximal exopolysaccharides for *G. lucidum* IBK 1900 and *T. versicolor* IBK 353 submerged cultivation on sugar beet vinasse was obtained on the tenth and seventh days, 11.26±0.1 g/L and 7.07±0.5 g/L, respectively. For the comparison, different strains of *T. versicolor* synthesized 8.9–10.6 g/L biomass and 1.13–1.15 g/L exopolysaccharides in submerged cultivation on milk permeate supplemented with glucose, yeast extract and salt (Cui et al., 2007). *G. lucidum* produced more biomass, 14.3–17.0 and 20.1 g/L at optimized temperature and pH, but less exopolysaccharides (1.2 g/L) in submerged cultivation on the cheese whey (Lee et al., 2003).

9.7.2 Production of Exopolysaccharides

The highest exopolysaccharides accumulation was observed in culture liquids of *C. comatus* IBK 137 and *G. lucidum* IBK 1900 after cultivation on the crude sugar beet vinasse (Table 9.3).

TABLE 9.3 EXOPOLYSACCHARIDE CONCENTRATION, G/L, IN THE CULTURE LIQUID OF FUNGAL STRAINS AFTER CULTIVATION ON SUGAR BEET VINASSE

	Exopolysaccharide Concentration, g/L	
Fungal Strains	**Crude Beet Vinasse**	**Beet Vinasse Diluted (1:1)**
C. comatus IBK 137	6.86±0.31	3.55±0.16
C. militaris IBK 1862	4.23±0.20	1.71±0.09
O. sinensis IBK 1928	5.79±0.25	2.66±0.11
F. fomentarius IBK 355	5.29±0.24	1.88±0.10
G. lucidum IBK 1900	6.72±0.05	3.61±0.15
I. obliquus IBK 1877	3.72±0.15	0.85±0.03
P. eryngii IBK 2015	3.93±0.16	2.49±0.11
S. commune IBK 1768	5.43±0.27	3.06±0.14
T. versicolor IBK 353	4.65±0.29	0.96±0.05

Exopolysaccharides synthesis was significantly ($p < 0.05$) lower after cultivation on the diluted sugar beet vinasse (Table 9.1). On average, fungal mycelia synthesized 5.18 g/L of exopolysaccharides on crude sugar beet vinasse, 4.10 g/L on corn stillage and 0.35 g/L on sorghum bagasse. Overexpression of homologous phosphomannomutase gene in *Ganoderma* increased the production of exopolysaccharides by *Ganoderma* by 1.41 times but only up to 1.53 g/L (Zhao et al., 2022). It was also shown that cocultivation of *G. lucidum* with *P. ostreatus* on potato dextrose broth increased exopolysaccharides synthesis from 1.43 to 3.35 g/L under optimized conditions (Asadi et al., 2021). *C. militaris* accumulated 2.34 g/L exopolysaccharides in a glucose–potato–peptone medium but this concentration increased up to 7.96 g/L after the addition of coix seed powder and optimization of cultivation conditions (Zhang et al., 2021). *F. fomentarius* synthetized only 0.81 g/L of exopolysaccharides with antiproliferative activity toward human gastric cancer cells on glucose–peptone medium but the concentration reached up to 3.64 g/L under optimized medium and conditions (Chen et al., 2008).

Therefore, cultivation of medicinal mycelial mushrooms can be an alternative way to utilize wastes of bioethanol production. Among investigated wastes (sorghum bagasse, molasses and corn stillage), the most efficient for biomass (up to 16 g/L for *G. lucidum* IBK 1900, surface cultivation) and exopolysaccharides accumulation (up to 11 g/L for *G. lucidum* IBK 1900, submerged cultivation) is liquid medium composed of crude sugar beet vinasse, which was patented in

Ukraine (Ivanova et al., 2019). A possible advantage of this substrate is the uniform distribution of nutrients in the liquid medium, which makes them more accessible for assimilation by fungal mycelium.

9.8 CONCLUSIONS

Vinasses with high concentration of soluble solids are generated in large quantities as the wastewaters of bioethanol production. The dark color, high organic load, acidic nature, unpleasant smell, toxic compounds and heavy metals of such vinasses may pose a threat to the environment if disposed untreated on the industrial scale. Fertirrigation with vinasse is a common, simple and inexpensive practice, but it can lead to long-term negative environmental effects. Biodigestion increases the value of vinasse as a fertilizer and improves the energy efficiency of vinasse treatment. One of the advanced technologies for the production of fertilizers from vinasse is the creation of a slow-releasing matrix which hinders the nutrients leaching and further eutrophication of the surface waters. Concentration of vinasse can make the transportation and storage of wastewater more efficient for further use as a fertilizer, biogas substrate or for combustion. The production of energy in the form of biogas or heat after incineration is the most explored option on the industrial scale after direct fertirrigation. Addition of cosubstrate and microelements, dilution and concentration had been applied for the optimization of vinasses anaerobic digestion. The combination of physicochemical, thermal and biological treatment methods for vinasses is not the easiest way, but it may provide a spectrum of valuable products with minimal effluents discharged to the environment.

Aerobic treatment of vinasse aims at the color and organic load removal with the obtaining of biomass and some targeted products using bacteria, fungi and algae. *Ganoderma lucidum* IBK 1900 appears to be the most efficient among 14 medicinal mushrooms for biomass (up to 16 g/L in surface cultivation) and exopolysaccharide (up to 11 g/L in submerged cultivation) production on the crude sugar beet vinasse medium. The obtained biomass and exopolysaccharides after further evaluation of medicinal properties can be potentially used for the creation of dietary supplements and functional foods with immunomodulatory properties.

In brief, this chapter describes different types of strategies for using effluents with strong potential for environmental pollution and at the same time containing compounds that can be useful for generating value-added products. The development of technologies friendly, or friendlier, with nature is a non-avoidable requisite for the future which is already here.

ACKNOWLEDGMENTS

The experimental research was supported by the National Academy of Sciences of Ukraine (partially by each of the projects: No. 0114U002171, No. 0118U005321, No. 0119U101434, No. 0120U101651).

REFERENCES

APHA. 2005. *Standard methods for the examination of water and wastewater*, 21st ed. American Public Health Association, Washington, DC.

Arslanoğlu, H. & Tümen, F. (2021). Potassium struvite (slow release fertilizer) and activated carbon production: Resource recovery from vinasse and grape marc organic waste using thermal processing. *Process Safety and Environmental Protection, 147*, 1077–1087. https://doi.org/10.1016/j.psep.2021.01.025

Asadi, F., Barshan-tashnizi, M., Hatamian-Zarmi, A., Davoodi-Dehaghani, F. & Ebrahimi-Hosseinzadeh, B. (2021). Enhancement of exopolysaccharide production from *Ganoderma lucidum* using a novel submerged volatile co-culture system. *Fungal Biology, 125*(1), 25-31. https://doi.org/10.1016/j.funbio.2020.09.010

Babitskaya, V. G., Scherba, V. V., Mitropolskaya, N. Y. & Bisko, N. A. (2000). Exopolysaccharides of some medicinal mushrooms production and composition. *International Journal of Medicinal Mushrooms, 2*(1), 51-54. https://doi.org/10.1615/IntJMedMushr.v2.i1.50

Babu, J., Alappat, B. J. & Rane, V. (1995). An algorithm to calculate the performance details of an RCFB incinerator with heat recovery for the treatment of distillery spent wash. *International Journal of Energy Research, 19*(4), 329–336.

Bisko, N. A., Lomberg, M. L., Mytropolska, N. Y. & Mykchaylova, O. B. (2016). *The IBK mushroom culture collection*. Alterpres, M. G. Kholodny Institute of Botany, National Academy of Sciences of Ukraine, Kyiv.

Braun, R. (2007). Anaerobic digestion – A multi-faceted process for energy, environmental management and rural development. In: P. Ranalli (Ed.), *Improvement of crop plants for industrial end uses* (pp. 335–416). Springer, Dordrecht. https://doi.org/10.1007/978-1-4020-5486-0_13

Buděšinský, B. & Krumlová, L. (1967). Determination of sulphur and sulphate by titration with barium perchlorate: Comparison of various colour indictors. *Analytica Chimica Acta, 39*, 375–381. https://doi.org/10.1016/S0003-2670(01)80517-1

Camargo, J. M. O., Gallego Rios, J. M., Antonio, G. C. & Leite, J. T. C. (2021). Physicochemical properties of sugarcane industry residues aiming at their use in energy processes. In: M. S. Khan (Ed.) *Sugarcane – Biotechnology for biofuels*. IntechOpen, London. https://doi.org/10.5772/intechopen.95936

Carrilho, E. N. V. M., Labuto, G. & Kamogawa, M. Y. (2016). Destination of vinasse, a residue from alcohol industry. In: M. N. V. Prasad & K. Shih (Eds.), *Environmental materials and waste* (pp. 21–43). Academic Press, Elsevier. https://doi.org/10.1016/b978-0-12-803837-6.00002-0

Carvajal-Zarrabal, O., Nolasco-Hipólito, C., Barradas-Dermitz, D. M., Hayward-Jones, P. M., Aguilar-Uscanga, M. G. & Bujang, K. (2012). Treatment of vinasse from tequila production using polyglutamic acid. *Journal of Environmental Management, 95*, S66–S70. https://doi.org/10.1016/j.jenvman.2011.05.001

Cerri, B. C., Borelli, L. M., Stelutti, I. M., Soares, M. R. & da Silva, M. A. (2020). Evaluation of new environmental friendly particulate soil fertilizers based on agroindustry wastes biopolymers and sugarcane vinasse. *Waste Management, 108*(1), 144–153. https://doi.org/10.1016/j.wasman.2020.04.038

Chen, W., Zhao, Z., Chen, S. F. & Li, Y. Q. (2008). Optimization for the production of exopolysaccharide from *Fomes fomentarius* in submerged culture and its antitumor effect *in vitro*. *Bioresource Technology, 99*(8), 3187–3194. https://doi.org/10.1016/j.biortech.2007.05.049

Choudhary, S. (2020). Characterization and applications of mushroom exopolysaccharides. In: J. Singh & P. Gehlot (Eds.), *New and future developments in microbial biotechnology and bioengineering* (pp. 171–181). Elsevier. https://doi.org/10.1016/B978-0-12-821007-9.00014-0

Christofoletti, C. A., Escher, J. P., Correia, J. E., Marinho, J. F. U. & Fontanetti, C. S. (2013). Sugarcane vinasse: Environmental implications of its use. *Waste Management, 33*(12), 2752–2761. https://doi.org/10.1016/j.wasman.2013.09.005

Crespi, M. S., Martins, Q. V., Almeida, S., Barud, H. S., Kobelnik, M. & Ribeiro, C. A. (2011). Characterization and thermal behavior of residues from industrial sugarcane processing. *Journal of Thermic Analytical Calorimetry, 106*, 753–757. https://doi.org/10.1007/s10973-011-1397-9

Cui, J., Tha Goh, K. K., Archer, R. & Singh, H. (2007). Characterisation and bioactivity of protein-bound polysaccharides from submerged-culture fermentation of *Coriolus versicolor* Wr-74 and ATCC-20545. *Journal of Industrial Microbiology & Biotechnology, 34*, 393–402. https://doi.org/10.1007/s10295-007-0209-5

España-Gamboa, E., Mijangos-Cortes, J., Barahona-Perez, L., Dominguez-Maldonado, J., Hernández-Zarate, J. & Alzate-Gaviria, L. (2011). Vinasses: Characterization and treatments. *Waste Management & Research, 29*(12), 1235–1250. https://doi.org/10.1177/0734242X10387313

Fermoso, F. G., Bartacek, J., Jansen, S. & Lens, P. N. L. (2009). Metal supplementation to UASB bioreactors: From cell-metal interactions to full-scale application. *Science of the Total Environment, 407*, 3652–3667. https://doi.org/10.1016/j.scitotenv.2008.10.043

Formann, S., Hahn, A., Janke, L., Stinner, W., Sträuber, H., Logroño, W. & Nikolausz, M. (2020). Beyond sugar and ethanol production: Value generation opportunities through sugarcane residues. *Frontiers in Energy Research, 8*, 579577. https://doi.org/10.3389/fenrg.2020.579577

Fuess, L. T., Rodrigues, I. J. and Garcia, M. L. (2017). Fertirrigation with sugarcane vinasse: Foreseeing potential impacts on soil and water resources through vinasse characterization. *Journal of Environmental Science and Health, Part A, 52*(11), 1063–1072. https://doi.org/10.1080/10934529.2017.1338892

Gebreeyessus, G. D., Mekonen, A. & Alemayehu, E. (2019). A review on progresses and performances in distillery stillage management. *Journal of Cleaner Production, 232*, 295–307. https://doi.org/10.1016/j.jclepro.2019.05.383

Ghernaout, D., Elboughdiri, N. & Ghareba, S. (2020). Fenton technology for wastewater treatment: Dares and trends. *Open Access Library Journal, 7*, 1–26. https://doi.org/10.4236/oalib.1106045

Gurgel, M. N. D. A., Correa, S. T. R., Dourado Neto, D. & Paula Júnior, D. R. D. (2015). Technology for sugarcane agroindustry waste reuse as granulated organomineral fertilizer. *Agricultural Engineering, 35*(1), 63–75. https://doi.org/10.1590/1809-4430-Eng.Agric.v35n1p63-75/2015

Hakika, D. C., Sarto, S., Mindaryani, A. & Hidayat, M. (2019). Decreasing COD in sugarcane vinasse using the Fenton reaction: The effect of processing parameters. *Catalysts*, *9*(11), 881. https://doi.org/10.3390/catal9110881

Hamidi, M., Okoro, O. V., Milan, P. B., Khalili, M. R., Samadian, H., Nie, L. & Shavandi, A. (2022). Fungal exopolysaccharides: Properties, sources, modifications, and biomedical applications. *Carbohydrate Polymers*, *284*, 119152. https://doi.org/10.1016/j.carbpol.2022.119152

Hoang, T. D. & Nghiem, N. (2021). Recent developments and current status of commercial production of fuel ethanol. *Fermentation*, *7*, 314. https://doi.org/10.3390/fermentation7040314

Hoarau, J., Caro, Y., Grondin, I. & Petit, T. (2018). Sugarcane vinasse processing: Toward a status shift from waste to valuable resource. A review. *Journal of Water Process Engineering*, *24*, 11–25. https://doi.org/10.1016/j.jwpe.2018.05.003

Huang, C. W., Hung, Y. C., Chen, L. Y., Asatiani, M., Elisashvili, V., Klarsfeld, G., Melamed, D., Fares, B., Wasser, S. P. & Maua, J. L. (2021). Chemical composition and antioxidant properties of different combinations of submerged cultured mycelia of medicinal mushrooms. *International Journal of Medicinal Mushrooms*, *23*(8), 1–24. https://doi.org/10.1615/IntJMedMushrooms.2021039339

Ivanov, V., Shevchenko, O., Marynin, A., Stabnikov, V., Gubenia, O., Stabnikova, O., Shevchenko, A., Gavva, O. & Saliuk, A. (2021). Trends and expected benefits of the breaking edge food technologies in 2021–2030. *Ukrainian Food Journal*, *10*(1), 7–36. https://doi.org/10.24263/2304-974X-2021-10-1-3

Ivanova, T. S., Krupodorova, T. A., Barshteyn, V. Y., Artamonova, A. B. & Shlyakhovenko, V. A. (2014). Anticancer substances of mushroom origin. *Experimental Oncology*, *36*(1), 1–9.

Ivanova, T. S., Kulichkova, G. I., Savytska, N. A., Volodko, O. I., Lukashevych, K. M., Syvak, V. O. & Tsygankov, S. P. (2021). Sugar beet vinasse into biogas solution. In: German Biogas and Bioenergy Society (Ed.), *Progress in biogas V – Science meets practice* (pp. 166–167). IBBK Fachgruppe Biogas GmbH, Stuttgart.

Ivanova, T. S., Titova, L. O., Bisko, N. A., Klechak, I. R., Novak, A. G. & Tsygankov, S. P. (2019). *Nutrient medium for fungal cultivation which contains molasses stillage*. Patent of Ukraine 118997, Bul. 7, class C05F5/00, A01G18/20.

Janke, L., Leite, A. F., Batista, K., Silva, W., Nikolausz, M., Nelles, M. & Stinner, W. (2016). Enhancing biogas production from vinasse in sugarcane biorefineries: Effects of urea and trace elements supplementation on process performance and stability. *Bioresource Technology*, *217*, 10–20. https://doi.org/10.1016/j.biortech.2016.01.110

Kharayat, Y. (2012). Distillery wastewater: Bioremediation approaches. *Journal of Integrative Environmental Sciences*, *9*(2), 69–91. https://doi.org/10.1080/1943815X.2012.688056

Kulichkova, G. I., Ivanova, T. S., Köttner, M., Volodko, O. I., Spivak, S. I., Tsygankov, S. P. & Blume, Ya. B. (2020). Plant feedstock and their biogas production potentials. *The Open Agriculture Journal*, *14*, 219–234. https://doi.org/10.2174/1874331502014010219

Lachman, J., Baláš, M., Lisý, M., Lisá, H., Milčák, P. & Elbl, P. (2021). An overview of slagging and fouling indicators and their applicability to biomass fuels. *Fuel Processing Technology*, *217*, 106804. https://doi.org/10.1016/j.fuproc.2021.106804

Larsson, E. & Tengberg, T. (2014). *Evaporation of vinasse: Pilot plant investigation and preliminary process design*. Göteborg, Sweden: Chalmers University of Technology.

Lee, H., Song, M., Yu, Y. & Hwang, S. (2003). Production of *Ganoderma lucidum* mycelium using cheese whey as an alternative substrate: Response surface analysis and biokinetics. *Biochemical Engineering Journal, 15*, 93–99. https://doi.org/10.1016/S1369-703X(02)00211-5

Li, L. Q., Song, A. X., Yin, J. Y., Siu, K. C., Wong, W. & Wu, J. Y. (2020). Anti-inflammation activity of exopolysaccharides produced by a medicinal fungus *Cordyceps sinensis* Cs-HK1 in cell and animal models. *International Journal of Biological Macromolecules, 149*, 1042–1050. https://doi.org/10.1016/j.ijbiomac.2020.02.022

Liu, L., Feng, J., Gao, K., Zhou, Sh., Yan, M., Tang, Ch., Zhou, J., Liu, Y. & Zhang, J. (2022). Influence of carbon and nitrogen sources on structural features and immunomodulatory activity of exopolysaccharides from *Ganoderma lucidum*. *Process Biochemistry, 119*, 96–105. https://doi.org/10.1016/j.procbio.2022.05.016

López-López, A., Davila-Vazquez, G., León-Becerril, E., Villegas-García, E. & Gallardo-Valdez, J. (2010). Tequila vinasses: Generation and full scale treatment processes. *Reviews in Environmental Science and Bio/Technology, 9*(2), 109–116. https://doi.org/10.1007/s11157-010-9204-9

Mahmoudabadi, Z. S., Rashidi, A. & Maklavany, D. M. (2022). Optimizing treatment of alcohol vinasse using a combination of advanced oxidation with porous α-Fe_2O_3 nanoparticles and coagulation-flocculation. *Ecotoxicology and Environmental Safety, 234*, 113354. https://doi.org/10.1016/j.ecoenv.2022.113354

Mane, J. D., Modi, S., Nagawade, S., Phadnis, S. P. & Bhandari, V. M. (2006). Treatment of spent wash using chemically modified bagasse and color removal studies. *Bioresource Technology, 97*(14), 1752–1755. https://doi.org/10.1016/j.biortech.2005.10.016

Maurya, R., Tirkey, S. R., Rajapitamahuni, S., Ghosh, A. & Mishra, S. (2019). Recent advances and future prospective of biogas production. In M. Hosseini (Ed.), *Advances in feedstock conversion technologies for alternative fuels and bioproducts* (pp. 159–178). https://doi.org/10.1016/B978-0-12-817937-6.00009-6

Mikucka, W. & Zielińska, M. (2020). Distillery stillage: Characteristics, treatment, and valorization. *Applied Biochemistry and Biotechnology, 192*, 770–793. https://doi.org/10.1007/s12010-020-03343-5

Moraes, B. S., Junqueira, T. L., Pavanello, L. G., Cavalett, O., Mantelatto, P. E., Bonomi, A. & Zaiat, M. (2014). Anaerobic digestion of vinasse from sugarcane biorefineries in Brazil from energy, environmental, and economic perspectives: Profit or expense? *Applied Energy, 113*, 825–835. https://doi.org/10.1016/j.apenergy.2013.07.018

Moraes, B. S., Triolo, J. M., Lecona, V. P., Zaiat, M. & Sommer, S. G. (2015). Biogas production within the bioethanol production chain: Use of co-substrates for anaerobic digestion of sugar beet vinasse. *Bioresource Technology, 190*, 227–234. https://doi.org/10.1016/j.biortech.2015.04.089

Morozova, I., Oechsner, H., Roik, M., Hülsemann, B. & Lemmer, A. (2020). Assessment of areal methane yields from energy crops in Ukraine, best practices. *Applied Sciences, 10*, 4431. https://doi.org/10.3390/app10134431

Nakashima, R. N. & de Oliveira Junior, S. (2020). Comparative exergy assessment of vinasse disposal alternatives: Concentration, anaerobic digestion and fertirrigation. *Renewable Energy, 147*, 1969–1978. https://doi.org/10.1016/j.renene.2019.09.124

Pant, D. & Adholeya, A. (2007). Biological approaches for treatment of distillery wastewater: A review. *Bioresource Technology, 98*(12), 2321–2334. https://doi.org/10.1016/j.biortech.2006.09.027

Park, H. D., Chang, I. S. & Lee, K. J. (2015). *Principles of membrane bioreactors for wastewater treatment*. CRC Press, Boca Raton, FL.

Parsaee, M., Kiani Deh Kiani, M. & Karimi, K. (2019). A review of biogas production from sugarcane vinasse. *Biomass and Bioenergy, 122*, 117–125. https://doi.org/10.1016/j.biombioe.2019.01.034

Peiter, F. S., Hankins, N. P. & Pires, E. C. (2019). Evaluation of concentration technologies in the design of biorefineries for the recovery of resources from vinasse. *Water Research, 157*, 483–497. http://doi.org/10.1016/j.watres.2019.04.003

Ramos, L. R., Lovato, G., Rodrigues, J. A. D. & Silva, E. L. (2022). Scale-up and energy estimations of single- and two-stage vinasse anaerobic digestion systems for hydrogen and methane production. *Journal of Cleaner Production, 349*, 131459. https://doi.org/10.1016/j.jclepro.2022.131459

Rathore, H., Prasad, S. & Sharma, S. (2017). Mushroom nutraceuticals for improved nutrition and better human health: A review. *PharmaNutrition, 5*(2), 35–46. http://doi.org/10.1016/j.phanu.2017.02.001

Reis, C. E. R., Bento, H. B. S., Alves, T. M., Carvalho, A. K. F. & de Castro, H. F. (2019). Vinasse treatment within the sugarcane-ethanol industry using ozone combined with anaerobic and aerobic microbial processes. *Environments, 6*, 5. https://doi.org/10.3390/environments6010005

Reis, C. E. R. & Hu, B. (2017). Vinasse from sugarcane ethanol production: Better treatment or better utilization? *Frontiers in Energy Research, 5*, 7. https://doi.org/10.3389/fenrg.2017.00007

Reyes-Cabrera, J., Leon, R. G., Erickson, J. E., Rowland, D. L., Silveira, M. L. & Morgan, K. T. (2017). Differences in biomass and water dynamics between a cotton-peanut rotation and a sweet sorghum bioenergy crop with and without biochar and vinasse as soil amendments. *Field Crops Research, 214*, 123–130. https://doi.org/10.1016/j.fcr.2017.09.012

Romaniuk, W., Rogovskii, I, Polishchuk, V., Titova, L., Borek, K., Shvorov, S., Roman, K., Solomka, O., Tarasenko, S., Didur, V. & Biletskii, V. (2022). Study of technological process of fermentation of molasses vinasse in biogas plants. *Processes, 10*(10), 2011. https://doi.org/10.3390/pr10102011

Satyawali, Y. & Balakrishnan, M. (2007). Removal of color from biomethanated distillery spentwash by treatment with activated carbons. *Bioresource Technology, 98*(14), 2629–2635. https://doi.org/10.1016/j.biortech.2006.09.016

Satyawali, Y. & Balakrishnan, M. (2008). Wastewater treatment in molasses based alcohol distilleries for COD and color removal: A review. *Journal of Environmental Management, 86*(3), 481–497. https://doi.org/10.1016/j.jenvman.2006.12.024

Scarpari, M., Reverberi, M., Parroni, A., Scala, V., Fanelli, C., Pietricola, Ch., Zjalic, S., Maresca, V., Tafuri, A., Ricciardi, M. R., Licchetta, R., Mirabilii, S., Sveronis, A., Cescutti, P. & Rizzo, R. (2017). Tramesan, a novel polysaccharide from *Trametes versicolor*. Structural characterization and biological effects. *PLoS One, 12*(8), e0171412. https://doi.org/10.1371/journal.pone.0171412

Silva, A. F. R., Lebron, Y. A. R., Brasil, Y. L., Lange, L. C. & Amaral, M. C. S. (2022). Effect of electrolyte solution recycling on the potassium recovery from vinasse by integrated electrodialysis and K-struvite precipitation processes. *Chemical Engineering Journal, 450*(1), 137975. https://doi.org/10.1016/j.cej.2022.137975

Singdevsachan, S. K., Auroshree, P., Mishra, J., Baliyarsingh, B., Tayung, K. & Thatoi, H. (2016). Mushroom polysaccharides as potential prebiotics with their antitumor and immunomodulating properties: A review. *Bioactive Carbohydrates and Dietary Fibre*, *7*(1), 1–14. https://doi.org/10.1016/j.bcdf.2015.11.001

Soković, M., Glamočlija, J., Ćirić, A., Petrović, J. & Stojković, D. (2018). Mushrooms as sources of therapeutic foods. In: A. M. Holban & A. M. Grumezescu (Eds.), *Handbook of food bioengineering series* (pp. 141–178). Academic Press. https://doi.org/10.1016/B978-0-12-811517-6.00005-2

Stabnikov, V. P. & Ivanov, V. N. (2006). The effect of various iron hydroxide concentrations on the anaerobic fermentation of sulfate-containing model wastewater. *Applied Biochemistry and Microbiology*, *42*(3), 284–288. https://doi.org/10.1134/S0003683806030112

Sun, H. Q., Yu, X. F., Li, T. & Zhu, Z. Y. (2021). Structure and hypoglycemic activity of a novel exopolysaccharide of *Cordyceps militaris*. *International Journal of Biological Macromolecules*, *166*, 496–508. https://doi.org/10.1016/j.ijbiomac.2020.10.207

Syaichurrozi, I., Sarto, S., Sediawan, W. B. & Hidayat, M. (2020). Mechanistic models of electrocoagulation kinetics of pollutant removal in vinasse waste: Effect of voltage. *Journal of Water Process Engineering*, *36*, 101312. https://doi.org/10.1016/j.jwpe.2020.101312

Tejada, M., Gonzalez, J. L., García-Martínez, A. M. & Parrado, J. (2008). Application of a green manure and green manure composted with beet vinasse on soil restoration: Effects on soil properties. *Bioresource Technology*, *99*(11), 4949–4957. https://doi.org/10.1016/j.biortech.2007.09.026

Tena, M., Perez, M. & Solera, R. (2021). Effect of hydraulic retention time on the methanogenic step of a two-stage anaerobic digestion system from sewage sludge and wine vinasse: Microbial and kinetic evaluation. *Fuel*, *296*, 120674. https://doi.org/10.1016/j.fuel.2021.120674

Toledo-Cervantes, A., Villafán-Carranza, F., Arreola-Vargas, J., Razo-Flores, E. & Oscar Méndez-Acosta, H. (2020). Comparative evaluation of the mesophilic and thermophilic biohydrogen production at optimized conditions using tequila vinasses as substrate. *International Journal of Hydrogen Energy*, *45*(19), 11000–11010. https://doi.org/10.1016/j.ijhydene.2020.02.051

Torroba, A. (2021). *Liquid biofuels atlas 2020–2021*. Inter-American Institute for Cooperation on Agriculture, San Jose, Costa Rica.

Valverde, M. E., Hernández-Pérez, T. & Paredes-López, O. (2015). Edible mushrooms – Improving human health and promoting quality life. *International Journal of Microbiology*, Article ID 376387. https://doi.org/10.1155/2015/376387

Volodko, O. I., Ivanova, T. S., Kulichkova, G. I., Lukashevych, K. M., Blume, Y. B. & Tsygankov, S. P. (2020). Fermentation of sweet sorghum syrup under reduced pressure for bioethanol production. *The Open Agriculture Journal*, *14*, 235–245. https://doi.org/10.2174/1874331502014010235

Wasser, S. P. & Weis, A. L. (1999). Medicinal properties of substances occurring in higher basidiomycetes mushrooms: Current perspectives. *International Journal of Medicinal Mushrooms*, *1*(1), 31–62. https://doi.org/10.1615/IntJMedMushrooms.v1.i1.30

Xiang, Y., Xu, X. & Li, J. (2012). Chemical properties and antioxidant activity of exopolysaccharides fractions from mycelial culture of *Inonotus obliquus* in a ground corn stover medium. *Food Chemistry, 134*(4), 1899–1905. https://doi.org/10.1016/j.foodchem.2012.03.121

Zhang, C., Jiang, L. & Wang, Z. (2021). Effect of coix seed on exopolysaccharide production of *Cordyceps militaris* in liquid culture. *Arabian Journal of Chemistry, 14*(3), 102999. https://doi.org/10.1016/j.arabjc.2021.102999

Zhang, P. J., Zhao, Z. G., Yu, S. J., Guan, Y. G., Li, D. & He, X. (2012). Using strong acid–cation exchange resin to reduce potassium level in molasses vinasses. *Desalination, 286*, 210–216. https://doi.org/10.1016/j.desal.2011.11.024

Zhang, Y., Kong, H., Fang, Y., Nishinari, K. & Phillips, G. O. (2013). Schizophyllan: A review on its structure, properties, bioactivities and recent developments. *Bioactive Carbohydrates and Dietary Fibre, 1*(1), 53–71. https://doi.org/10.1016/j.bcdf.2013.01.002

Zhao, L. N., Cao, Y. B., Luo, Q., Xu, Y. L., Li, N., Wang, C. X. & Xu, J. W. (2022). Overexpression of phosphomannomutase increases the production and bioactivities of *Ganoderma* exopolysaccharides. *Carbohydrate Polymers, 294*, 119828. https://doi.org/10.1016/j.carbpol.2022.119828

Chapter 10

Microbial Surfactants Production from Industrial Waste

Tetyana Pirog
National University of Food Technologies
Institute of Microbiology and Virology

Viktor Stabnikov and Olena Stabnikova
National University of Food Technologies

CONTENTS

DOI: 10.1201/9781003329671-10

10.1 INTRODUCTION

Surfactants of microbial origin possess emulsifying, antimicrobial, antiadhesive and biofilm-disrupting properties, and are promising for use in the oil and gas industries, food industry, medicine, household and personal care, and environmental technologies (Chong & Li, 2017; Parthipan et al., 2017; Paulino et al., 2016; Sałek et al., 2022). In addition, microbial surfactants are physically and chemically stable substances, whose application is very attractive due to biodegradability and nontoxicity compared with their synthetic counterparts (Gayathiri et al., 2022). Positive aspect is that the application of biosurfactants leads to the damage of the cytoplasmic membrane of microbial cells causing the reduction of the possibility of microbial resistant forms occurrence (Otzen, 2017). Meanwhile, currently, the industrial production of microbial surfactants is performed by a limited number of companies (Sekhon Randhawa & Rahman, 2014). The high costs of the biosynthesis process, as well as the low concentration of synthesized surfactants in cultural media, are the reasons for the insufficient efficiency of microbial technologies for surfactant production. It was estimated that up to 50% of the total cost comes from the preparation of the medium for biosurfactant production, and high raw material costs are essential points that make the commercial production of biosurfactants still a challenge (Ebadipour et al., 2016). One of the approaches to reduce the cost of microbial surfactants is the use of industrial waste as substrates (Banat et al., 2014; Bjerk et al., 2021).

Used cooking oil (UCO) is generated during the frying of food products both in the food processing industry (commercial, industrial and institutional wastes) and in households using vegetable oil (De Feo et al., 2020). The amount of disposed UCO in the world is estimated at about 15 million tons annually (Wang et al., 2019). Inappropriate disposal of this waste could have a serious negative environmental impact, meanwhile, recovery rate of UCO at present time is low. The promising way for utilization of UCO is its biotechnological conversion in value-added by-products such as biodiesel, plasticizer, polyurethane, detergents, and surfactants (Awogbemi et al., 2021). However, heated cooking oils contain various toxic substances (Wang et al., 2021), which can be potential inhibitors of the growth of microorganisms and the synthesis of the target product.

Production of biodiesel as a biodegradable and nontoxic substitute for petroleum diesel by the transesterification of UCO is a common way for its utilization (da Silva et al., 2022; Hossain et al., 2015; Wang et al., 2019) and has both environmental and economic benefits. The main advantages of biodiesel production from UCO include the utilization of toxic wastes and reduction of agricultural land occupied by the cultivation of plants for the production of biodiesel. However, despite the environmental advantages, this technology is characterized by high cost, a short storage time of biodiesel and formation of technical (crude) glycerol (Panadare & Rathod, 2015). Produced crude glycerol accounts for 10% by weight of biodiesel, and it also contains toxic methanol residues, high concentrations of salts and free fatty acids. To be applied in such traditional areas as pharmaceutical, cosmetics and food industry, crude glycerol should come through costly processes of purification to achieve the required level of purity (≥ 99%). There are some studies to evaluate the potential bioconversion of crude glycerol in value-added products such as biohydrogen (Rodrigues et al., 2018), 1,3-propanediol and citric acid (Mitrea et al., 2017), 1,3-propanediol and lactate (Wang et al., 2019), butanol (Yadav et al., 2014), poly(hydroxyalkanoates) (Cavalheiro et al., 2009) and ethanol (Chilakamarry et al., 2021; Liu et al., 2012).

The aim of the present research was to study the possibilities of using such waste streams as used cooking oil and crude glycerol for the production of microbial surfactants.

10.2 BIOCONVERSION OF INDUSTRIAL WASTES INTO SURFACE-ACTIVE SUBSTANCES USING BACTERIAL STRAINS *ACINETOBACTER CALCOACETICUS* IMB B-7241, *RHODOCOCCUS ERYTHROPOLIS* IMB AC-5017 AND *NOCARDIA VACCINII* IMB B-7405

The strains of hydrocarbon-oxidizing bacteria isolated from an oil-contaminated soil samples *Acinetobacter calcoaceticus* IMV B-7241, *Rhodococcus erythropolis* IMV Ac-5017 and *Nocardia vaccinii* IMV B-7405 from the Collection of Microorganisms of the D.K. Zabolotny Institute of Microbiology and Virology of the National Academy of Sciences of Ukraine were used in this study. All these strains growing in media with glucose or ethanol as sources of carbon were able to synthesize extracellular surface-active substances (SAS) (Pirog et al., 2005; 2020; 2023). As a source of carbon and energy in the present study, such wastes were used: (a) crude glycerol, which is a waste of biodiesel production (Biofuel plant, Zaporizhzhia, Ukraine); and (b) oil after frying of potatoes and meat (McDonald's fast food restaurant chain, Kyiv, Ukraine).

TABLE 10.1 SYNTHESIS OF SURFACE-ACTIVE SUBSTANCES (SAS) BY *ACINETOBACTER CALCOACETICUS* IMV B-241, *RHODOCOCCUS ERYTHROPOLIS* IMV AC-5017 AND *NOCARDIA VACCINII* IMV B-7405 IN MEDIA WITH DIFFERENT TYPES OF GLYCEROL

Bacteria	Concentration of SAS, g/L in the Medium with Glycerol		
	I (Pure)	II (Modified)	III (Crude)
A. calcoaceticus	2.4±0.12	3.2±0.16	4.7±0.23
R. erythropolis	0.5±0.03	0.7±0.03	1.0±0.05
N. vaccinii	1.8±0.09	2.5±0.12	3.5±0.18

Compositions of the base media for bacteria cultivation were the following: (a) for *A. calcoaceticus*, g/L: $(NH_2)_2CO$, 0.35; $MgSO_4{\cdot}7H_2O$, 0.1; NaCl, 1.0; Na_2HPO_4, 0.6; KH_2PO_4, 0.14; yeast extract, 5 mL; solution of trace elements, 1 mL; deionized water, 1L; pH 6.8–7.0; (b) for *R. erythropolis*, g/L: $NaNO_3$, 1.3; $MgSO_4{\cdot}7H_2O$, 0.1; NaCl, 1.0; Na_2HPO_4, 0.6; KH_2PO_4, 0.14; $FeSO_4{\cdot}7H_2O$, 0.001; deionized water, 1L; pH 6.8–7.0; (b) for *N. vaccinii*, g/L: $NaNO_3$, 0.5; $MgSO_4{\cdot}7H_2O$, 0.1; $CaCl_2{\cdot}2H_2O$, 0.1; KH_2PO_4, 0.1; $FeSO_4{\cdot}7H_2O$, 0.001; yeast extract, 5 mL; deionized water, 1L; 6.8–7.0. Trace element solution contains, g/100 mL: $ZnSO_4{\cdot}7H_2O$, 1.1; $MnSO_4{\cdot}H_2O$, 0.6; $FeSO_4{\cdot}7H_2O$, 0.1; $CuSO_4{\cdot}5H_2O$, 0.004; $CoSO_4{\cdot}7H_2O$, 0.03; H_3BO_3, 0.006; KI, 0.0001; and EDTA (Trilon B), 0.5 (Pirog et al., 2023). Inoculum was produced by the cultivation of the strains in the appropriate medium with 0.5% (v/v) of the corresponding substrate. The cultural liquid with a concentration of bacterial cells 10^4 to 10^5 cells/mL was taken from the exponential growth phase and added to the medium for the strain cultivation, 10% (v/v). Cultivation of each bacterial strain was conducted under aeration at 30°C for 120 h. Results of synthesis of SAS by strains in the correspondent medium with addition as a source of carbon pure glycerol, 1.0% v/v (I); "modified" glycerol, 1.0% v/v, and added with sodium chloride, 2.5%, and methanol, 0.3% v/v (II), and crude glycerol, 2.2% (III) are shown in Table 10.1.

The concentration of carbon in medium III with crude glycerol was the same as in medium I with pure glycerol. The amounts of SAS produced by all tested bacterial strains grown in medium II with pure glycerol added with the components which are usually present in crude glycerol, particularly, sodium chloride and methanol, were higher by 11%–68% in comparison with medium I (pure glycerol). The amounts of synthesized SAS in the medium with crude glycerol were almost twice higher than those in medium I with pure glycerol.

Increasing the amount of the inoculum to 10%–15% and doubling (compared to the base medium) the concentrations of the nitrogen-containing components in media allowed to conduct cultivation of bacterial strains at a concentration of crude glycerol up to 7%–8% (v/v) (Figure 10.1).

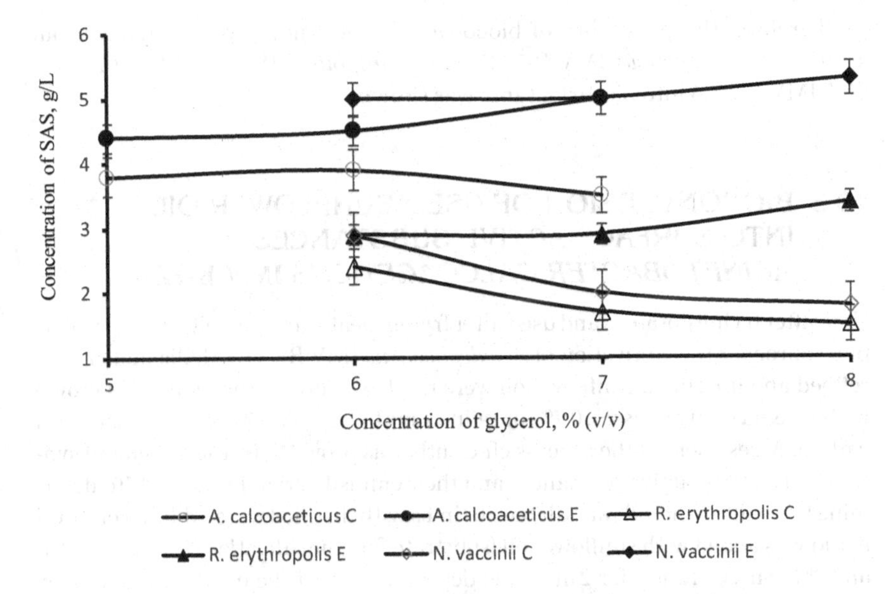

Figure 10.1 Effect of the concentration of nitrogen on the synthesis of surface-active substances (SAS) by *Acinetobacter calcoaceticus* IMV B-7241, *Rhodococcus erythropolis* IMV Ac-5017 and *Nocardia vaccinii* IMV B-7405: C, control—single dose of nitrogen source; E, experiment—double dose of nitrogen source.

The amount of SAS synthesized by the studied strains in the media with increased content of nitrogen compound varied from 2.9 to 5.3 g/L which is 1.4–3.0 times higher than in the base medium with a similar crude glycerol concentration.

According to literature data, the amounts of surfactants synthesized by various bacterial strains using crude glycerol are usually low. Thus, *Bacillus subtilis* LSFM cultivated in a medium with 5% crude glycerol synthesized 1.36 g/L of surfactin (de Faria et al., 2011); *Pseudomonas aeruginosa* MSIC02 in a medium with 5% prehydrolyzed (treated with sulfuric acid) crude glycerol synthesized only 1.27 g/L of rhamnolipids (de Sousa et al, 2011) and *Pseudomonas aeruginosa* RS6 in the medium with 1% biodiesel waste plus glycerol and 0.2 M sodium nitrate produced 2.73 g/L of rhamnolipids (Baskaran et al., 2021). Wherein, in most studies the concentration of crude glycerol in the medium for cultivation of surfactant producers was not higher than 5% (v/v) (de Faria et al., 2011; de Sousa et al, 2011; Liu et al., 2011). An increase in the concentration of crude glycerol to 8% (v/v) was accompanied by a two-fold decrease in the amount of synthesized glycolipids by the basidiomycete *Ustilago maydis* (Liu et al., 2011). Thus, the selected strains are not inferior and even exceed those described in the literature in terms of the amount of synthesized exopolysaccharides using crude glycerol as a source of carbon and energy.

Therefore, the possibility of bioconversion of crude glycerol by bacterial strains *A. calcoaceticus* IMV B-7241, *R. erythropolis* IMV Ac-5017 and *N. vaccinii* IMV B-7405 into biosurfactants was shown.

10.3 BIOCONVERSION OF USED SUNFLOWER OIL INTO SURFACE-ACTIVE SUBSTANCES BY *ACINETOBACTER CALCOACETICUS* IMV B-7241

Used after frying potatoes and used after frying meat sunflower oil served as a carbon source for the cultivation of *A. calcoaceticus* IMV B-7241. Media added with refined and unrefined sunflower oil were used as controls. Inoculum was grown in the media with molasses, 0.7%, or refined sunflower oil, 0.5% (v/v), as a source of carbon. Assessment of the process efficiently was provided by the amount of synthesized surface-active substances and their emulsification index, E24. To determine E24, culture liquid was diluted 50 times with distilled water; diluted cultural liquid was added with sunflower oil (substrate for emulsification) in a ratio of 1:1, and shaken vigorously for 2 min. The determination of the emulsification index was carried out after 24 h as the ratio of the height of the emulsion layer to the total height of the liquid in the test tube and expressed as a percentage (Table 10.2).

TABLE 10.2 SYNTHESIS OF SURFACE-ACTIVE SUBSTANCES (SAS) BY *ACINETOBACTER CALCOACETICUS* IMV B-7241 IN THE MEDIA WITH 4 % (V/V) OF SUNFLOWER OIL

Carbon Source in Media		Characteristics	
For Inoculum Preparation	For SAS Synthesis	SAS (g/L)	E_{24} (%)
Molasses	Refined sunflower oil	4.0±0.20	56±4.1
	Unrefined sunflower oil	2.3±0.12	50±3.3
	UCO[a] after potato frying	1.5±0.08	45±3.4
	UCO after meat frying	2.8±0.14	49±4.0
Refined sunflower oil	Refined sunflower oil	3.4±0.17	51±4.9
	Unrefined sunflower oil	3.3±0.16	47±3.0
	UCO after potato frying	3.9±0.19	52±3.9
	UCO after meat frying	4.5±0.22	54±4.2
UCO after potato frying	UCO after potato frying	5.0±0.25	50±3.4
UCO after meat frying	UCO after meat frying	8.5±0.42	54±4.0

[a] UCO, used sunflower oil.

It was shown that the use of molasses as a carbon source for inoculum preparation resulted in a decrease in the amount of SAS synthesized in the media with unrefined and used sunflower oil in 1.7–2.7 times compared to these values for the medium with the refined substrate (Table 10.2). However, in the case of using molasses in the medium for preparation of inoculum for seeding of the medium with refined sunflower oil, an increase in the synthesis of microbial surfactants was observed. When inoculum was prepared in the medium with used sunflower oil, the concentrations of synthesized surfactants were highest, 5.0 and 8.5 g/L. There were no significant changes in the emulsification index for all variants.

Applications of refined oils or fat-and-oil production waste for the synthesis of microbial surfactants were proposed (Banat et al., 2014; Bhardwaj et al., 2013). Meanwhile, not much research has been published regarding the use of UCO for biosurfactant synthesis, which may be due to the low amounts of SAS obtained. Thus, *Pseudomonas fluorescence* MFS03 synthesized 4.2 g/L of surfactant in medium with 2% of used vegetable oil (Govindammal & Parthasarathi, 2013) and cultivation of *Pseudomonas aeruginosa* PB3A in a medium with 1% of used oil was accompanied by the production of 0.3–0.6 g/L surfactant (Saravanan & Subramaniyan, 2014). However, when *Pseudomonas aeruginosa* ATCC 9027 was grown on overcooked sunflower oil (initial concentration 15 mL/L followed by addition of 20 mL/L at 72 h of growth), the concentration of rhamnolipids was 8.5 g/L (Luo et al., 2013).

Thus, the possibility of replacing refined sunflower oil with used one for the synthesis of surfactants by *A. calcoaceticus* IMV B-7241 was shown.

10.4 BIOTRANSFORMATION OF MIXED INDUSTRIAL WASTE INTO SURFACE-ACTIVE SUBSTANCES BY *NOCARDIA VACCINII* IMV B-7405

The possibility to increase the synthesis of biosurfactants in the medium with a mixture of crude glycerol and used sunflower oil by *N. vaccinii* IMV B-7405 was studied. Concentrations of substrates in media for cultivation were, % (v/v): refined oil, 0.7-2; UCO, 0.7-4; pure glycerol, 2–3.3; crude glycerol, 1–4. Concentration of sodium nitrate was 0.5 g/L. As inoculum, a culture from exponential phase growth in the medium containing as a carbon source, % (v/v): monosubstrate (refined or used sunflower oil, purified or crude glycerin), 0.5; a mixture of used sunflower oil, 0.25, or crude glycerol, 0.25.

Effect of the different carbon source in the medium used for the inoculum preparation on the synthesis of surfactants by *N. vaccinii* on a mixture of crude glycerol and used sunflower oil was studied (Table 10.3).

TABLE 10.3 SYNTHESIS OF SURFACE-ACTIVE SUBSTANCES (SAS) BY *NOCARDIA VACCINII* IMV B-7405 ON A MIXTURE OF CRUDE GLYCEROL AND USED SUNFLOWER OIL, DEPENDING ON THE CARBON SOURCE USED FOR INOCULUM PREPARATION

Carbon Source in Media		
For SAS Synthesis	**For Inoculum Preparation**	**SAS (g/L)**
Used sunflower oil, 4%	Used sunflower oil, 0.5%	2.1±0.11
Crude glycerol, 4%	Crude glycerol, 0.5%	3.0±0.15
Used sunflower oil, 2%+crude glycerol, 2%	Used sunflower oil, 0.5%	2.8±0.14
	Crude glycerol, 0.5%	3.3±0.16
	Used sunflower oil, 0.5%+crude glycerol, 0.5%	3.2±0.16

The highest concentrations of surfactants, 3.2–3.3 g/L, synthesized by *N. vaccinii* in the medium with a mixture of UCO and crude glycerol were when the inoculum was prepared using also the same mixed substrate or crude glycerol. The amounts of surfactants synthesized on the mixed substrate were 1.5–1.6 times higher than that on the used sunflower oil.

Effect of concentrations of crude glycerol and used sunflower oil in the mixture on the synthesis of surface-active substances is shown in Figure 10.2.

The concentration of sodium nitrate in the medium was 0.5 g/L. Inoculum was prepared in the medium with crude glycerol, 0.5% (v/v). The results showed that with an increase in the concentration of monosubstrates in the mixture from 1.0% to 2.5%, the amount of synthesized surfactants increased by 1.5 times. At the same time, during the cultivation of *N. vaccinii* in a medium with crude glycerol, 3.0%, and used sunflower oil, 3%, the amount of surfactant produced decreased. However, the emulsification index of the culture liquid remained almost constant, 40%–42%, for all variants of bacteria cultivation.

Cultivation of *N. vaccinii* in a medium with used sunflower oil, 1.0% (v/v) and crude glycerol, 3.0% (v/v) and concentration of $NaNO_3$, 0.75 g/L, allowed to increase the synthesis of SAS up to 5.4 g/L (Table 10.4).

It was shown that using as a carbon source mixture of used sunflower oil, 1% (v/v) and crude glycerol, 3% (v/v) (molar ratio of monosubstrate concentrations was 0.078:1) the amounts of synthesized SAS were 5.1 and 5.4 g/L at a concentration of $NaNO_3$, 0.5 and 0.75 g/L, respectively, that was higher ones in the medium with monosubstrates (Table 10.4).

Cultivation of microorganisms on a mixture of substrates required for growth avoids unproductive consumption of carbon and energy, which takes place when monosubstrates are used, ensures an increase in the efficiency of

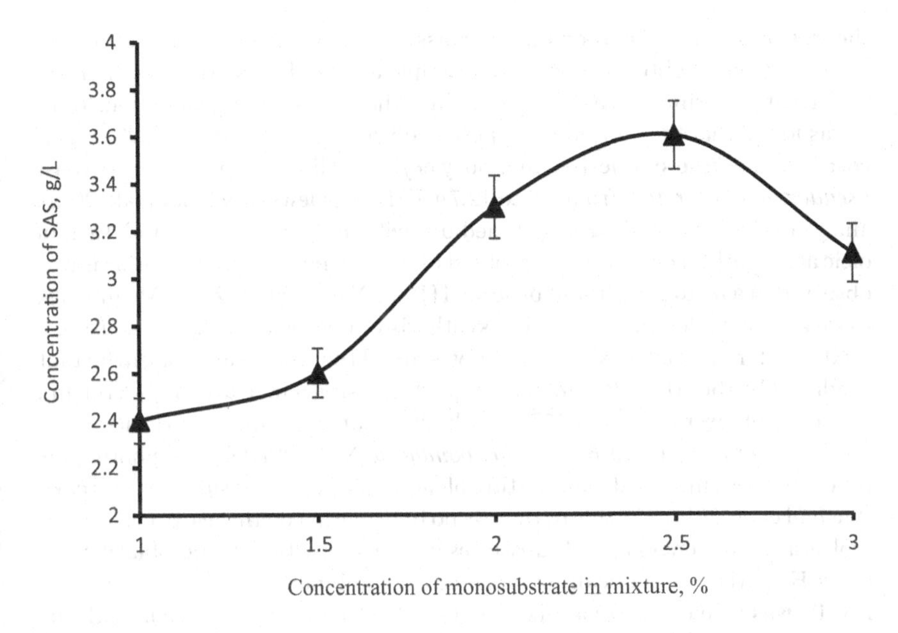

Figure 10.2 Effect of the concentration of monosubstrates (used sunflower oil or crude glycerol) in the mixture on the synthesis of surface-active substances (SAS) by *Nocardia vaccinii* IMV B-7405.

TABLE 10.4 SYNTHESIS OF SURFACE-ACTIVE SUBSTANCES (SAS) BY *NOCARDIA VACCINII* IMV B-7405 IN MEDIA WITH DIFFERENT CONCENTRATIONS OF USED SUNFLOWER OIL, CRUDE GLYCEROL AND $NANO_3$

Concentration of Carbon Sources, % (v/v)		SAS, g/L, at Concentration $NaNO_3$, g/L	
Used Sunflower Oil	**Crude Glycerol**	**0.5**	**0.75**
1.0	n/d[a]	2.1±0.09	n/d
0	3.0	2.3±0.13	n/d
2.0	2.0	3.3±0.17	3.6±0.18
1.0	2.5	3.4±0.17	3.1±0.16
1.0	3.0	5.1±0.25	5.4±0.27
0.75	3.25	4.4±0.22	4.5±0.23

[a] n/d—not determined.

the transformation of carbon into biomass, as well as intensify the synthesis of secondary metabolites. Recently, this simple approach has attracted increasing attention, primarily, as a way to reduce the cost of their production. Thus, it was found that the addition of rapeseed oil, 5%, to a medium with 2% of glucose increased the synthesis of mannosylerythritol lipids by phyllosphere yeast *Pseudozyma antarctica* from 2.2 to 12.7 g/L (Dziegielewska & Adamzak, 2013). The yeast *Pichia caribbica* in the medium with 100 g/L of xylose and 4 g/L of oleic acid synthesized 7.5 g/L of xylolipids; however, no surfactant synthesis was observed in a medium without oleic acid (Joshi-Navare et al., 2014). Meanwhile, there are only a few reports on the synthesis of microbial surfactants using a mixture of industrial wastes. Thus, it was found that the yield of biosurfactant produced by the strain *Pseudomonas* sp. PS-17 using a mixture of glycerol and cooked sunflower oil was by 15%–20% higher than when monosubstrates were used (Karpenko et al., 2016). *Candida bombicola* NRRL Y-17069, a sophorolipids producer, was cultivated on a mixture of motor oil waste and sunflower oil cake (Rashad et al., 2014). Currently, there is no information on the use of crude glycerol as a component of mixed substrates for the production of microbial surfactants. However, metabolically engineered a recombinant strain *Escherichia coli* KNSP1 was able to utilize a mixture of crude glycerol and fatty acid with the production of succinate and polyhydroxyalkanoates (Kang et al., 2011).

The application of used sunflower oil and crude glycerin for the production of microbial surfactants will simultaneously solve several important problems: reduce the cost of the target product, dispose of toxic industrial waste and increase the profitability of biodiesel production.

10.5 PROPERTIES OF SURFACE-ACTIVE SUBSTANCES SYNTHESIZED ON INDUSTRIAL WASTE

10.5.1 Antimicrobial and Antiadhesion Activity of Surface-Active Substances Synthesized by *Acinetobacter calcoaceticus* IMV B-7241 on Waste from Biodiesel Production

Microbial surfactants are multifunctional preparations, since, in addition to surface-active and emulsifying properties, they have antimicrobial and antiadhesive effects including the ability to destroy biofilms (Tiso et al., 2017). One of the reasons for the growing interest in surfactants of microbial origin as antimicrobial agents is an increase in the emergence of antibiotic-resistant pathogenic bacteria. The presence of toxic components in the crude glycerol can affect the quality of synthesized target products, particularly, biosurfactants. The strain of bacteria *Acinetobacter calcoaceticus* IMV B-7241 synthesizes extracellular surface-active substances which are a complex of glyco- (trehalose mono- and dimycolates, trehalose mono- and diacelates) and aminolipids (Pirog et al., 2023). Comparison of biological properties such as antimicrobial and antiadhesive activity, as well as ability to destroy biofilms of surfactants synthesized by *A. calcoaceticus* IMV B-7241 in the medium with pure or crude glycerol was conducted.

Pure or crude glycerol (Komsomolsk Biofuel Plant, Poltava, Ukraine) was used as a carbon source. To achieve equimolar quantities of pure or crude glycerol by carbon in media, their volume fractions were 3% and 5%, respectively. Surfactants were extracted from the culture liquid with a Folch mixture (chloroform and methanol, 2:1) (Pirog et al., 2021). The antimicrobial activity of surfactants was determined by minimum inhibitory concentration (MIC). The determination of MIC was carried out by the method of two-fold serial dilutions in meat-peptone broth (MPB) for bacteria and liquid wort for yeast and fungi (Chebbi et al., 2017). *Escherichia coli* IEM1, *Bacillus subtilis* BT2 (vegetative cells and spores), *Staphylococcus aureus* BMS1, *Enterobacter cloacae* C8, *Leuconostoc mesenteroides* PB7; yeast *Candida albicans* D6, *C. tropicalis* PE2, *C. utilis* EI8; micromycetes *Aspergillus niger* P3 and *Fusarium culmorum* T7 from the collection of microorganisms of the Department of Biotechnology and Microbiology of the National University of Food Technologies were used as test cultures for determining the biological properties of surfactants.

The antiadhesion activity was determined by the method as reported by Rufino et al. (2011) and Pirog et al. (2021). The antimicrobial activity of surfactants was estimated by MIC. Determination of the antiadhesive properties of surfactants was carried out as described in Janek et al. (2012). The number of adherent cells (degree of adhesion) was determined by the spectrophotometric method as the ratio of the optical density of the suspension obtained from materials treated with surfactant preparations (tile, steel, linoleum) to the optical density of control samples (without surfactant treatment) and expressed as a percentage. Study of biofilm disruption using SAS was done by the method by Gomes and Nitschke (2012).

The antimicrobial activities of surfactants synthesized by *A. calcoaceticus* on purified and crude glycerol estimated by minimum inhibitory concentration of SAS are shown in Table 10.5.

The antimicrobial activities of surfactant synthesized on crude glycerol against bacteria and yeast were two to eight times lower than that for SAS produced in pure glycerol. There was a much significant difference in MIC of these surfactants against fungi: the antimicrobial activity of surfactants synthesized on crude glycerol was more than 60 times lower than that of surfactants produced in the medium with pure glycerol. However, values of MIC of surfactants synthesized by *A. calcoaceticus* in crude glycerol, 0.96–15.2 μg/mL, are quite low and demonstrate a high SAS antimicrobial activity (Table 10.5). These

TABLE 10.5 THE ANTIMICROBIAL ACTIVITY OF SURFACE-ACTIVE SUBSTANCES (SAS) SYNTHESIZED BY *ACINETOBACTER CALCOACETICUS* IMV B-7241 IN THE MEDIUM WITH PURE OR CRUDE GLYCEROL

	MIC[a], μg/L of SAS Produced in the Medium with Glycerol	
Tested Microorganisms	**Pure**	**Crude**
E. coli IEM1	0.96	3.80
B. subtilis BT2 (vegetative cells)	0.24	0.96
B. subtilis BT2 (spores)	3.80	15.2
S. aureus BMS1	3.80	7.60
E. cloacae C8	0.48	3.80
C. albicans D6	1.90	15.20
C. tropicalis PE2	3.80	7.60
C. utilis EI8	1.90	7.60
A. niger P3	0.06	3.80
F. culmorum T7	0.03	1.90

[a] MIC—minimum inhibitory concentration.

values are comparable with the activity of aminolipids, the most active antimicrobial agents among surfactants of microbial origin. Thus, the MICs of surfactin and iturin against different strains of *Escherichia coli* were 15.6 and 300 μg/mL, respectively; MICs of polypeptin and octapeptin against *Proteus vulgaris* were 50–100 and 6.3 μg/mL, respectively; MICs of iturin, polypeptin and octapeptin against *S. aureus* were 400; 6.3 and 50 μg/mL, respectively (Cochrane & Vederas, 2016). The minimum inhibitory concentration of the aminolipids synthesized by *Streptomyces amritsarensis* sp. nov. against *B. subtilis* MTCC 619 did not exceed 10 μg/mL, and against *Staphylococcus epidermidis* MTCC 435 it was 15 μg/mL (Sharma et al., 2014).

Antiadhesion activity of surfactants was determined by the spectrophotometric method (Pirog et al., 2019), as the ratio of absorbance of cell suspension obtained from the abiotic materials treated with SAS to absorbance of suspension from control samples (without treatment with SAS) and expresses as a percentage. It was found that surfactants synthesized on crude glycerol had much lower antiadhesive activity compared to SAS produced on a pure substrate (Table 10.6).

Thus, the percent of adhered cells of *L. mesenteroides*, *E. cloaceae* and *C. albicans* to the surfaces of such abiotic materials as ceramic-tile, steel, and linoleum treated with solutions of SAS which were synthesized on crude glycerol was in

TABLE 10.6 THE ANTIADHESION ACTIVITY OF SURFACE-ACTIVE SUBSTANCES (SAS) SYNTHESIZED BY *ACINETOBACTER CALCOACETICUS* IMV B-7241 IN THE MEDIUM WITH PURE OR CRUDE GLYCEROL

	Adhesion of Microorganisms, %, after Treatment of Materials with SAS Solutions[a] Produced in the Medium with Glycerol					
	Ceramic-Tile		Steel		Linoleum	
Microorganisms	**Pure**	**Crude**	**Pure**	**Crude**	**Pure**	**Crude**
L. mesenteroides PB7	21	85	23	80	26	54
E. cloaceae C8	46	68	39	80	35	87
C. albicans D6	21	34	61	76	69	86

[a] Concentration of SAS in solutions was 5 µg/mL.

the range from 54% to 87%; meanwhile, after their treatment with SAS obtained using pure glycerol it varied from 21% to 46%. The values of antiadhesive activity remained practically unchanged when the concentration of surfactants was reduced to 2.5 µg/mL. Thus, the antiadhesive activity of surfactants synthesized by *A. calcoaceticus* on crude glycerol was not high enough. Perhaps this is due to application of surfactants at low concentrations 2.5–5 µg/mL. However, it is known that the maximum antiadhesive effect of microbial aminolipids was achieved at their average concentration of 2–50 µg/mL (Sriram et al., 2011).

Destruction of biofilms formed by *E. cloacae* and *L. mesenteroides* after their treatment with SAS synthesized by *A. calcoaceticus* in the medium with pure or crude glycerol is shown in Figure 10.3.

The maximum destructions of biofilm formed by *E. cloaceae,* due to the treatment with surfactants formed in pure or crude glycerol, 55% and 43%, respectively, were observed at SAS concentration of 124 µg/mL. The maximum destruction of biofilm formed by *L. mesenteroides* due to the treatment with surfactant formed in crude glycerol, 58%, was achieved at SAS concentration 124 µg/mL; meanwhile, a concentration of SAS, 1.9 µg/mL, synthesized on pure substrate destruction of biofilm was 80%.

SAS synthesized by *A. calcoaceticus* IMV B-7241 includes aminolipids, which are considered to possess the highest antimicrobial and antiadhesive activity among microbial surfactants (Cochrane and Vederas, 2016). Lower biological activity of SAS produced in the medium in which pure glycerol was replaced with crude glycerol may be caused by the decrease of the content of aminolipids in the composition of the synthesized surfactant complex. It was shown, for example, that aminolipids produced by *Bacillus amylofaciens* AR2 grown in the medium with dextrose, sucrose or glycerol, were presented by a complex

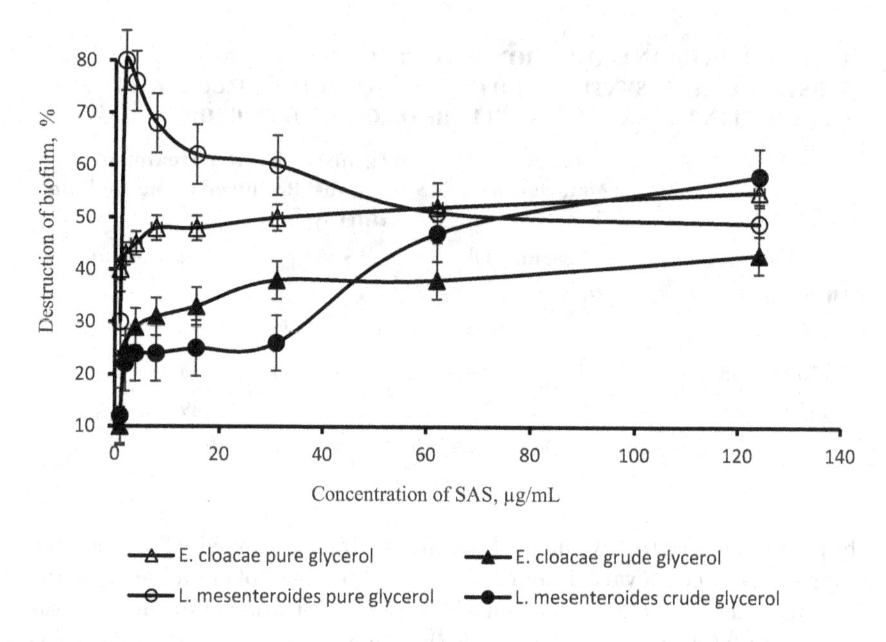

Figure 10.3 Destruction of biofilms formed by *Enterobacter cloacae* C8 and *Leuconostoc mesenteroides* PB7 after their treatment with surface-active substances (SAS) synthesized by *Acinetobacter calcoaceticus* IMV B-7241 in the medium with pure or crude glycerol.

of surfactin, iturin and fengycin; however, only iturin was produced in case of cultivation in the medium containing maltose, lactose and sorbitol, and SAS synthesized on sucrose and dextroge possessed the highest antifungal activity (Singh et al., 2014).

In turn, the inhibition of the synthesis of aminolipids in the medium with crude glycerol may be due to the presence of a high concentration of potassium or sodium cations (Garlapati et al., 2015), which could be potential inhibitors of nicotinamide adenine dinucleotide phosphate ($NADP^+$)-dependent glutamate dehydrogenase (NADP-GDH), which is a key enzyme in aminolipid biosynthesis in these bacteria. It was confirmed that the addition of 50 and 100 mM sodium or potassium cations in the cell-free extract of *A. calcoaceticus* IMV B-7241, decreased $NADP^+$-dependent glutamate dehydrogenase activity by 45% and 29%, respectively, compared to control without addition of cations (Figure 10.4).

Thus, surfactants synthesized by *A. calcoaceticus* IMV B-7241 on biodiesel production waste had antimicrobial activity and antiadhesive properties (MIC 0.96–15.2 μg/mL, destruction of 50% of the biofilm at a concentration of 124 μg/mL), which are comparable to those of known microbial surfactant aminolipids.

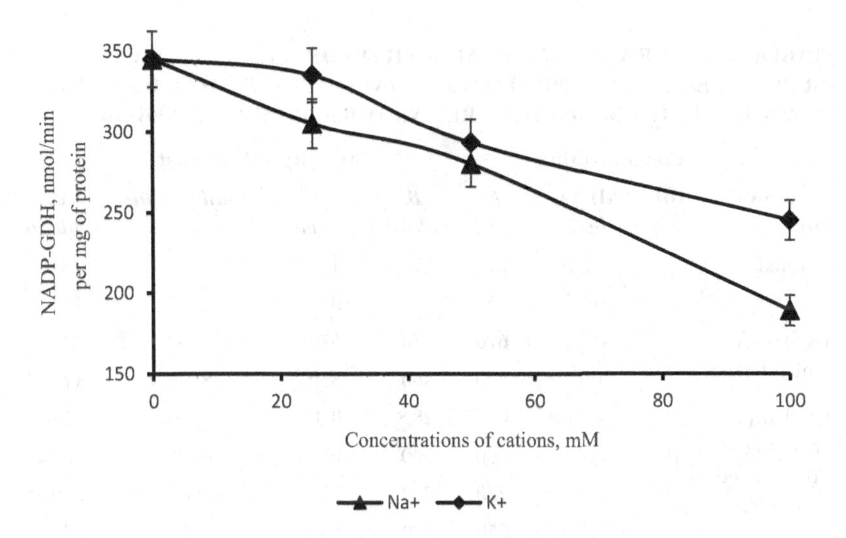

Figure 10.4 Activity of NADP-dependent glutamate dehydrogenase at different concentrations of sodium or potassium cations.

10.5.2 Antimicrobial and Antiadhesion Activity of Surface-Active Substances Synthesized by *Acinetobacter calcoaceticus* IMV B-7241 on Used Cooking Oil

It was shown that *Acinetobacter calcoaceticus* IMV B-7241 produces biosurfactants in the medium with used sunflower oil (see Section 10.3). However, the composition of the oil changes greatly during long-term processing at high temperatures. For example, due to the passage of iron cations from meat into cooked oil, its oxidation and thermal degradation are observed. Oil used for frying meat contains toxic heterocyclic amines, which may inhibit the activity of $NADP^+$-dependent glutamate dehydrogenase (Choe & Min, 2007). So, the antimicrobial and antiadhesive activities as well as the ability to destroy biofilms of SAS synthesized by *A. calcoaceticus* IMV B-7241 in the medium with refined and used sunflower oil were compared.

The antimicrobial activities of SAS synthesized by *A. calcoaceticus* IMV B-7241 during cultivation in the media with refined and used sunflower oil and different amounts of nitrogen are shown in Table 10.7.

An increase in the concentration of used sunflower oil to 4%–5% in the medium for *A. calcoaceticus* cultivation resulted in production of surfactants with extremely low antimicrobial activity: MICs for the bacterial test cultures were higher than 600–900, and for yeast higher than 1,200–1,700 μg/mL (Table 10.7). The content of the nitrogen source in the medium increased with increase in the concentration of the carbon source to maintain the carbon/nitrogen ratio at the

TABLE 10.7 THE ANTIMICROBIAL ACTIVITY OF SURFACE-ACTIVE SUBSTANCES (SAS) SYNTHESIZED BY *ACINETOBACTER CALCOACETICUS* IMV B-7241 IN THE MEDIA WITH REFINED OR USED SUNFLOWER OIL

	Concentration		MIC[a], µg/mL against				
Sunflower Oil	**Oil (%)**	**$(NH_2)_2CO$ (g/L)**	***E. coli***	***B. subtilis***	***S. aureus***	***Pseudomonas sp.***	***C. albicans***
Refined	2	0.35	0.8	51.2	1.6	0.8	25.6
	4	0.7	15	64	32	16	130
Used after meat frying	4	0.7	670	670	670	670	1340
	4	1.0	850	850	850	850	1700
Used after cooking French fried potatoes	2	0.35	0.9	28.8	0.45	0.45	57.6
	4	0.7	600	600	600	600	1,200
	4	1.0	750	750	750	750	n/d[b]
	5	0.7	850	850	850	850	n/d
	5	1.0	900	900	900	900	n/d

[a] MIC, minimum inhibitory concentration.
[b] n/d, not determined.

optimal level for surfactant synthesis. Antimicrobial activities of surfactants synthesized in the media with UCO were much lower than those with the same concentrations of refined oil, except for medium with 2% of oil after cooking French fried potatoes. This confirms that the used cooking oil contains inhibitors of the synthesis of surfactant components responsible for antimicrobial properties.

The number of research works studying the antimicrobial properties of microbial surfactants and their potential use in medicine is constantly increasing (Ceresa et al., 2021; Markande et al., 2021). However, there are only a few reports in the literature about the antimicrobial properties of surfactants synthesized from oil-containing industrial waste. Thus, rhamnolipid synthesized by *Thermus thermophilus* HB8 using sunflower seed oil inhibited the growth of *Micrococcus lysodeikticus* (Pantazaki et al., 2010). In the presence of Rufisan, 6–12 mg/L, synthesized by *Candida lipolytica* UCP0988 on soybean oil production waste, 6%, the growth of bacteria of the genus *Streptococcus* was inhibited by 35%–65%, and representatives of the genus *Staphylococcus* by only 15%–18%. However, Rufisan did not show an antimicrobial effect against *Candida albicans* (Rufino et al., 2011). Rhamnolipid synthesized by *Pseudomonas aeruginosa* LBI on soybean oil production waste inhibited the growth of *Bacillus cereus* and *Mucor miehei* at a concentration of 64 µg/mL, and growth of *Neurospora crassa*, *Staphylococcus aureus* and *Micrococcus luteus* at a concentration of 256 µg/mL (Nitschke et al., 2010). Meanwhile, after these publications, no new information about the antimicrobial activity of surfactants synthesized using UCO has appeared in the literature, and there are only few data on the biological properties of surfactants obtained from refined oils or a mixture of carbohydrates and refined oils. Thus, rhamnolipid synthesized by *Pseudomonas aeruginosa* DKB1 on olive oil had a high antimicrobial activity with MIC 8 μg/mL against *Bacillus subtilis* (Sanjivkumar et al., 2021). In the presence of 0.5–5 g/L of lipopeptides synthesized by *P. aeruginosa* MA-1 on olive oil, the inhibition zones of *Staphylococcus aureus* ATCC 43300 growth did not exceed 7–9.5 mm (Tazdaït et al., 2018). The minimum inhibitory concentrations of sophorolipids (synthesized by *Starmerella bombicola* on a medium with glucose and rapeseed oil) with rhamnolipid (provided by Jeneil Biotech Inc., Saukville, WI) against *Actinomyces naeslundii, Streptococcus oralis, S. mutans, S. sanguinis* and *Neisseria mucosa* ranged between 100 and 400 μg/mL (Elshikh et al., 2017).

Since surfactant synthesized by *A. calcoaceticus* in the medium with the concentration of used cooking oil 4%–5% had low antimicrobial activity, the antiadhesive properties of this SAS synthesized in a medium with 2% of oil after cooking French fried potatoes were compared to SAS produced in the medium with the same amount of refined oil (Table 10.8).

TABLE 10.8 THE ANTIADHESION ACTIVITY OF SURFACE-ACTIVE SUBSTANCES (SAS) SYNTHESIZED BY *ACINETOBACTER CALCOACETICUS* IMV B-7241 IN THE MEDIA WITH REFINED OR USED SUNFLOWER OIL

			Adhesion of microorganisms, %, after treatment of materials with SAS solutions			
Microorganism	**Sunflower Oil**	**Concentration of SAS, μg/mL**	**polystyrene**	**ceramic-tile**	**steel**	**linoleum**
B. subtilis BT2	Refined	1.25	27	47	48	41
		2.5	34	36	41	33
	Used	1.25	29	49	44	44
		2.5	32	34	48	34
S. aureus BMS1	Refined	1.25	46	81	56	36
		2.5	57	58	47	25
	Used	1.25	53	79	51	37
		2.5	56	54	45	25
C. albicans D6	Refined	1.25	42	49	47	65
		2.5	46	52	34	47
	Used	1.25	n/d	48	43	51
		2.5	51	51	40	44
A. niger P3	Refined	1.25	77	n/d	n/d	n/d
		2.5	60	n/d	n/d	n/d
	Used	1.25	51	n/d	n/d	n/d
		2.5	54	n/d	n/d	n/d

According to the results, the adhesion of the test microbial cultures to abiotic materials treated with SAS solutions, synthesized both on 2% refined or used sunflower oil, was almost the same. Thus, the number of cells of *B. subtilis* BT-2, *S. aureus* BMS1 and *C. albicans* D6 attached to different surfaces was 27%–48%, 25%–58% and 40%–65%, respectively. It is important that a significant decrease in the adhesion of test cultures to surfactant-treated materials was achieved at very low concentrations of SAS 1.25–2.5 μg/mL. There is limited research on the antiadhesive properties of surfactants synthesized on oil-containing substrates. It was shown that Rufisan synthesized by *C. lipolytica* UCP0988 on soybean oil production waste at the concentration of 0.75 mg/L reduced the adhesion of cells of *Streptococcus* and *Lactobacillus* to polystyrene to 61%–91%; and at a concentration of 12 mg/L reduced the number of attached *E. coli* and *C. albicans* cells to 21%–51% (Rufino et al., 2011). Probiotic strain *Propionibacterium freudenreichii* PTCC 1674 grown in various substrates, including used sunflower oil, synthesized a surfactant that at a concentration of 10 mg/mL reduced the number of *E. coli* cells attached to plastic by 13%, and *S. aureus* by 37 (Hajfarajollah et al., 2014). Biosurfactants synthesized by *A. calcoaceticus* IMV B-7241 in refined and used sunflower oil reduced the adhesion of test microbial cultures by 35%–75% at a concentration of 1.25-2.5 μg/L, so they were more effective antiadhesive agents than SAS reported in the literature.

Destruction of bacterial biofilms by surfactants synthesized by *A. calcoaceticus* IMV B-7241 in the media with refined or used sunflower oil was studied (Table 10.9).

TABLE 10.9 THE ANTIADHESION ACTIVITY OF SURFACE-ACTIVE SUBSTANCES (SAS) SYNTHESIZED BY *ACINETOBACTER CALCOACETICUS* IMV B-7241 IN THE MEDIA WITH REFINED OR USED SUNFLOWER OIL

Microorganism	**Sunflower Oil**	**Destruction of Biofilm, %, after Treatment with Solutions of SAS at Concentration, μg/mL**			
		29±1.16	58±0.22	116±5.11	233±4.80
B. subtilis BT2	Refined	5±0.12	7±0.02	37±2.24	48±1.11
	Used	11±0.09	19±0.08	47±2.01	63±3.14
S. aureus BMS1	Refined	52±0.20	53±0.25	53±2.6	55±3.21
	Used	36±1.52	38±1.95	57±2.45	66±3.02
E. coli IEM1	Refined	50±2.34	54±2.33	67±3.22	75±3.7
	Used	83±3.25	42±1.78	33±1.43	25±1.00
E. cloacae C8	Refined	24±1.20	29 ±1.11	52 ±2.32	54±2.54
	Used	32±1.22	47±1.87	53±1.87	55±2.41

Destruction of bacterial biofilms in the presence of SAS depended on the quality of the substances used as the oil substrate (refined or used), concentration of surfactant and test bacterial culture. Thus, the destructions of the *B. subtilis* biofilm using all studied concentrations of surfactant (29, 58, 116 and 233 μg/mL) obtained from media with used sunflower oil were by 6%–15% higher compared to those using SAS synthesized in the media with the same concentrations of refined oil. The destructions of the *S. aureus* biofilm in the presence of surfactants formed on used sunflower oil were by 4%–11% higher compared to results using SAS synthesized on the same concentrations of refined oil but only at concentrations of SAS 116 and 233 μg/mL. Meanwhile, the maximum destruction of the *E. coli* biofilm, 83%, was achieved at a concentration of SAS synthesized on used oil 29 μg/mL. The destructions of the *E. cloacae* biofilm in the presence of SAS synthesized either on refined or used oils in concentrations of 116 or 233 μg/mL were almost the same, 52 and 55, respectively. However, at lower concentrations, 29 and 58 μg/mL, surfactant obtained from used oil destroyed the *E. cloacae* biofilm more effectively. Thus, the replacement in the medium used for the cultivation of *A. calcoaceticus* refined oil with the used oil resulted in the synthesis of SAS, which effectively destroyed bacterial biofilms at low concentrations. Although there is much research on the destruction of biofilms by microbial surfactants, the effect of surfactants obtained in oil-containing substrates has not been studied enough.

Biosurfactants can be used as dispersants for bioremediation of oil-polluted soil or for combating oil spills because they are stable, highly efficient and biodegradable so do not cause secondary contamination of environment (Nikolova & Gutierrez, 2021; Patowary et al., 2018). The use of biosurfactants can also enhance oil degradation by activating indigenous oil-oxidizing microorganisms (Das & Chandran, 2011). However, the high antimicrobial activity of microbial surfactants can become a significant obstacle to their use in environmental technologies. For example, in some cases, at high concentrations, surfactants can reduce the rate of the purification process, adversely affecting degrading microorganisms. Thus, Whang and coauthors (2008) found that surfactin at a concentration of 40 mg/L had a negative effect on soil remediation from diesel fuel, while at a concentration of 400 mg/L remediation was completely inhibited. The effect of the surfactant *A. calcoaceticus* IMV B-7241 with different antimicrobial activity on the decomposition of oil in water was studied (Table 10.10).

It was shown that destruction of petroleum, %, and the total number of microorganisms in the presence of SAS, synthesized in a medium with 4% used oil, were higher than those when SAS obtained in a medium with 2% UCO was used. This can be explained by the fact that an increase in the concentration of UCO from 2% to 4% in the medium resulted in the production of biosurfactant with lower antimicrobial activity (Table 10.7).

TABLE 10.10 PETROLEUM DESTRUCTION AND CONCENTRATION OF BACTERIA AT THE PRESENCE OF SURFACE-ACTIVE SUBSTANCES (SAS) SYNTHESIZED BY *ACINETOBACTER CALCOACETICUS* IMV B-7241 USING REFINED OR USED SUNFLOWER OIL

Oil	Concentration of Oil (%)	Petroleum Destruction, %, at its Initial Concentration, g/L		Microorganisms, CFU/mL, at Initial Concentration of Petroleum, g/L	
		3.0	6.0	3.0	6.0
Refined	2	71±6	70±5	$(1.9\pm0.09)\cdot10^6$	$(2.2\pm0.11)\cdot10^6$
Used	2	72±6	70±4	$(2.0\pm0.10)\cdot10^6$	$(2.2\pm0.11)\cdot10^6$
	4	88±7	80±7	$(3.1\pm0.15)\cdot10^6$	$(2.9\pm0.14)\cdot10^6$

10.5.3 Regulation of the Antimicrobial and Antiadhesion Activity of Surface-Active Substances (SAS) Synthesized by *Acinetobacter calcoaceticus* IMV B-7241 on Used Cooking Oil

It was shown (Section 10.5.1) that synthesis of surface-active substances by *Acinetobacter calcoaceticus* IMV B-7241 in the medium with crude glycerol could be inhibited by increasing of concentration of potassium or sodium cations to 50 and 100 mM in the medium. The addition of sodium or potassium cations in these quantities into the cell-free extract of *A. calcoaceticus* IMV B-7241 decreased $NADP^+$-dependent glutamate dehydrogenase activity. Meanwhile, this enzyme plays an important role in biosynthesis of aminolipids which are part of SAS synthesized by *A. calcoaceticus* IMV B-7241. So, the possibility to regulate the antimicrobial and antiadhesive activity of SAS synthesized by *A. calcoaceticus* IMV B-7241 changing the concentration of potassium or sodium cations was studied.

The basal medium with 2% of used sunflower oil after frying potatoes with NaCl, 1 g/L; basal without addition of NaCl (medium I); with addition of NaCl, 2 g/L (medium II), and with NaCl, 1g/L, and KCl, 1 g/L (medium III) were used for SAS production.

The lowest amount of SAS was in the basal medium and medium II, in which only sodium cations were present, and the highest amount of surfactant was observed in medium III that contained both sodium and potassium ions (Table 10.11).

The surface-active substances produced by *A. calcoaceticus* IMV B-7241 are a complex of aminolipids and glycolipids (Pirog et al., 2023). The activity of enzymes participated in aminolipids biosynthesis, $NADP^+$-dependent glutamate dehydrogenase, enzymes participated in glycolipids biosynthesis, phosphoenolpyruvate (PEP) synthase and phosphoenolpyruvate (PEP) carboxykinase) were determined in cell-free extract of *A. calcoaceticus* IMV B-7241 at different concentrations of monovalent cations in the reaction mixture.

TABLE 10.11 PRODUCTION OF SURFACE-ACTIVE SUBSTANCES (SAS) BY *ACINETOBACTER CALCOACETICUS* IMV B-7241 IN MEDIUM WITH DIFFERENT CONCENTRATIONS OF MONOVALENT CATIONS

Medium	NaCl (g/L)	KCl (g/L)	SAS (g/L)
Basal	1.0	0	3.6±0.18
I	0	0	6.1±0.30
II	2.0	0	2.4±0.13
III	1.0	1.0	7.7±0.35

The activity of PEP synthase (EC 2.7.9.2) was determined by the rate of pyruvate formation, which was analyzed by NADH oxidation at 340 nm in a conjugated reaction with lactate dehydrogenase; the activity of PEP carboxykinase (EC 4.1.1.49) was determined by the formation of PEP and pyruvate during NADH oxidation. When studying the effect of monovalent cations on $NADP^+$-dependent glutamate dehydrogenase activity, 5–100 mM K^+ or Na^+ was added to the cell-free extract as a solution of KCl and NaCl salts (Table. 10.12).

It was found that potassium and sodium cations at concentrations of 50 and 100 mM inhibited $NADP^+$-dependent glutamate dehydrogenase, but in the presence of lower concentrations (5 and 10 mM of Na^+, 20 and 30 mM of K^+), a two-fold increase in the activity of this enzyme was observed compared to that without cation addition. The introduction of cations in the cell-free extract at a concentration of 5–20 mM was accompanied by an increase in PEP-carboxykinase and PEP-synthase activity. Data on inhibitory of $NADP^+$-dependent glutamate dehydrogenase activity with potassium and sodium cations in relatively high concentrations over 50 mM are consistent with the literature data (Lee et al., 2002; Wakamatsu et al., 2013). Thus, in the presence of 1 M of potassium chloride, a 30% decrease in the activity of the purified enzyme of hyperthermophilic archaea *Thermococcus waiotapuensis* was observed (Lee et al., 2002), the activity of purified $NADP^+$-dependent glutamate dehydrogenase of *Pyrobaculum calidifontis* decreased by 50% in the presence of 100–200 mM of potassium chloride and 100–300 mM of sodium chloride (Wakamatsu et al., 2013).

The antimicrobial and antiadhesive activity of SAS synthesized by *A. calcoaceticus* IMV B-7241 in media with different content of monovalent cations was analyzed. These results showed that in the case of cultivation of *A. calcoaceticus* in media I, II and III, synthesized SAS had an antimicrobial activity which was significantly lower compared to the activity of surfactants obtained on the basal medium (Table 10.13).

The results showed that SAS synthesized by *A. calcoaceticus* IMV B-7241 in media I, II and III had lower antimicrobial activity against the studied microorganisms compared to the activity of SAS produced using the basal medium,

TABLE 10.12 EFFECT OF DIFFERENT CONCENTRATIONS OF MONOVALENT CATIONS ON ENZYMES ACTIVITY PARTICIPATED IN THE BIOSYNTHESIS OF AMINO- AND GLYCOLIPIDS

Cations	Concentration (Mm)	Enzyme Activity, nmoL/min per mg of Protein		
		NADP+-Dependent Glutamate Dehydrogenase[a]	PEP Carboxykinase[b]	PEP Synthase
Cation free		192±10	192±10	10962±548
Na^+	5	385±19	385±19	12308±615
	10	385±19	577±29	10000±500
	20	192±10	769±39	9808±490
	30	192±10	192±10	7500±375
	50	160±8	n/d[c]	n/d
	100	107±5	n/d	n/d
K^+	5	192±10	769±39	11538±576
	10	195±11	772±38	12500±625
	20	385±19	768±37	6154±308
	30	385±19	192±10	4200±210
	50	160±8	n/d	n/d
	100	137±7	n/d	n/d

[a] NADP+—nicotinamide adenine dinucleotide phosphate.
[b] PEP—phosphoenolpyruvate.
[c] n/d—not determined.

and MIC was in two to three orders of magnitude higher than those for the surfactants formed on the basal medium.

It was found that surfactants obtained in modified media I and II had lower antiadhesive activity than those synthesized in the basal medium (Table 10.14).

It is evident that the results demonstrated the possibility to regulate the antimicrobial and antiadhesive activity of surfactants by changing the content of cations in the medium for cultivation, which are inhibitors/activators of enzymes responsible for the synthesis of components of the SAS complex.

TABLE 10.13 ANTIMICROBIAL ACTIVITY OF SURFACE-ACTIVE SUBSTANCES (SAS) SYNTHESIZED BY *ACINETOBACTER CALCOACETICUS* IMV B-7241 IN THE MEDIA WITH DIFFERENT CONTENTS OF POTASSIUM AND SODIUM CATIONS

	Minimum Inhibitory Concentrations of SAS, μg/mL, Synthesized in the Medium			
Microorganism	**Basal**	**I**	**II**	**III**
Pseudomonas sp. MI2	14	384	18	60
S. aureus BMS1	0.88	24	576	480
E. coli IEM1	0.88	24	144	480
B. subtilis BT2	1.75	384	18	60
E. cloaceae C8	0.88	48	72	480
C. albicans D6	28	770	144	960
R. nigricans P1	56	384	1,150	960
A. niger P3	0.88	1.5	144	120
F. culmorum T7	3.50	48	1,150	240

TABLE 10.14 ANTIADHESION ACTIVITY OF SURFACE-ACTIVE SUBSTANCES (SAS) SYNTHESIZED BY *ACINETOBACTER CALCOACETICUS* IMV B-7241 IN THE MEDIA WITH DIFFERENT CONTENT OF POTASSIUM AND SODIUM CATIONS

		Adhesion[a] (%)			
Microorganism	**Medium**	**Glass**	**Ceramic-Tile**	**Steel**	**Polystyrene**
S. aureus BMS1	Basal	58	71	26	75
	I	58	92	50	90
	II	87	90	67	95
	III	62	70	36	78
B. subtilis BT2	Basal	68	38	26	70
	I	78	61	50	83
	II	75	62	61	85
	III	56	48	36	79
C. albicans D6	Basal	65	53	48	68
	I	78	63	68	81
	II	82	70	81	89
	III	58	59	57	72

(Continued)

TABLE 10.14 (*Continued*) ANTIADHESION ACTIVITY OF SURFACE-ACTIVE SUBSTANCES (SAS) SYNTHESIZED BY *ACINETOBACTER CALCOACETICUS* IMV B-7241 IN THE MEDIA WITH DIFFERENT CONTENT OF POTASSIUM AND SODIUM CATIONS

		Adhesion[a] (%)			
Microorganism	**Medium**	**Glass**	**Ceramic-Tile**	**Steel**	**Polystyrene**
A. niger P3	Basal	76	n/d[b]	n/d	60
	I	90	n/d	n/d	79
	II	86	n/d	n/d	75
	III	82	n/d	n/d	68

[a] The concentration of surfactant solutions was 0.12 μg/ml.
[b] n/d, not determined.

10.5.4 Antimicrobial Activity of Surface-Active Substances (SAS) Synthesized by *Acinetobacter calcoaceticus* IMV B-7241 in the Medium with Crude Glycerol in the Presence of Biological Inducers

In recent years, a lot of publications have appeared on the cocultivation of producers of antimicrobial compounds with competitive microorganisms or biological inducers. In response to the presence of such inducers, an increase in antimicrobial activity, and/or intensification of synthesis of the target products, and even synthesis of unusual metabolites are observed (Abdel-Wahab et al., 2019; Benitez et al., 2011; Stierle et al., 2017). Studies of the effect of the presence of competitive microorganisms on the synthesis of antimicrobial compounds can be divided into three groups: (a) both strains (the producer and the competitive microorganism) are introduced into the medium in a ratio of 1:1. This is so-called classical cocultivation of two microorganisms, the cultivation of artificial microbial associations (consortiums) (Matevosyan et al., 2019; Stierle et al., 2017); (b) living or inactivated cells of the inducer are introduced into the medium at a significantly lower concentration compared to the cells of the producer of the target metabolites (Rateb et al., 2013); and (c) supernatant (filtrate) after growing a competitive microorganism is used as an inductor (Rateb et al., 2013; Wang et al., 2013). Competitive microorganisms (or their supernatants) used in the second and third groups are called inducers or elicitors. It is considered that during cocultivation, natural conditions are imitated and ecological competition for the substrate is going among the basic culture, producer of second metabolites and other microorganisms present in this artificial ecosystem (Abdel-Wahab et al., 2019; Buijs et al., 2021). Many genes responsible for the synthesis of biologically active compounds remain "silent" in axenic culture of the producing strain, but in coculture new bioactive secondary metabolites could appear (Marmann et al., 2014). This relates to microorganisms isolated from similar habitats. For example, to increase the antimicrobial activity of antibiotics synthesized by soil streptomycetes, strain *B. subtilis*, which is typical inhabitant of soil, was chosen to play the role of competitive microorganism or inducer (Liang et al., 2019).

Antimicrobial activity of surfactants synthesized by *A. calcoaceticus* IMV B-7241 in the presence of inducers in a medium with pure or crude glycerol was studied. Living or inactivated cells of *B. subtilis* BT2, as well as the supernatant after cultivation of the strain, were used as inducers, which were added in an amount of 2.5%–10% (v/v) into the medium with pure or crude glycerol for cultivation of *A. calcoaceticus* IMV B-7241. Effect of addition of inducers on the synthesis of SAS by *A. calcoaceticus* IMV B-7241 on pure or crude glycerol was shown (Table 10.15).

It was shown that the presence of inducers in the medium in the form of inactivated cells of *B. subtilis* or the supernatant did not change the concentration of the synthesized SAS, which was 1.54 and 1.44 g/L on pure glycerol and 2.32 and 2.36 g/L on crude glycerol, respectively. The introduction of living cells of *B. subtilis* BT2 into the medium with pure glycerol resulted in a decrease in the amount of surfactant by 1.5 times to 0.96 g/L compared to control without inducer, 1.42 g/L, which may be due to the competition of the surfactant producer and inducer for the substrate. Meanwhile, the addition of living cells of *B. subtilis* BT2 in the medium with crude glycerol increased concentration of SAS by 1.4 times, 3.56 g/L. One of the reasons for this may be that toxic impurities in the composition of industrial waste inhibited the growth of the inducer, while methanol, ethanol, triglycerides and fatty acids served as additional sources of carbon for the surfactant producer.

TABLE 10.15 EFFECT OF ADDITION OF INDUCERS ON THE SYNTHESIS OF SURFACE-ACTIVE SUBSTANCES (SAS) BY *ACINETOBACTER CALCOACETICUS* IMV B-7241

Glycerol	Inducer	SAS (g/L)
Pure, 3% (v/v)	Control without inducer	1.42±0.07
	Living cells of *B. subtilis* BT2	0.96±0.04
	Inactivated cells of *B. subtilis* BT2	1.54±0.07
	Supernatant	1.44±0.07
Crude, 5% (v/v)	Control without inducer	2.52±0.12
	Living cells of *B. subtilis* BT2	3.56±0.17
	Inactivated cells of *B. subtilis* BT2	2.32±0.11
	Supernatant	2.36±0.11

Antimicrobial activity of SAS synthesized by *A. calcoaceticus* IMV B-7241 in the presence of inducers was determined against some bacterial cultures (Table 10.16).

The most effective among the used inducers, living cells of *B. subtilis* BT-2 were: their addition into the medium with either pure or crude glycerol resulted in the synthesis of SAS with the minimum inhibitory concentrations against studied bacterial strains in 3–23 times lower in comparison with SAS synthesized in the medium without inducing microorganisms. The least effective of the studied inducers was the supernatant: the antimicrobial activity of surfactants synthesized in its presence against the most bacterial test cultures was only two times lower than that of surfactants produced in control.

The effect of surfactants synthesized in the presence of inducers on yeast of the genus Candida is shown (Table. 10. 17).

The use of living cells of *B. subtilis* BT-2 as inducers resulted in the synthesis of surfactants with highest activity against yeasts of genus *Candida,* the inactivated cells of *B. subtilis* were less effective inducer, and the supernatant had no positive effect on SAS activity against yeasts *Candida* except its addition into the medium with crude glycerol, where SAS synthesized in the medium with inducer showed twice a higher activity against *C. tropicalis* PE2.

Results reported in literature on the efficiency of application of living or inactivated cells, or the corresponding supernatant as inducers differ greatly. Thus,

TABLE 10.16 THE ANTIBACTERIAL ACTIVITY OF SURFACE-ACTIVE SUBSTANCES (SAS) SYNTHESIZED BY *ACINETOBACTER CALCOACETICUS* IMV B-7241 IN THE MEDIUM WITH PURE OR CRUDE GLYCERIN IN THE PRESENCE OF INDUCERS

Glycerol	Inducer	Minimum Inhibitory Concentrations, µg/mL, Against			
		B. subtilis	***S. aureus***	***Proteus vulgaris***	***E. cloacae***
Pure	Control without inducer	2.8	2.8	5.6	5.6
	Living cells of *B. subtilis*	0.23	0.23	1.84	0.46
	Inactivated cells of *B. subtilis*	1.4	1.4	1.4	0.7
	Supernatant	1.4	2.8	2.8	1.4
Crude	Control without inducer	9.8	4.9	9.8	19.6
	Living cells of *B. subtilis*	0.85	0.85	1.7	0.85
	Inactivated cells of *B. subtilis*	2.2	2.2	2.2	4.4
	Supernatant	4.6	2.3	4.6	18.4

TABLE 10.17 THE ANTIFUNGAL ACTIVITY OF SURFACE-ACTIVE SUBSTANCES (SAS) SYNTHESIZED BY *ACINETOBACTER CALCOACETICUS* IMV B-7241 IN THE MEDIUM WITH PURE OR CRUDE GLYCERIN IN THE PRESENCE OF INDUCERS

Glycerol	Inducer	Minimum Inhibitory Concentrations, μg/mL, against	
		C. tropicalis BT-2	*C. albicans* D-6
Pure	Control without inducer	11.2	11.2
	Living cells of *B. subtilis*	1.87	3.75
	Inactivated cells of *B. subtilis*	2.8	5.6
	Supernatant	11.2	11.2
Crude	Control without inducer	19.7	19.7
	Living cells of *B. subtilis*	3.5	7.0
	Inactivated cells of *B. subtilis*	4.5	9.0
	Supernatant	9.2	18.4

it was shown that the spectrum of metabolites synthesized by actinobacteria in the presence of living inducer cells was wider than when inactivated with heat-treated cells were used (Liang et al., 2020). Meanwhile, the activity of phenazine synthesized by *Pseudomonas aeruginosa* was almost the same independent on living or inactivated cells of the *E. coli*, *B. subtilis* and *Saccharomyces cerevisiae* were used as inducers (Luti & Yonis, 2013). The use of supernatant of *Streptomyces bullii* C2 as an inducer did not have a positive effect on the synthesis of antimicrobial compounds by the fungus *Aspergillus fumigatus* MBC-F1–10, while in the presence of living cells of the inducer, the synthesis of nine new antimicrobial metabolites that were not synthesized by monoculture was observed (Rateb et al., 2013). The supernatant of fungal elicitor *Penicillium chrysogenum* AS 3.5163 showed higher inducing efficiency on biosynthesis of antibiotic natamycin by the *S. natalensis* HW-2 than inactivated fungal biomass (Wang et al., 2013). These data indicate various mechanisms for increasing the synthesis or activity of antimicrobial compounds synthesized under the presence of inducers.

10.6 CONCLUSIONS

1. Bioconversion of the waste from biodiesel production, crude glycerol, by bacterial strains *Acinetobacter calcoaceticus* IMV B-7241, *Rhodococcus erythropolis* IMV Ac-5017 and *Nocardia vaccinii* IMV B-7405 into biosurfactants was demonstrated.

2. Application of the mixture of large-scale wastes, crude glycerol after biodiesel production, and used cooking oil, as carbon sources, 3% (v/v) and 1% (v/v), respectively, allowed production of 5.4 g/L of surfactants by *N. vaccinii* IMV B-7405, which was higher than when monosubstrates were used separately.
3. It was shown that refined sunflower oil could be replaced with used cooking oil (UCO) for the synthesis of surfactants by *A. calcoaceticus* IMV B-7241. Production of surfactants was 5.0 and 8.5 g/L in the media with UCO after potato frying and after meat frying, with emulsification index, E24, 50 and 54, respectively.
4. Surfactants synthesized by *A. calcoaceticus* IMV B-7241 on crude glycerol, biodiesel production waste, had antimicrobial activity and antiadhesive properties that are comparable to those of known microbial surfactant aminolipids.
5. Surfactants synthesized by *A. calcoaceticus* IMV B-7241 in the medium with 2% of oil after cooking of French fried potatoes have the minimum inhibited concentration in the range from 0.9 to 58 μg/mL against *E. coli* IEM1, *B. subtilis* BT2, *S. aureus* BMS1, *Pseudomonas* sp. MI2 and *C. albicans* D6; surfactants possessed antiadhesion activity at low concentrations 1.25–2.5 μg/mL, and ensured destructions of the bacterial biofilms which were higher compared to those using SAS synthesized in the media with the same concentrations of refined oil.
6. The possibility to regulate biological activity of surface-active substances synthesized by *A. calcoaceticus* IMV B-7241 by the addition of the cells of inducing bacteria particularly *B. subtilis* BT-2 into the medium for cultivation was shown. SAS produced during cocultivation with inducing bacteria possessed significantly increased antimicrobial and antiadhesion activities, as well as increased ability to destroy bacterial and yeast biofilms.
7. Application of industrial wastes, such as crude glycerol from biodiesel production and used cooking oil for microbial surfactant synthesis, will solve several important problems, including reducing the cost of surfactants, utilization of toxic substances and decreasing the environmental impacts of economic activity.

REFERENCES

Abdel-Wahab, N.M., Scharf, S., Ozkaya, F.C., Kurtan, T., Mandi, A., Fouad, M.A., Kamel, M.S., Müller, W.E.G., Kalscheuer, R., Lin, W., Daletos, G., Ebrahim, W., Liu, Z. & Proksch, P. (2019). Induction of secondary metabolites from the marine-derived fungus *Aspergillus versicolor* through co-cultivation with *Bacillus subtilis*. *Planta Medica, 85*(6), 503–512. https://doi.org/10.1055/a-0835-2332

Awogbemi, O., Kallon, D.V., Aigbodion, V.S. & Panda, S. (2021). Advances in biotechnological applications of waste cooking oil. *Case Studies in Chemical and Environmental Engineering, 4*, 00158. https://doi.org/10.1016/j.cscee.2021.100158

Banat, I.M., Satpute, S.K., Cameotra, S.S., Patil, R. & Nyayanit, N.V. (2014). Cost effective technologies and renewable substrates for biosurfactants' production. *Frontiers in Microbiology, 5*, 697. https://doi.org/10.3389/fmicb.2014.00697

Baskaran, S.M., Zakaria, M.R., Mukhlis Ahmad Sabri, A.S., Mohamed, M.S., Wasoh, H., Toshinari, M., Hassan, M.A. & Banat, I.M. (2021). Valorization of biodiesel side stream waste glycerol for rhamnolipids production by *Pseudomonas aeruginosa* RS6. *Environmental Pollution, 276*, 116742. https://doi.org/10.1016/j.envpol.2021.116742

Benitez, L., Correa, A., Daroit, D. & Brandelli, A. (2011). Antimicrobial activity of *Bacillus amyloliquefaciens* LBM 5006 is enhanced in the presence of *Escherichia coli*. *Current Microbiology, 62*, 1017–1022. https://doi.org/10.1007/s00284-010-9814-z

Bhardwaj, G., Cameotra, S.S. & Chopra, H.K. (2013). Utilization of oleo-chemical industry by-products for biosurfactant production. *AMB Express, 3*, 68. https://doi.org/10.1186/2191-0855-3-68

Bjerk, T.R., Severino, P., Jain, S., Marques, C., Silva, A.M., Pashirova, T. & Souto, E.B. (2021). Biosurfactants: Properties and applications in drug delivery, biotechnology and ecotoxicology. *Bioengineering, 8*(8), 115. https://doi.org/10.3390/bioengineering8080115

Buijs, Y., Zhang, S.D., Jørgensen, K.M., Isbrandt, T., Larsen, T.O. & Gram, L. (2021). Enhancement of antibiotic production by co-cultivation of two antibiotic producing marine *Vibrionaceae* strains. *FEMS Microbiology Ecology, 97*(4), fiab041. https://doi.org/10.1093/femsec/fiab041

Cavalheiro, J.M.B.T., De Almeida, M., Grandfils, C. & Da Fonseca, M. (2009). Poly (3-hydroxybutyrate) production by *Cupriavidus necato* using waste glycerol. *Process Biochemistry, 44*(5), 509–515. https://doi.org/10.1016/j.procbio.2009.01.008

Ceresa, C., Fracchia, L., Fedeli, E., Porta, C. & Banat, I.M. (2021). Recent advances in biomedical, therapeutic and pharmaceutical applications of microbial surfactants. *Pharmaceutics, 13*(4), 466. https://doi.org/10.3390/pharmaceutics13040466

Chebbi, A., Elshikh, M., Haque, F., Dobbin, S., Marchant, R., Sayadi, S., Chamkha, M. & Banat, I.M. (2017). Rhamnolipids from *Pseudomonas aeruginosa* strain W10; as antibiofilm/antibiofouling products for metal protection. *Journal of Basic Microbiology, 57*, 364–375. https://doi.org/10.1002/jobm.201600658

Chilakamarry, C.R., Sakinah, A.M.M., Zularisam, A.W. & Pandey, A. (2021). Glycerol waste to value added products and its potential applications. *Systems Microbiology and Biomanufacturing, 1*, 378–396. https://doi.org/10.1007/s43393-021-00036-w

Choe, E. & Min, D.B. (2007). Chemistry of deep-fat frying oils. *Journal of Food Science, 72*(5), R77–86. https://doi.org/10.1111/j.1750-3841.2007.00352.x

Chong, H. & Li, Q. (2017). Microbial production of rhamnolipids: Opportunities, challenges and strategies. *Microbial Cell Factories, 16*(1), 137. https://doi.org/10.1186/s12934-017-0753-2

Cochrane, S.A. & Vederas, J.C. (2016). Lipopeptides from *Bacillus* and *Paenibacillus* spp.: A gold mine of antibiotic candidates. *Medicinal Research Reviews, 36*(1), 4–31. https://doi.org/10.1002/med.21321

Das, N. & Chandran, P. (2011). Microbial degradation of petroleum hydrocarbon contaminants: An overview. *Biotechnology Research International*, 941810. https://doi.org/10.4061/2011/941810

da Silva, C.A., dos Santos, R.N., Oliveira, G.G., de Souza Ferreira, T.P., de Souza, N.L.G.D., Soares, A.S., de Melo, J.F., Colares, C.J.G., de Souza, U.J.B., de Araújo-Filho, R.N., de Souza Aguiar, R.W., dos Santos, G.R., Gabev, E.E. & Campos, F.S. (2022). Biodiesel and bioplastic production from waste-cooking-oil transesterification: An environmentally friendly approach. *Energies, 15*(3), 1073. https://doi.org/10.3390/en15031073

de Faria, A.F., Teodoro-Martinez, D.S., de Oliveira Barbosa, G.N., Vaz, B.G., Silva, I.S., Garcia, J.S., Totola, M.R., Eberlin, M.N., Grossman, M., Alves, O.L. & Durrant, L.R. (2011). Production and structural characterization of surfactin (C14/Leu7) produced by *Bacillus subtilis* isolate LSFM-05 grown on raw glycerol from the biodiesel industry. *Process Biochemistry, 46*(10), 1951–1957. https://doi.org/10.1016/j.procbio.2011.07.001

de Feo, G., Di Domenico, A., Ferrara, C., Abate, S. & Sesti Osseo, L. (2020). Evolution of waste cooking oil collection in an area with long-standing waste management problems. *Sustainability, 12*(20), 8578. https://doi.org/10.3390/su12208578

de Sousa, J.R., da Costa Correia, J.A., de Almeida, J.G.L., Rodrigues, S., Pessoa, O.D.L., Melo, V.M.M. & Goncalves, L.R.B. (2011). Evaluation of a co-product of biodiesel production as carbon source in the production of biosurfactant by *Pseudomonas aeruginosa* MSIC02. *Process Biochemistry, 46*(9), 1831–1839. https://doi.org/10.1016/j.procbio.2011.06.016

Dziegielewska, E. & Adamzak, M. (2013). Free fatty acids and a high initial amount of biomass in the medium increase the synthesis of mannosylerythritol lipids by *Pseudozyma antarctica*. *Environmental Biotechnology, 9*(1), 14–18.

Ebadipour, N., Lotfabad, T.B., Yaghmaei, S., & RoostAazad, R. (2016). Optimization of low-cost biosurfactant production from agricultural residues through response surface methodology. *Preparative Biochemistry and Biotechnology, 46*(1), 30–38. https://doi.org/10.1080/10826068.2014.979204

Elshikh, M., Moya-Ramírez, I., Moens, H., Roelants, S., Soetaert, W., Marchant, R., & Banat, I.M. (2017). Rhamnolipids and lactonic sophorolipids: Natural antimicrobial surfactants for oral hygiene. *Journal of Applied Microbiology, 123*(5), 1111–1123. https://doi.org/10.1111/jam.13550

Garlapati, V.K., Shankar, U. & Budhiraja, A. (2015). Bioconversion technologies of crude glycerol to value added industrial products. *Biotechnology Reports, 9*, 9–14. https://doi.org/10.1016/j.btre.2015.11.002

Gayathiri, E., Prakash, P., Karmegam, N., Varjani, S., Awasthi, M.K. & Ravindran, B. (2022). Biosurfactants: Potential and eco-friendly material for sustainable agriculture and environmental safety — A review. *Agronomy, 12*(3), 662. https://doi.org/10.3390/agronomy12030662

Gomes, M.Z.V. & Nitschke, M. (2012). Evaluation of rhamnolipids surfactants as agents to reduce the adhesion of *Staphylococcus aureus* to polystyrene surfaces. *Letters in Applied Microbiology, 49*(1), 960–965. https://doi.org/10.1016/j.foodcont.2011.11.025

Govindammal, M. & Parthasarathi, R. (2013). Production and characterization of biosurfactant using renewable substrates by *Pseudomonas fluorescence* isolated from mangrove ecosystem. *Journal of Applied Chemistry, 2*(1), 55–62.

Hajfarajollah, H., Mokhtarani, B. & Noghabi, K.A. (2014). Newly antibacterial and antiadhesive lipopeptide biosurfactant secreted by a probiotic strain, *Propionibacterium freudenreichii*. *Applied Biochemistry and Biotechnology, 74*(8), 2725–2740. https://doi.org/10.1007/s12010-014-1221-7

Hossain, A.B.M.S. & AlEissa, M.S. (2015). Biodiesel fuel production from palm, sunflower waste cooking oil and fish byproduct waste as renewable energy and environmental recycling process. *Biotechnology Journal International, 10*(4), 1–9. https://doi.org/10.9734/BBJ/2016/22338

Janek, T., Lukaszewicz, M., & Krasowska, A. (2012). Antiadhesive activity of the biosurfactant pseudofactin II secreted by the Arctic bacterium *Pseudomonas fluorescens* BD5. *BMC Microbiology, 12*, 24. https://doi.org/10.1186/1471-2180-12-24

Joshi-Navare, K., Singh, P.K. & Prabhune, A.A. (2014). New yeast isolate *Pichia caribbica* synthesizes xylolipid biosurfactant with enhanced functionality. *European Journal of Lipid Science and Technology, 116*(8), 1070–1079. https://doi.org/10.1002/ejlt.201300363

Kang, Z., Du, L., Kang, J., Wang, Y., Wang, Q., Liang, Q. & Qi, Q. (2011). Production of succinate and polyhydroxyalkanoate from substrate mixture by metabolically engineered *Escherichia coli*. *Bioresource Technology, 102*(11), 6600–6004. https://doi.org/10.1016/j.biortech.2011.03.070

Karpenko, I., Midyana, G., Karpenko, O. & Novikov, V. (2016). Influence of food industry wastes as substrates on the yield of biosurfactants of the strain *Pseudomonas* sp. PS-17. *Ecological Engineering and Environment Protection, 1*, 44–51.

Lee, M.K., González, J.M. & Robb, F.T. (2002). Extremely thermostable glutamate dehydrogenase (GDH) from the freshwater archaeon *Thermococcus waiotapuensis*: Cloning and comparison with two marine hyperthermophilic GDHs. *Extremophiles, 6*(2), 151–159. https://doi.org/10.1007/s007920100238

Liang, L., Sproule, A., Haltli, B., Marchbank, D.H., Berrué, F., Overy, D.P., McQuillan, K., Lanteigne, M., Duncan, N., Correa, H. & Kerr, R.G. (2019). Discovery of a new natural product and a deactivation of a quorum sensing system by culturing a "producer" bacterium with a heat-killed "inducer" culture. *Frontiers in Microbiology, 9*, 3351. https://doi.org/10.3389/fmicb.2018.03351

Liang, L., Wang, G., Haltli, B., Marchbank, D.H., Stryhn, H., Correa, H. & Kerr, R.G. (2020). Metabolomic comparison and assessment of co-cultivation and a heat-killed inducer strategy in activation of cryptic biosynthetic pathways. *Journal of Natural Products, 83*(9), 2696–2705. https://doi.org/10.1021/acs.jnatprod.0c00621

Liu, X., Jensen, P.R. & Workman, M. (2012). Bioconversion of crude glycerol feedstocks into ethanol by *Pachysolen tannophilus*. *Bioresource Technology, 104*, 579–586. https://doi.org/10.1016/j.biortech.2011.10.065

Liu, Y., Koh, C.M. & Ji, L. (2011). Bioconversion of crude glycerol to glycolipids in *Ustilago maydis*. *Bioresource Technology, 102*(4), 3927–3933. https://doi.org/10.1016/j.biortech.2010.11.115

Luo, Z., Yuan, X., Zhong, H., Zeng, G., Liu, Z., Ma, X. & Zhu, Y. (2013). Optimizing rhamnolipid production by *Pseudomonas aeruginosa* ATCC 9027 grown on waste frying oil using response surface method and batch-fed fermentation. *Journal of Central South University of Technology, 20*, 1015–1021. https://doi.org/10.1007/s11771-013-1578-8

Luti, K.J.K. & Yonis, R.W. (2013). Elicitation of *Pseudomonas aeruginosa* with live and dead microbial cells enhances phenazine production. *Romanian Biotechnological Letters, 18*(6), 8769–8778.

Markande, A.R., Patel, D. & Varjani, S. (2021). A review on biosurfactants: Properties, applications and current developments. *Bioresource Technology, 330*, 124963. https://doi.org/10.1016/j.biortech.2021.124963

Marmann, A., Aly, A.H., Lin, W., Wang, B. & Proksch, P. (2014). Co-cultivation: A powerful emerging tool for enhancing the chemical diversity of microorganisms. *Marine Drugs, 12*(2), 1043–1065. https://doi.org/10.3390/md12021043

Matevosyan, L., Bazukya, I. & Trchounian, A. (2019). Antifungal and antibacterial effects of newly created lactic acid bacteria associations depending on cultivation media and duration of cultivation. *BMC Microbiology, 19*(1), 46–63. https://doi.org/10.1186/s12866-019-1475-x

Mitrea, L., Trif, M., Catoi, A.F. & Vodnar, D.C. (2017). Utilization of biodiesel derived-glycerol for 1,3-PD and citric acid production. *Microbial Cell Factories, 16*, 190. https://doi.org/10.1186/s12934-017-0807-5

Nikolova, C. & Gutierrez, T. (2021). Biosurfactants and their applications in the oil and gas industry: Current state of knowledge and future perspectives. *Frontiers in Bioengineering and Biotechnology, 9*, 626639. https://doi.org/10.3389/fbioe.2021.626639

Nitschke, M., Costa, S.G. & Contiero, J. (2010). Structure and applications of a rhamnolipid surfactant produced in soybean oil waste. *Applied Biochemistry and Biotechnology, 160*(7), 2066–2074. https://doi.org/10.1007/s12010-009-8707-8

Otzen, D.E. (2017). Biosurfactants and surfactants interacting with membranes and proteins: Same but different? *Biochimica et Biophysica Acta - Biomembranes, 1859*(4), 639–649. https://doi.org/10.1016/j.bbamem.2016.09.024

Panadare, D.C. & Rathod, V.K. (2015). Applications of waste cooking oil other than biodiesel: A review. *Iranian Journal of Chemical Engineering, 12*(3), 55–76. http://www.ijche.com/article_11253.html

Pantazak, A.A., Dimopoulou, M.I., Simou, O.M. & Pritsa, A.A. (2010). Sunflower seed oil and oleic acid utilization for the production of rhamnolipids by *Thermus thermophilus* HB8. *Applied Microbiology and Biotechnology, 88*(4), 939–951. https://doi.org/10.1007/s00253-010-2802-1

Parthipan, P., Preetham, E., Machuca, L.L., Rahman, P.K., Murugan, K. & Rajasekar, A. (2017). Biosurfactant and degradative enzymes mediated crude oil degradation by bacterium *Bacillus subtilis* A1. *Frontiers in Microbiology, 8*, 193. https://doi.org/10.3389/fmicb.2017.00193. 2017

Patowary, R., Patowary, K., Kalita, M.C. & Deka, S. (2018). Application of biosurfactant for enhancement of bioremediation process of crude oil contaminated soil. *International Biodeterioration & Biodegradation, 129*, 50–60. https://doi.org/10.1016/j.ibiod.2018.01.004

Paulino, B.N., Pessôa, M.G., Mano, M.C., Molina, G., Neri-Numa, I.A. & Pastore, G.M. (2016). Current status in biotechnological production and applications of glycolipid biosurfactants. *Applied Microbiology and Biotechnology, 100*(24), 10265–10293. https://doi.org/10.1007/s00253-016-7980-z

Pirog, T., Kluchka, L., Lytsai, D. & Stabnikov, V. (2021). Factors affecting antibiofilm properties of microbial surfactants. *Scientific Study & Research: Chemistry & Chemical Engineering, Biotechnology, Food Industry, 21*(1), 27–37.

Pirog, T., Kluchka, L., Skrotska, O. & Stabnikov, V. (2020). The effect of co-cultivation of *Rhodococcus erythropolis* with other bacterial strains on biological activity of synthesized surface-active substances. *Enzyme and Microbial Technology, 142*, 109677. https://doi.org/10.1016/j.enzmictec.2020.109677

Pirog, T., Shevchuk, T., Voloshina, I. & Grechirchak, N. (2005). Use of claydite-immobilized oil-oxidizing microbial cells for purification of water from oil. *Applied Biochemistry and Microbiology, 41*(1), 58–63. https://doi.org/10.1007/s10438-005-0010-z

Pirog, T., Stabnikov, V. & Stabnikova, O. (2023). Bacterial microbial surface-active substances in food-processing industry. In O. Paredes-López, O. Shevchenko, V. Stabnikov & V. Ivanov (Eds.), *Bioenhancement and fortification of foods for a healthy diet* (pp. 271–294). CRC Press, Boca Raton, London. https://doi.org/10.1201/9781003225287-6

Rashad, M.M., Al-Kashef, A.S., Nooman, M.U. & El-din-Mahmoud, A.E. (2014). Co-utilization of motor oil waste and sunflower oil cake on the production of new sophorolipids by *Candida bombicola* NRRL Y-17069. *Research Journal of Pharmaceutical, Biological and Chemical Science, 5*(4), 1515–1528.

Rateb, M.E., Hallyburton, I., Houssen, W.E., Bull, A.T., Goodfellow, M., Santhanam, R., Jaspars, M. & Ebel, R. (2013). Induction of diverse secondary metabolites in *Aspergillus fumigatus* by microbial co-culture. *RSC Advances, 3*(34), 14444. https://doi.org/10.1039/c3ra42378f

Rodrigues, C.V., Nespeca, M.G., Sakamoto, I.K., Oliveira, J.E.D., Varesche, M.B.A. & Maintinguer, S.I. (2018). Bioconversion of crude glycerol from waste cooking oils into hydrogen by sub-tropical mixed and pure cultures. *International Journal of Hydrogen Energy, 44*(1), 144–154. https://doi.org/10.1016/j.ijhydene.2018.02.174

Rufino, R.D., Luna, J.M., Sarubbo, L.A., Rodrigues, L.R., Teixeira, J.A. & Campos-Takaki, G.M. (2011). Antimicrobial and anti-adhesive potential of a biosurfactant Rufisan produced by *Candida lipolytica* UCP 0988. *Colloids and Surfaces B: Biointerfaces, 84*, 1–5. https://doi.org/10.1016/j.colsurfb.2010.10.045

Sałek, K., Euston, S.R. & Janek, T. (2022). Phase behaviour, functionality, and physicochemical characteristics of glycolipid surfactants of microbial origin. *Frontiers in Bioengineering and Biotechnology, 10*, 816613. https://doi.org/10.3389/fbioe.2022.816613

Sanjivkumar, M., Deivakumari, M. & Immanuel, G. (2021). Investigation on spectral and biomedical characterization of rhamnolipid from a marine associated bacterium *Pseudomonas aeruginosa* (DKB1). *Archives of Microbiology, 203*(5), 2297–2314. https://doi.org/10.1007/s00203-021-02220-x

Saravanan, V. & Subramaniyan, V. (2014). Production of biosurfactant by *Pseudomonas aeruginosa* PB3A using agro-industrial wastes as a carbon source. *Malaysian Journal of Microbiology, 10*(1), 57–62.

Sekhon Randhawa, K.K. & Rahman, P.K. (2014). Rhamnolipid biosurfactants-past, present, and future scenario of global market. *Frontiers in Microbiology, 5*, 454. https://doi.org/10.3389/fmicb.2014.00454

Sharma, D., Mandal, S.M. & Manhas, R.K. (2014). Purification and characterization of a novel lipopeptide from *Streptomyces amritsarensis* sp. nov. active against methicillin-resistant *Staphylococcus aureus*. *AMB Express, 4*, 50. https://doi.org/10.1186/s13568-014-0050-y

Singh, A.K., Rautela, R. & Cameotra, S.S. (2014). Substrate dependent in vitro antifungal activity of *Bacillus* sp. strain AR2. *Microbial Cell Factories, 13*, 67. https://doi.org 10.1186/1475-2859-13-67

Sriram, M.I., Kalishwaralal, K., Deepak, V., Gracerosepat, R., Srisakthi, K. & Gurunathan, S. (2011). Biofilm *Bacillus cereus* inhibition and antimicrobial action of lipopeptide biosurfactant produced by heavy metal tolerant strain NK1. *Colloids and Surfaces B: Biointerfaces, 85*(2), 174–181. https://doi.org/10.1016/j.colsurfb.2011.02.026

Stierle, A.A., Stierle, D.B., Decato, D., Priestley, N.D., Alverson, J.B., Hoody, J., McGrath, K. & Klepacki, D. (2017). The berkeleylactones, antibiotic macrolides from fungal coculture. *Journal of Natural Products, 80*, 1150–1160. https://doi.org/10.1021/acs.jnatprod.7b00133

Tazdaït, D., Salah, R. & Mouffok, S. (2018). Preliminary evaluation of a new low-cost substrate (amurca) in production of biosurfactant by *Pseudomonas aeruginosa* isolated from fuel-contaminated soil. *Journal of Materials and Environmental Science, 9*(3), 964–970. https://www.jmaterenvironsci.com/Document/vol9/vol9_N3/107-JMES-3338-Tazdait.pdf

Tiso, T., Zauter, R., Tulke, H., Leuchtle, B., Li, W.J., Behrens, B., Wittgens, A., Rosenau, F., Hayen, H. & Blank, L.M. (2017). Designer rhamnolipids by reduction of congener diversity: Production and characterization. *Microbial Cell Factories, 16*(1), 225. https://doi.org/10.1186/s12934-017-0838-y

Wakamatsu, T., Higashi, C., Ohmori, T., Doi, K. & Ohshima, T. (2013). Biochemical characterization of two glutamate dehydrogenases with different cofactor specificities from a hyperthermophilic archaeon *Pyrobaculum calidifontis. Extremophiles, 17*(3), 379–389. https://doi.org/10.1007/s00792-013-0527-7

Wang, D., Yuan, J., Gu, S. & Shi, Q. (2013). Influence of fungal elicitors on biosynthesis of natamycin by *Streptomyces natalensis* HW-2. *Applied Microbiology and Biotechnology, 97*(12), 5527–5534. https://doi.org/10.1007/s00253-013-4786-0

Wang, Q., Hengel, M.J. & Shibamoto, T. (2021). Investigation of toxic α-dicarbonyl compounds formed in the headspace of various heated cooking oils. *Food and Nutrition Research, 4*(3), 1–7.

Wang, X.L., Zhou, J.J., Sun, Y.Q. & Xiu, Z.L. (2019). Bioconversion of raw glycerol from waste cooking-oil-based biodiesel production to 1,3-propanediol and lactate by a microbial consortium. *Frontiers in Bioengineering and Biotechnology, 7*, article 14. https://www.frontiersin.org/article/10.3389/fbioe.2019.00014

Whang, L.M., Liu, P.W., Ma, C.C. & Cheng, S.S. (2008). Application of biosurfactants, rhamnolipid, and surfactin, for enhanced biodegradation of diesel-contaminated water and soil. *Journal of Hazardous Materials, 151*(1), 155–163. https://doi.org/10.1016/j.jhazmat.2007.05.063

Yadav, S., Rawat, G., Tripathi, P. & Saxena, R.K. (2014). A novel approach for biobutanol production by *Clostridium acetobutylicum* using glycerol: A low cost substrate. *Renewable Energy, 71*, 37–42. https://doi.org/10.1016/j.renene.2014.05.004

Chapter 11

Biosensors Practical Application

Lyudmyla Shkotova, Oleksandr Soldatkin, Valentyna Arkhypova, Viktoriya Peshkova, Olga Saiapina, Tetyana Sergeyeva, Alexei Soldatkin, and Sergei Dzyadevych
Institute of Molecular Biology and Genetics of the National Academy of Sciences of Ukraine

CONTENTS

DOI: 10.1201/9781003329671-11

11.1 INTRODUCTION

To date, the use of biosensors related to analytical biotechnologies is one of the world's key exponential technologies, the parameters of which are improving by tens and even hundreds of percent per year at the same cost level. And if, at the same time, they allow sharing with other technologies, such as nanotechnology, a programmed technological explosion occurs. For example, for 15 years since 2007, the sensor market has grown a thousand-fold, while the unit cost of such a system given its functionality has decreased also thousand times. It is the best demonstration of the progress in this industry. Interest in biosensors is due to their certain advantages over classical methods of analysis. Today, electrochemical biosensors are considered a successful alternative to traditional methods of analysis given the existing advances in the development of their laboratory prototypes and the entry into the market of a number of competitive measuring instruments. First, the use of biosensors eliminates the need for expensive and cumbersome equipment. Second, biosensor analysis is simpler and more user-friendly, so no special skills are required. Thus, the use of biosensor analysis, in general, is cheaper and more convenient. As screening tools, biosensors can help to select a certain number of suspicious samples, which will be further analyzed using classical methods, thereby reducing the cost and time of analysis.

To date, a number of reviews (Perumal & Hashim, 2014; Rotariu et al., 2016; Sin et al., 2014) and monographs (Cosnier, 2015; Dzyadevych & Soldatkin, 2008; Marks et al., 2007) have been published as well as numerous experimental works describing different types of biosensors have been done. According to the classical IUPAC definition, "a biosensor is an integrated device based on a receptor and a transducer, which is able to provide quantitative or semiquantitative analysis using a biological recognition element" (Thévenot et al., 1999). As shown in Figure 11.1, several events occur during the sample analysis by the biosensor: the biomaterial is in contact with the sample; the target molecule reacts with the biomaterial; the processes in the biomaterial change the

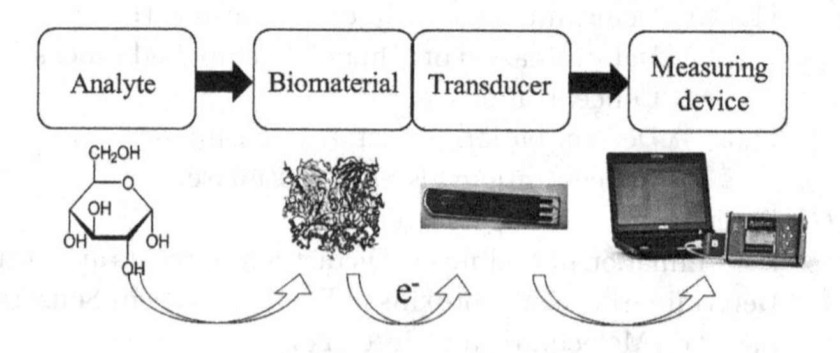

Figure 11.1 General scheme of biosensor operation using the glucose biosensor as an example.

physical and/or chemical properties of the surface of the transducer (for example, the pH changes); these changes are proportional to the concentration of the target analyte and are measured by a transducer, which transmits the information to the measuring device, and the latter usually transmits the information to the computer.

Biosensors are analytical devices, the sensor systems of which are of a biochemical nature and based on reactions involving biomolecular or supramolecular structures. Biosensors consist of three components: a bioreceptor is a detector layer of an immobilized biomaterial; a physicochemical transducer capable of transforming a biological response into a measurable signal; and an electronic system for signal amplification and recording.

11.2 STRUCTURES OF BIOSENSORS AND DEVICES FOR OPERATION WITH THEM

11.2.1 Potentiometric Biosensors

We used the transducers with a differential pair of *p*-channel transistors on a single crystal with a total area of 8 mm × 8 mm (Figure 11.2).

The crystal included two identical transistors separated by a protective 50 μm wide *n*+—region with contact to the substrate, *p*+—diffusion buses brought to the edge of the chip with contacts to the drain source, output to the built-in reference microelectrode, as well as two test MDS transistors with a metal gate, designed to check the electrical parameters of the produced crystals. The

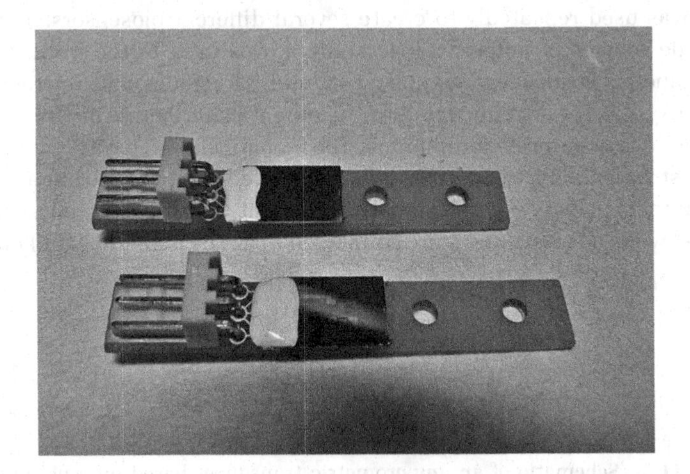

Figure 11.2 General view of p-channel pH-sensitive field-effect transistors manufactured at the V.Ye. Lashkarev Institute of Semiconductor Physics, NAS of Ukraine.

ion-selective properties of the transistor are due to the gate dielectric layer consisting of a thermally oxidized SiO_2 film 50 nm thick and a Si_3N_4 film 70 nm thick deposited in the reduced pressure reactor (Kukla et al., 2008). The measurements were performed by using a portable device operating according to the scheme of direct current measurement in the transistor channel with active load. The sensitivity of the device was about 25 μA/pH, which corresponded to a conditional pH sensitivity of about 80 mV/pH. The portable device and the sensor crystals were developed and manufactured at the V.Ye. Lashkarev Institute of Semiconductor Physics, National Academy of Sciences of Ukraine.

11.2.2 Amperometric Biosensors

As amperometric transducers, the sensors based on platinum disk electrodes, which were produced in our department, were used. Schematic representation and appearance of such transducers are presented in Figure 11.3.

When creating platinum disk electrodes, first a platinum wire, 0.4 mm in diameter and 3 mm long, was sealed in the end part of the glass capillary with an outer diameter of 3.5 mm. The open end of the wire extended above the working surface of the transducer. The platinum wire was then connected to a silver conductor inside a capillary using Wood's low-melting alloy. At the other end of the conductor, a copper contact pad was placed to connect the measuring device. The working surface of the electrodes was obtained by grinding with alumina powder (particles 0.1 μm and 0.05 μm); before immobilization of the bioselective element the electrodes were wiped with alcohol. Periodically, the electrode surface was renewed by similar grinding. Thus, the same transducer was used repeatedly to create several different biosensors. The three-electrode scheme of amperometric analysis was used in the work. Working amperometric electrodes, auxiliary platinum electrodes and reference electrodes (Ag/AgCl) were connected to the potentiostat PalmSens (Netherlands) or the analog of domestic production. The 8-channel device (CH-8 multiplexer, PalmInstruments BV, the Netherlands) connected to a potentiostat allowed us to simultaneously receive the signals from eight working electrodes, but usually, 2–3 working electrodes were connected to it. The potentiostat of domestic

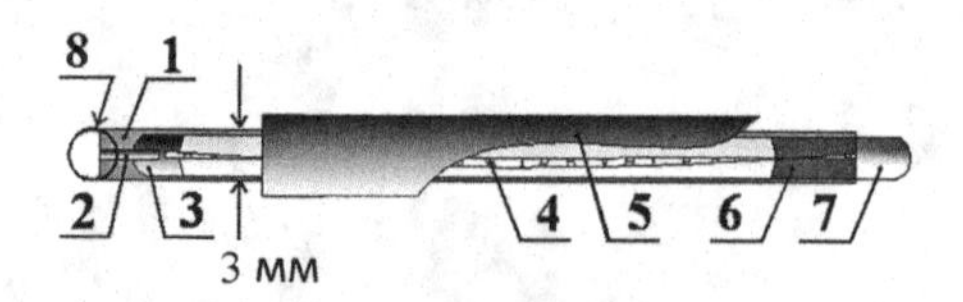

Figure 11.3 Schematic of an amperometric transducer based on a platinum disk electrode: 1, glass tube; 2, platinum wire; 3, wood alloy; 4, silver wire; 5, protective coating; 6, epoxy resin; 7, contact pad; 8, sensitive area of the transducer.

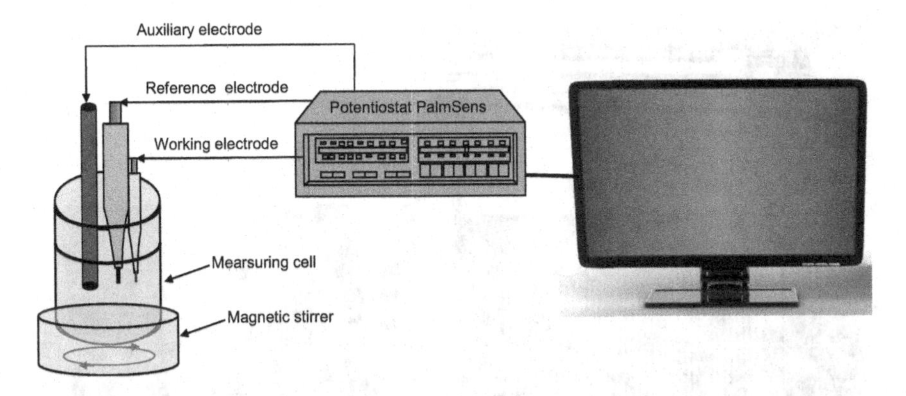

Figure 11.4 Scheme of the experimental setup for amperometric measurements.

production did not need a multiplexer, since it could work on four channels simultaneously. The distance between the auxiliary platinum electrode and all working biosensors during the measurement was the same, approximately 5 mm. The scheme of the measuring setup for amperometric measurements is shown in Figure 11.4.

11.2.3 Conductometric Biosensors

During the working stage, we used the transducers manufactured according to our recommendations at the V.Ye. Lashkarev Institute of Semiconductor Physics, NAS of Ukraine (Kyiv, Ukraine). They were 5 mm × 30 mm in size and consist of two identical pairs of thin-layer planar gold electrodes, produced by the technology of vacuum deposition of gold on a nonconductive ceramic substrate. To improve the adhesive properties of gold, a 50-nm-thick layer of chromium was applied on a ceramic substrate. Each system consisted of 20 pairs of interdigitated electrodes, a digit width and a gap between them of 20 μm, with a total sensitive surface of about 2 mm^2 (Dzyadevych, 2005) (Figure 11.5).

Figure 11.6 presents a diagram of the experimental setup based on a portable analyzer for carrying out measurements with biosensors. A schematic view of the biosensor is shown on the left. A working membrane was applied to one pair of electrodes (1) of this biosensor, a reference membrane (4)—to the second pair of electrodes (2). The portable conductometer "MCP-3" (5) was developed according to our recommendations and manufactured at the Institute of Electrodynamics, NAS of Ukraine. The conductometric circuit also included a holder for the biosensor (6) and a tripod (7). When performing the measurements, the working cell (8) with the test solution (9) was placed on the base of the tripod set, and the entire sensor unit is installed on a magnetic stirrer (10).

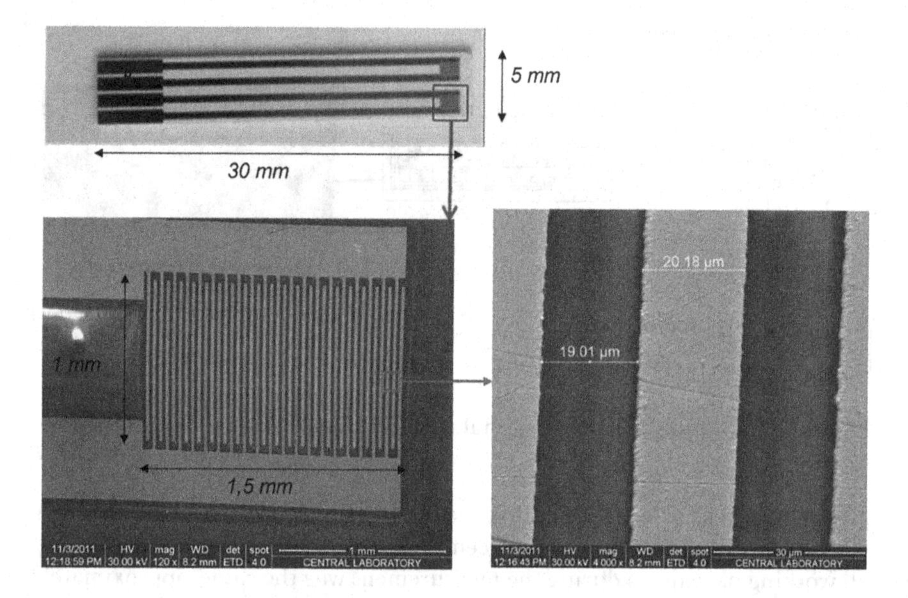

Figure 11.5 General view of conductometric transducers with a differential pair of gold interdigitated electrodes deposited on a sital substrate.

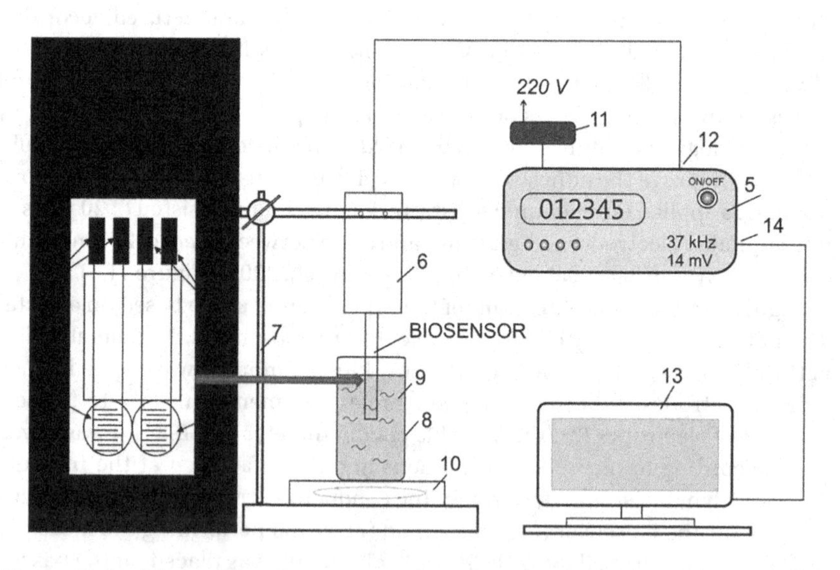

Figure 11.6 Scheme of conductometric system based on portable analyzer "MCP-3".

"MCP-3" is connected to the electrical supply network via the adapter (11), to the biosensor—by connecting wires via contact (12), and to the personal computer (13) with the installed software package—via contact (14). The measurements were performed at a current frequency of 37 kHz and an amplitude of 14 mV (Figure 11.6).

11.3 PRACTICAL APPLICATION OF BIOSENSORS

11.3.1 Determination of Glycoalkaloids in Samples of Potatoes and Tomatoes

Cultivated potatoes are one of the main agricultural crops consumed daily by millions of people. Potatoes are grown in almost 80% of all countries, and its world production reaches 350 million tons annually. The initial level of the total content of alkaloids in potatoes is genetically determined and differs significantly depending on the potato varieties and the area where they are grown. It is known that glycoalkaloids are involved in some mechanisms of potato resistance to diseases and insects. Commercial potato varieties consumed in Europe and the USA contain about 20–250 mg of glycoalkaloids per 1 kg of unpeeled tubers. Various factors—such as annual and regional variations, exposure to light (green potatoes) or mechanical damage during collection and storage—can cause a significant increase in the primary concentration of glycoalkaloids, for example, in green or infected potatoes the level of glycoalkaloids can increase up to 5,00 mg per 1 kg of unpeeled tubers (Baup, 1826; Friedman & McDonald, 1997).

Alfa-solanine and α-chaconine amount up to 95% of the total content of glycoalkaloids in potatoes, other types of glycoalkaloids are present only in trace concentrations. On the other hand, it does not make much sense to consider the toxicity of individual glycoalkaloids, since in real potatoes, glycoalkaloids exist in combination. The human body is exposed to the combined action of all glycoalkaloids; therefore, an assessment of the total concentration of glycoalkaloids in potato samples is necessary. The same is true for glycoalkaloids in tomatoes and in any edible food sample. The glycoalkaloid known as tomatine was first isolated by Fontaine et al. (Johnson and Hellenas, 1983). It is less toxic compared to solanine and chaconine. In young green tomatoes, its content sometimes reaches 500 mg per 1 kg of fresh weight. When ripe, the level of tomatine decreases and for mature red tomatoes it can be about 5 mg per 1 kg of fresh weight. The development of a measurement protocol for real potato samples requires precalibration of the biosensor, i.e., it is necessary to determine the type and ratio of the glycoalkaloids to be determined using a biosensor. Typically, this is the ratio between α-solanine and α-chaconine in potatoes,

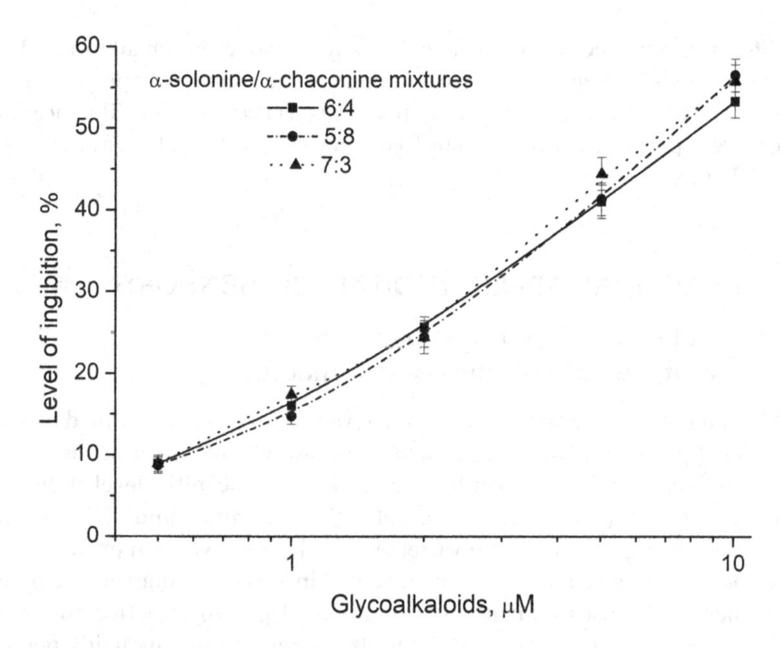

Figure 11.7 Calibration curves for the determination of different α-solanine/α-chaconine mixtures. Measurements were performed with 1 mM BuChE in 5 mM phosphate buffer, pH 7.2.

which is 6:4, although other options (5:5 and 7:3) are also possible. Figure 11.7 shows the dependence of the inhibition of immobilized BuChE on the concentration of α-solanine and α-chaconine and their ratio in the mixture.

Since the ratio may differ slightly, the above-mentioned ratio of 6:4 was used in subsequent experiments for the sensor calibration and further determination of the total concentration of glycoalkaloids in potato juice.

It is known from the literature that the genetically determined composition of glycoalkaloids may change at storage, transportation and mechanical damage of the harvested crop. Therefore, to improve the analytical methodology, it was important to investigate whether these factors affect the entire crop, or influence individual tubers in different ways. The level of BuChE inhibition by juice obtained from different tubers of the same potato variety was investigated. The results showed that the average content of glycoalkaloids in tubers of one variety is the same, but sometimes there is a deviation when the concentration of alkaloids increases by 1.5–2 times, which is most likely due to external factors (Figure 11.8). Therefore in the future, to minimize the difference in experimental responses, at least five different potato tubers of the same variety were used for homogenization.

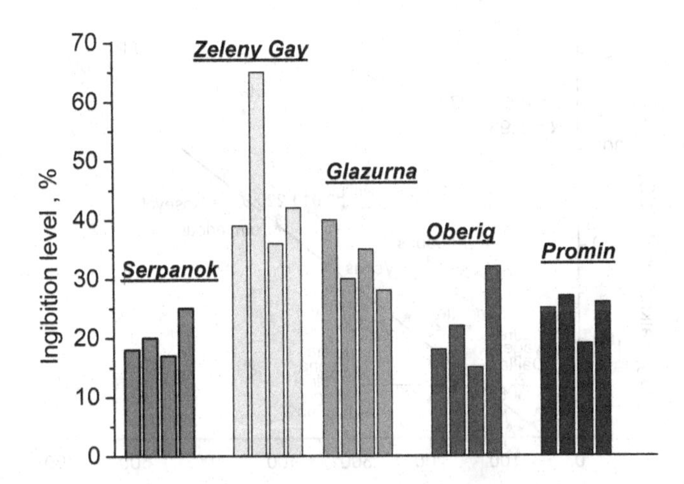

Figure 11.8 The level of BuChE inhibition by juice obtained from individual tubers of different potatoes varieties. Measurements were performed in 5 mM phosphate buffer, pH 7.5, substrate concentration 1 mM.

Potatoes of different varieties were grown at the experimental potato station Arvalis of the Plant Institute (Boigneville, France). These experimental cultivars (*Elkana, Monalisa, Pompadour, Starter, Pollux, Roseval, Phoebus, Venus, Charlotte, Redlaure, Voyager, Daifla, Albane, Lucie, Epona, Virginia, Hinga, Kaptah V, 91 F220 9, 209, 309, 204, 310, 303, 207, 201, 308, 306, 302*) and two commercial varieties (*Caesar* and *Agata*), in which the main part of glycoalkaloids are α-solanine and α-chaconine, were used for analysis. Control measurements of potato glycoalkaloids were performed by HPLC (Dzyadevych et al., 2004; Hellenäs & Branzell, 1997). Tubers were washed, dried, homogenized, and then the glycoalkaloids were extracted with acetic acid. The extracts were concentrated and purified using a Sep-PackPlus C_{18} cartridge. Separation and quantification of α-solanine and α-chaconine were performed on a Zorbax Extend C_{18} column (3.5 μm, 150 × 4.6 mm) at a wavelength of 202 μm. The comparison of the results of measurement of the total content of glycoalkaloids in the potato samples of 31 experimental cultivars is shown in Figure 11.9.

The results were obtained using the biosensor and the HPLC method with complex pretreatment of the samples. The figure shows a good correlation between the results. Certain disagreements can be explained, firstly, by the difference in the procedure of samples preparing, and secondly, by the fact that the analyses were carried out in different places and at different times. Additionally, the potatoes varieties bred by traditional selection at the Institute of Potato of the National Academy of Agrarian Sciences of Ukraine (NAASU) were studied, including the early (Povin', Serpanok, Dnipryanka, Zagadka,

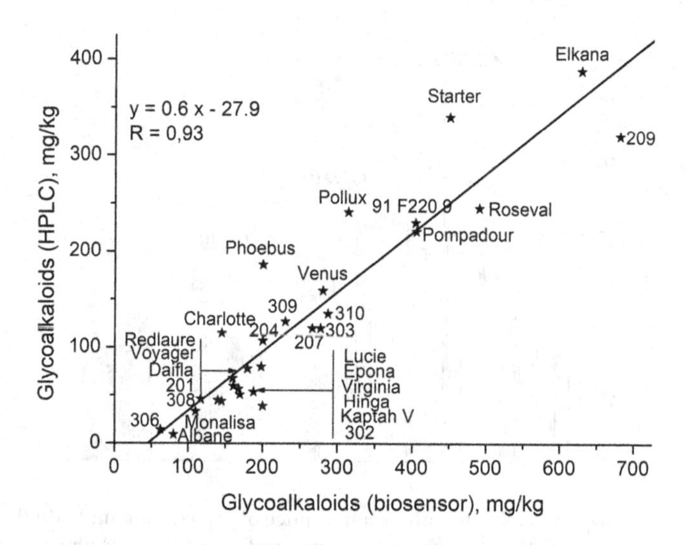

Figure 11.9 Comparative analysis of the glycoalkaloids content in different potatoes varieties grown at the experimental potato station ARVALIS of Institute of Plants (Boigneville, France). Results were obtained using a biosensor and HPLC method.

Podolyanka, Skarbnytsa, Glazurna, Svyatkova), mid-early (Levada, Svitanok Kievsky, Zelenyi Gai, Oberig, Zabava), mid-season (Dovira, Bylyna, Slovyanka, Yavir, Mandrivnytsya, Vernisazh) and mid-late (Promin', Chervona Ruta, Poliske Dzherelo) harvested in 2007 and 2008. The results of biosensor determination of the glycoalkaloids concentration in these samples are presented in Table. 11.1. Other properties of potatoes were also analyzed, namely, yield, starch and ripening time. As seen, the main factor is the genetically determined amount of glycoalkaloids in each sample.

Different commercial tomato varieties were used for tomatine analysis. The samples were washed by hand in running water and rubbed through a sieve to obtain juice. The juice samples were analyzed using a biosensor. Two biosensors methods of analysis such as by using a calibration curve and the standard addition method were used. The results of biosensor determination of the glycoalkaloids concentration in different commercial varieties of tomatoes are presented in Figure 11.10. The obtained total amount of glycoalkaloids in tomatoes ranged from 30 mg/kg to 70 mg/kg of fresh weight, which corresponds to the typical values. Biosensors demonstrated good operational stability during 20 days of work.

The figure clearly shows that the degree of inhibition of immobilized BuChE in the biosensor composition differs essentially for juice samples from various fruits and vegetables. In the case of potatoes, aubergine and tomatoes,

TABLE 11.1 YIELD, STARCH CONTENT, RIPENING TIME AND TOTAL CONCENTRATION OF GLYCOALKALOIDS DETERMINED BY THE BIOSENSOR FOR DIFFERENT VARIETIES OF POTATOES

	Variety	Glycoalkaloid Concentration (mg/kg Potato)		Ripening Time	Starch Content (%)	Yield (100 kg/ha)
		Yield 2007	Yield 2008			
1	Zaviya	-	120	II	-	-
2	Tyras	-	126	I	-	-
3	Zabava	230	150	II	15	430
4	Zvizdal'	-	150	III	-	-
5	Podolyanka	220	160	I	14.5	430
6	Vodograi	-	175	II	-	-
7	Oberig	230	180	II	14	480
8	Kalynivska	-	180	III	-	-
9	Dorogin'	-	200	IV	-	-
10	Polis'ka Yuvileina	-	206	IV	-	-
11	Dniprianka	200	215	I	14.5	430
12	Levada	280	230	II	17.5	440
13	Lugovska	-	230	III	-	-
14	Yavir	180	238	III	17.5	475
15	Zagadka	315	270	I	13.5	410
16	Serpanok	275	290	I	13.5	460
17	Slovyanka	265	300	III	13.5	500

(Continued)

TABLE 11.1 (*Continued*) YIELD, STARCH CONTENT, RIPENING TIME AND TOTAL CONCENTRATION OF GLYCOALKALOIDS DETERMINED BY THE BIOSENSOR FOR DIFFERENT VARIETIES OF POTATOES

	Variety	Glycoalkaloid Concentration (mg/kg Potato)		Ripening Time	Starch Content (%)	Yield (100 kg/ha)
		Yield 2007	Yield 2008			
18	Poliske Dzherelo	350	300	IV	16.5	450
19	Chervona Ruta	370	330	IV	19.5	455
20	Promin'	280	335	IV	15.5	465
21	Skarbnytsa	440	400	I	14	455
22	Glazurna	465	430	I	14.5	465
23	Dovira	370	450	III	17	440
24	Povin'	410	450	I	15.5	455
25	Shchedryk	-	450	I	-	-
26	Rakurs	-	480	IV	-	-
27	Zelenyi Gai	530	500	II	14	460
28	Svitanok Kievsky	510	650	II	19	420
29	Mangrivnytsa	715	670	III	17	440
30	Bylyna	1400	1,000	III	15.5	445
31	Svyatkova	270	-	I	14.5	465
32	Vernisazh	225	-	III	15	450

I, *Early;* **II**, *Middle Early;* **III**, *Medium-Mature;* **IV**, *Middle Late.*
All samples are cultivars.

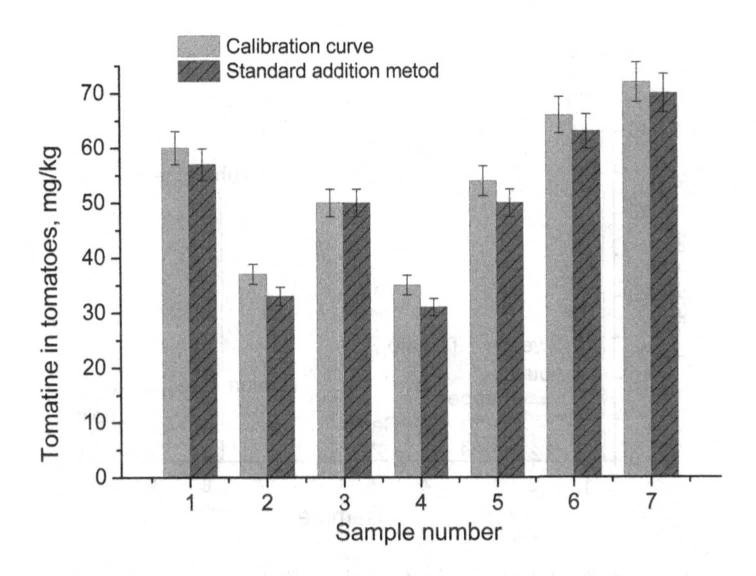

Figure 11.10 The results of glycoalkaloids analysis in different varieties of tomatoes by two methods. Measurements were performed with 1 mM BuCh in 5 mM phosphate buffer, pH 7.2.

inhibition is significantly larger. This is understandable because they belong to the same class of nightshade crops, which, unlike others, may contain glycoalkaloids. The analysis is simple, fast and does not require significant time.

11.3.2 Determination of Aflatoxins in Cereal Samples

Analysis of mycotoxins is a difficult task because these molecules are present in complex matrices at low concentrations; they can occur in various combinations and, at the same time, be produced by one or more species of fungi. To control the level of aflatoxins (AF), many countries have implemented rules governing their content. Acceptable limits of aflatoxins depend on the type of agricultural product. In the European Union, maximum levels are set at 8.0 ng$\times$g^{-1} for AFB1 and 15.0 ng$\times$g^{-1} for total aflatoxins in peanuts and other oilseeds to be sorted or otherwise physically processed prior to human consumption or use as food ingredients. These limits are much lower (2.0 ng$\times$g^{-1} for AFB1 and 4.0 ng$\times$g^{-1} for total aflatoxins) for peanuts and other oilseeds and processed products intended for direct human consumption. The same restrictions are given for all cereals and all products derived from cereals, including processed cereals. The maximum levels for corn and rice that will undergo grading or other physical processing prior to human consumption or using as food ingredients are 5.0 ng$\times$g^{-1} for AFB1 and 10.0 ng$\times$g^{-1} for the total amount of aflatoxins

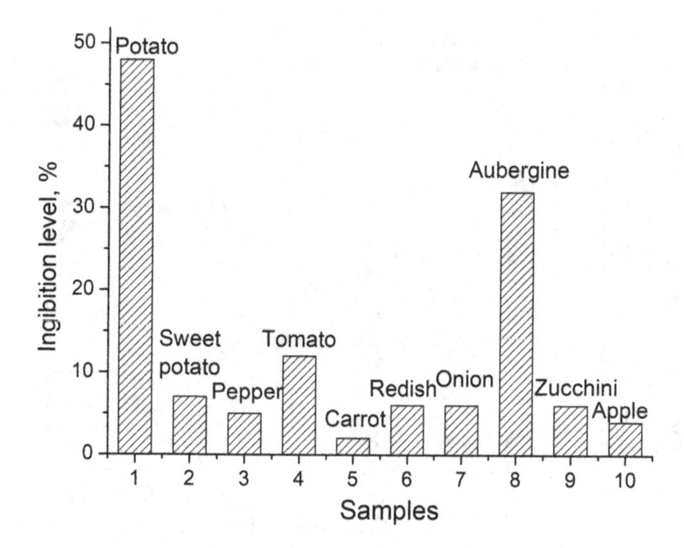

Figure 11.11 Results of the analysis of the glycoalkaloids content in various vegetables and fruits.

(European Commission, 2014). Wheat, oats, corn and peanuts specifically infected with *Aspergillus* were used as real samples to test the developed biosensor. The samples were specially prepared at the D.K. Zabolotny Institute of Microbiology and Virology, NAS of Ukraine. Aflatoxin B1 producer was grown on substrates (wheat, oats, corn and peanuts) for 21 days. Figure 11.12 shows photographs of the growth process of the fungus *Aspergillus* on the 15th day.

To analyze the composition of aflatoxins in contaminated cereal samples, measurement protocols were first developed. As a control, we used the extracts obtained by the same method from the same substrates, but not infected with the *Aspergillus*. The measurements were carried out as follows: first, several test responses to 4 mM substrate were received. Then, the calibration was performed by the volume of the control sample. Next, the biosensor was thoroughly washed in the working buffer until the response was restored, and the volume calibration for the infected sample was performed. Figure 11.13 presents the results of the experiment for the control sample and infected wheat.

The figure clearly shows that inhibition of AChE in the bioselective membrane by the control sample practically does not occur, whereas the infected extract inhibits the immobilized AChE rather strongly. According to the same algorithm, the measurements were performed for the samples of oats infected with the fungus *Aspergillus* and its control extract.

Figure 11.14 presents the kinetics of the effect of the control sample of oat extract and the sequential inhibition of the bioselective element by the extract from the infected sample.

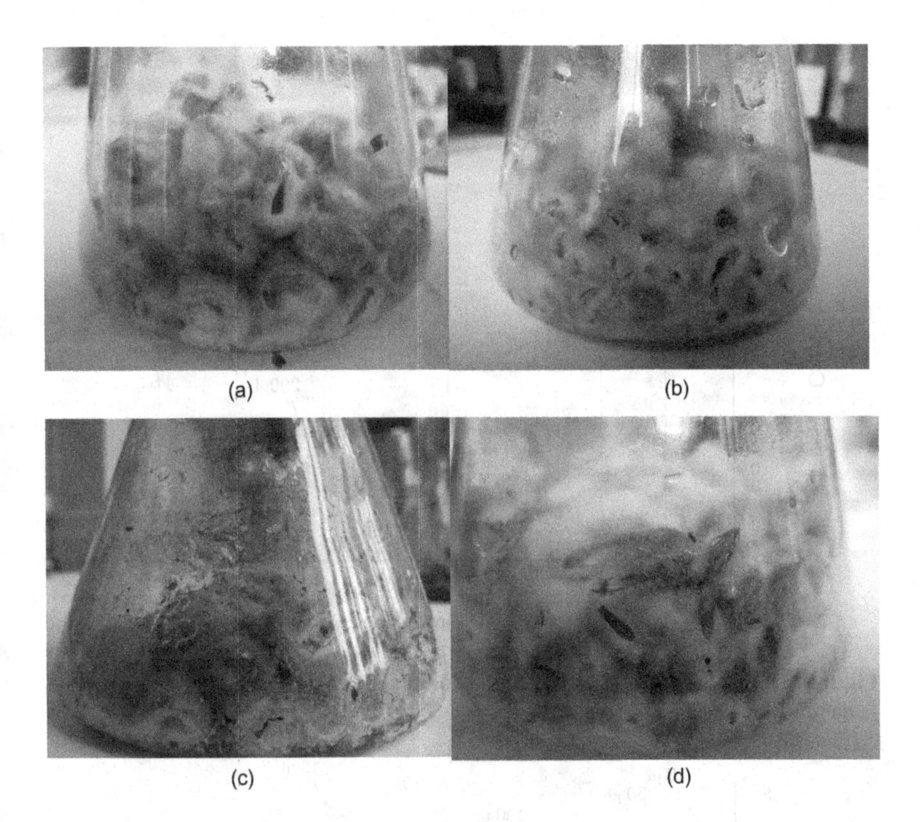

Figure 11.12 Photos of crops infected with the fungus *Aspergillus flavus* on the 15th day of cultivation (a, maize, b, wheat, c, peanuts, d, oats).

There is a slight influence of the control sample on the selective element, which can be associated with the "matrix" effect, whereas noticeable inhibition of the enzyme is observed for the studied sample. Thus, it is possible to draw a conclusion about rather good sensitivity of the developed biosensor to AFB1 in real samples.

The next series of experiments was performed with the samples obtained from corn. The measurements were performed on both infected and control samples. To be sure that the response reduced not due to the effect of the solvent, a series of responses to the same volumes of pure methanol was obtained after the responses to the control sample were measured (Figure 11.15).

The results confirmed that methanol (in the concentrations used in the experiment) does not affect the biosensor operation; there is a slight "matrix effect" for the control sample and a strong enough effect on the response of the infected sample. It was practically impossible to carry out measurements with samples obtained from infected nuts and a control lot due to the very high fat content of the nut extract. Therefore, the extracts turned out to be very oily,

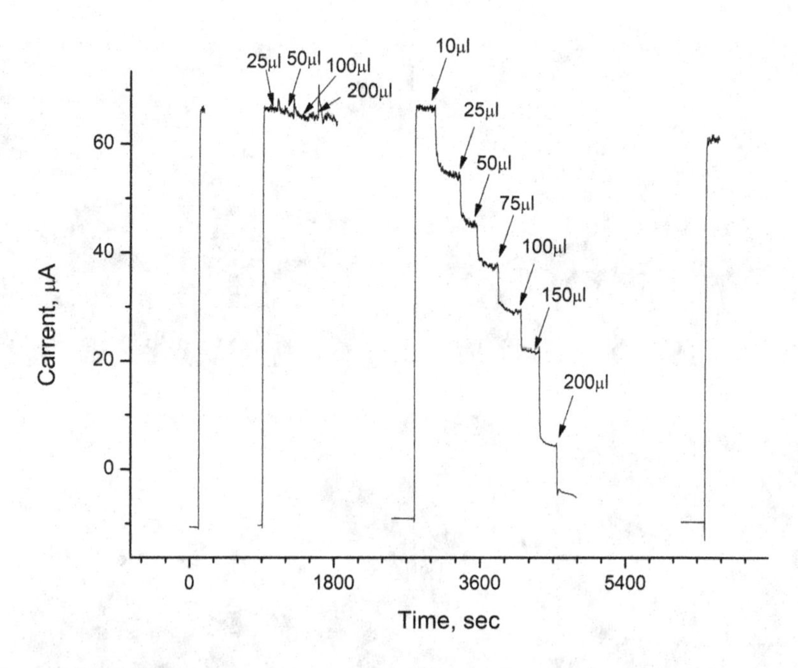

Figure 11.13 Biosensor responses to the addition of a control sample and an extract from infected wheat.

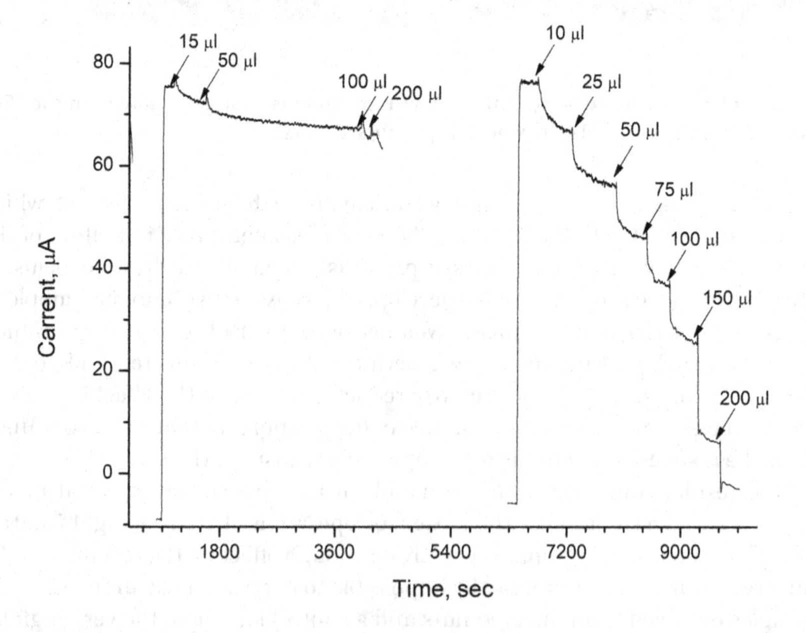

Figure 11.14 Biosensor responses to the addition of a control sample and an extract from infected oats.

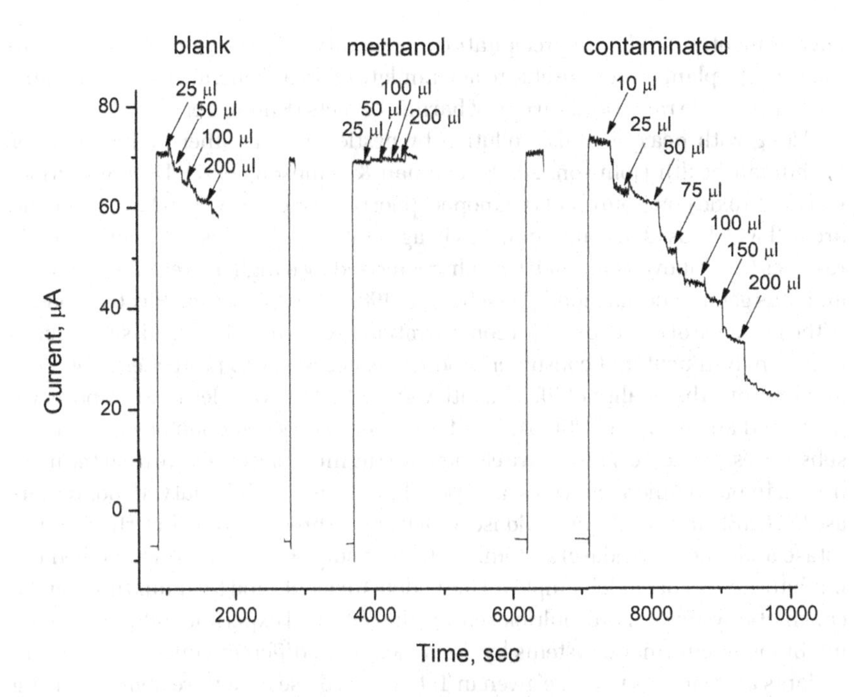

Figure 11.15 Biosensor responses to the addition of different volumes of a control sample, corresponding volume of methanol solvent and infected maize.

which led to the formation of films both on the measuring surface of the cell and on the surface of the selective element. The access of active substances to the membrane was sharply limited, and we did not even get reproducible responses to the AcChCl substrate. Therefore, the measurement of aflatoxins in nuts using biosensors is impossible, at least so far.

11.3.3 Determination of Heavy Metal Ions and Pesticides in Aqueous Samples of the Environment

Every year, environmental monitoring becomes more and more important all over the world (Yatsenko, 1996). This is due to the significant development of the chemical industry, the intensive use of chemicals in agriculture and the increasing utilization of a variety of chemical products in other branches of human activity. These chemicals, often toxic, in turn get into the air, land, water, thereby polluting large areas and the food of humans and farm animals. This, in turn, leads to deterioration of human health and the emergence of various diseases (Bundesgesetzblatt, 1986). It is well known that pesticides and heavy metals have a special place among the toxic substances polluting the environment. Heavy metals and their compounds are characterized by relatively high resistance to degradation in the environment, solubility in precipitation, the ability to sorption by soil and accumulation by plants. They are able to accumulate in organisms, are toxic to humans and have a wide range and variety of harmful effects (Kumara et al., 2017).

Along with heavy metals, pollution by pesticides is another high-risk factor for human health (Johnson, 2000; Schuman & Simpson, 1997). Decomposition-resistant toxic compounds of organophosphorus pesticides, which have been and are still widely used in some countries in agriculture, as well as their not less toxic residues (Arkhipova at al., 2001) are characterized by a high degree of penetration and thus get into human food (Verscheuren, 1983). Thus, constant effective control of the presence of these toxins (in concentrations exceeding the permissible values) in the environment and consumer products is necessary to protect environment and improve the quality of life (Dzantiev et al., 2004). Considering the above, we developed an enzyme multibiosensor for the inhibitory detection of various toxic substances. To create bioselective elements of the multibiosensor, we used the most useful, in our opinion, enzymes: acetylcholinesterase (AChE), butyrylcholinesterase (BuChE), urease, glucose oxidase (GOD) and a three-enzyme (invertase, mutarotase and glucose oxidase) system. In the first stage of this work, we studied the inhibitory effect of model samples of individual toxicants and their mixtures on the enzymatic systems of the multibiosensor. The obtained experimental values of the inhibition of enzymatic systems by the toxicants in different concentrations and variants of their mixtures are given in Table 11.2. These data were analyzed using the methods of mathematical statistics to upgrade the approaches of semiquantitative analysis of toxicants when working with real samples of the environment. The enzymes used in the experiment have a pronounced selectivity at least to some inhibitors.

Table 11.3 shows the conditional levels of the enzyme affinity to individual inhibitors. These levels are divided into three categories according to the following principle: "++"—the value of response of the bioselective element to the 1 mM inhibitor exceeds 80%, "+"—the response value is less than 80% but more than 20%, and "–"—the response value is less than 20%. As can be seen from the table, only two of the five enzymes are inhibited by pesticides, BuHE has a pronounced selectivity to pesticides and is much less inhibited by heavy metal ions. In general, this selectivity is also manifested at the action of mixtures of inhibitors: the enzymes selective to pesticides demonstrate higher responses to the mixtures with higher concentrations of pesticides.

To assess the possibilities of qualitative analysis of unknown mixtures using the considered set of enzymes, the experimental data were determined by the principal component analysis (PCA) (Figure 11.16).

In the obtained diagram, we observe a certain correlation between the location of the label corresponding to the mixture of inhibitors and the labels corresponding to the individual components of this mixture. Thus, a preliminary conclusion can be made about the possibility of qualitative analysis of the composition of inhibitor mixtures according to the principle of relative content of heavy metal ions or pesticides ("more–less") using this set of enzymes.

TABLE 11.2 INHIBITORY EFFECT OF TOXIC SUBSTANCES AND THEIR MIXTURES ON THE BIOSELECTIVE ELEMENTS OF THE MULTIBIOSENSOR (100% CORRESPONDS TO COMPLETE INHIBITION)

Inhibitor	Degree of Inhibition of Bioselective Elements (%)				
	Urease	BuChE	AChE	GOD	Three-Enzyme System
1 μM trichlorophone	0	15	0	0	0
10 μM trichlorophone	0	50	5	0	0
50 μM trichlorophone	0	70	25	0	0
1 mM trichlorophone	0	100	85	0	0
10 mM trichlorophone	0	100	100	0	0
1 μM carbofuran	0	25	5	0	0
10 μM carbofuran	0	70	25	0	0
100 μM carbofuran	0	100	50	0	0
2 mM carbofuran	0	100	100	0	10
1 μM Ag^+	0	0	5	15	11
10 μM Ag^+	0	3	25	60	65
50 μM Ag^+	10	7	70	100	99
0.2 μM Hg^{2+}	0	0	0	0	5
1 μM Hg^{2+}	4	0	0	10	22
10 μM Hg^{2+}	25	3	10	50	70
50 μM Hg^{2+}	65	7	70	90	100
10 μM Cu^{2+}	10	0	0	0	0
50 μM Cu^{2+}	30	0	0	0	5
200 μM Cu^{2+}	70	0	15	10	30
10 μM Cd^{2+}	12	0	0	0	5
50 μM Cd^{2+}	65	0	15	10	30
200 μM Cd^{2+}	100	0	40	40	70
Mixture 1	12	85	80	90	100
Mixture 2	20	100	80	70	75
Mixture 3	0	100	60	40	50
Mixture 4	50	60	30	45	80
Mixture 5	40	80	40	30	55

(*Continued*)

TABLE 11.2 (*Continued*) INHIBITORY EFFECT OF TOXIC SUBSTANCES AND THEIR MIXTURES ON THE BIOSELECTIVE ELEMENTS OF THE MULTIBIOSENSOR (100% CORRESPONDS TO COMPLETE INHIBITION)

Inhibitor	Degree of Inhibition of Bioselective Elements (%)				
	Urease	BuChE	AChE	GOD	Three-Enzyme System
Mixture 6	0	100	40	5	15
Mixture 7	40	60	7	15	30
Mixture 8	45	5	35	70	70
Mixture 9	30	100	100	100	100
Mixture 10	0	75	40	30	25
Mixture 11	0	100	40	15	10
Mixture 12	0	60	20	5	5
Mixture 13	0	60	10	0	0
Mixture 14	5	75	20	35	45
Mixture 15	5	50	15	35	50

Notes: Mixture 1, 6 μM Hg^{2+} + 8 μM Ag^{+} + 53 μM trichlorophone; Mixture 2, 9 μM Cd^{2+} + 12 μM Ag++ 50 μM carbofuran; Mixture 3, 5 μM Ag^{+} + 85 μM carbofuran + 13 μM trichlorophone; Mixture 4, 23 μM Cd^{2+} + 5 μM Hg^{2+} + 10 μM trichlorophone; Mixture 5, 25 μM Cu^{2+} + 5 μM Hg^{2+} + 150 μM trichlorophone; Mixture 6, 0.5 μM Hg^{2+} + 10 μM carbofuran + 10 μM trichlorophone; Mixture 7, 50 μM Cu^{2+} + 1 μM Hg^{2+} 5 μM trichlorophone; Mixture 8, 15 μM Cd^{2+} + 15 μM Cu^{2+} + 15 μM $^{Ag+}$; Mixture 9, 25 μM Hg^{2+} + 25 μM Ag^{+} + 50 μM trichlorophone + 50 μM carbofuran; Mixture 10, 2 μm Cd^{2+} + 2.5 μM Ag^{+} + 10 μM carbofuran; Mixture 11, 1 μM Ag^{+} + 17 μM carbofuran + 2.5 μM trichlorophone; Mixture 12, 0.25 μM Ag^{+} + 4 μM carbofuran + 0.75 μM trichlorophone; Mixture 13, 0.1 μM Hg^{2+} + 2 μM carbofuran + 2 μM trichlorophone; Mixture 14, 1.25 μM Hg^{2+} + 1.25 μM Ag^{+} + 2.5 μM trichlorophone + 2.5 μM carbofuran; Mixture 15, 1.25 μM Hg^{2+} + 1.6 μM Ag^{+} + 10 μM trichlorophone.

To verify the efficiency of the developed multibiosensor, natural water samples for the presence of toxic substances were analyzed. The samples were taken from a number of water reservoirs in the Kyiv city in the places of the most common contact of the population with water, namely on the city beaches of two districts, Obolon and Osokorki. The list of tested samples included water samples from lakes "Vyrlytsia," "Sunny," "Minister," "Opechen," "Opechen Nyzhnya," from the Dnieper River near the North and South bridges, and the Obolon Bay. We added a known amount of toxic substances to several samples to determine an increase in the levels of inhibition of the bioselective elements of the multibiosensor and its accordance with the actual values of the

TABLE 11.3 CONDITIONAL LEVELS OF THE ENZYME SYSTEMS AFFINITY TO TOXICANTS

Enzyme System	Trichlorfon	Carbofuran	Ag^+	Hg^{2+}	Cu^{2+}	Cd^{2+}
Urease	–	–	++	++	++	++
BuChE	++	++	+	+	–	–
AChE	++	++	++	++	+	++
GOD	–	–	++	++	+	++
Three-enzyme system	–	–	++	++	++	++

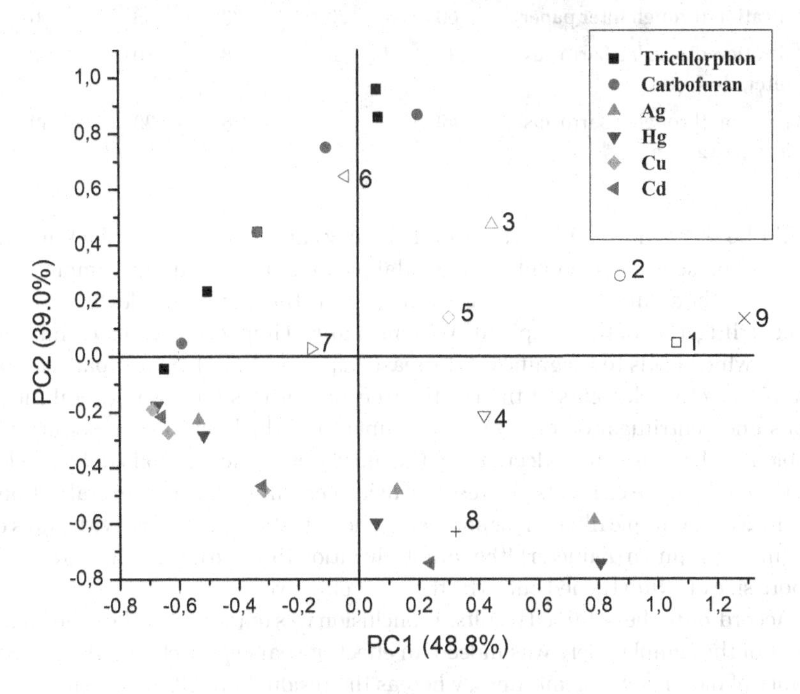

Figure 11.16 Graphical representation of experimental data in the plane of two main components. The numbers indicate the test mixtures of inhibitors shown in Table 11.1.

concentrations of toxic substances. Furthermore, we tested the multibiosensor in the analysis of multicomponent complex samples. For this purpose, the water samples were taken from the solid domestic waste landfill (SDW) No.5 in the village of Pidhirtsi, Obukhiv district, Kyiv region. It is during the inhibitory analysis of the sample from the solid domestic waste landfill (SDW) No. 5 that the main problems arose. After the multibiosensor incubation in a sample from the landfill, the sensitive elements of the multibiosensor completely lost the ability to respond to the introduction of appropriate substrates. The reason

TABLE 11.4 INHIBITION OF BIOSELECTIVE ELEMENTS OF THE MULTIBIOSENSOR SAMPLES FROM THE LANDFILL AFTER VARIOUS PRESAMPLE PREPARATION PROCEDURES

	Level of Inhibition of Bioselective Elements (%)				
Sample Preparation Procedure	**Urease**	**BuChE**	**AChE**	**GOD**	**Three-Enzyme System**
Without preparation	100	100	100	100	100
Centrifugation	75	45	50	70	75
Filtration through filter paper	60	20	30	55	60
Filtration through a Sartorius filter, d=45 μm	45	5	8	30	40
Filtration through a Sartorius filter, d=20 μm	40	3	8	20	30

could be either 100% inhibition of all the enzymes used or/and blocking the pores of bioselective elements by colloidal particles present in the sample.

To test the assumption, it was necessary to get rid of coarse particles by filtration or centrifugation of the sample to avoid mechanical impact (blocking membrane pores, which leads to a significant decrease in permeability) of large particles on the bioselective elements of the multibiosensor. For this purpose, several filtrations and centrifugation of the water sample from the landfill were performed. Table 11.4 demonstrates a decrease in the interference effect of colloidal particles on the bioselective elements as a result of using centrifugation and several options of filtration (a simple filter paper and two standard filters "Sartorius" with pores of 20 μm and 45 μm in diameter). The sample filtration through a Sartorius filter with a pore size of 20 μm turned out to be the most effective.

According to the obtained results, a conclusion was made that without pretreatment of the sample, there was indeed an effect of coarse particles on the permeability of bioselective membranes, whereas the residual inhibitory activity of the sample is a result of the presence of toxic substances in the sample. The obtained levels of inhibition of bioselective elements indicate that it is the heavy metal ions, which are present in the sample. It is demonstrated just by the inhibition of bioselective elements based on urease, GOD and three-enzyme system (Soldatkin et al., 2008). In parallel with the biosensor analysis, all samples used in the work were tested at the L.I. Medved Institute of Ecohygiene and Toxicology for the presence of toxic substances. The research was conducted using traditional methods for the toxicant determination (atomic absorption spectroscopy, thin-layer chromatography and atomic absorption analyzer of mercury). The results of the comparison of traditional and multibiosensory methods are presented in Table 11.5.

TABLE 11.5 COMPARISON OF DIFFERENT METHODS OF ANALYSIS OF REAL ENVIRONMENTAL SAMPLES

Place of the Samples Selection	Traditional Methods (mg/l)			Multibiosensor Method		
	$AAAHg^{2+}$	AAC	TLC			
Lake Vyrlytsia (Pozniaky, Kyiv)	-	-	-	-	-	-
Lake Vyrlytsia + 400 нM Hg^{2+}	0.079	-	-	+	-	-
Dnipro river (Osokorky, Kyiv)	-	-	-	-	-	-
Dnipro river + 10 µM trichlorophone	-	-	2.57	-	-	+
Lake Sonyachne (Osokorki, Kyiv)	-	-	-.	-	-	-
Lake Sonyachne + 5 мкM Cu^{+2}	-	0.321(Cu^{2+})	-	-	+	-
Lake Ministerske (Obolon, Kyiv)	-	-	-	-	-	-
Lake Opechen (Obolon, Kyiv)	-	-	-	-	-	-
Lake Opechen Nyzhnya (Obolon, Kyiv)	-	-	-	-	-	-
Lake Verbne (Obolon, Kyiv)	-	-	-	-	-	-
Dnieper, Obolon Bay, Kyiv	-	-	-	-	-	-
Dnipro, (Obolon, Kyiv)	-	-	-	-	-	-
Solid domestic waste landfill (SDW) No. 5 (Pidhirtsi village, Obukhov district, Kyiv region)	-	0.317(Cu^{2+}), 0.034(Co^{2+}), 1.471(Zn^{2+}), 0.988(Cr^{2+})	-	-	++	-

Notes: $AAAHg^{2+}$, atomic absorption analyzer of mercury; AAS, atomic absorption spectroscopy; TLC, thin layer chromatography; "+," exceeding maximum permissible concentrations (MPC); "++," exceeding the MPC by several orders of magnitude.

Rather a high correlation of the data obtained using traditional methods with the results obtained by the multibiosensor is observed. As expected, exceeding the permissible concentrations was found in the samples, to which appropriate aliquots of toxicants were intentionally added, namely, mercury excess (line 2), copper excess (line 6) and trichlorophone excess (line 4). Using the multibiosensor, it was also shown that the rest of samples from Kyiv reservoirs contained no toxicants in dangerous concentrations. Using traditional methods, we registered the exceeding of the maximum permissible concentrations for copper, cobalt, zinc and chromium, and the absence of mercury and pesticides (line 13), which is completely consistent with the data obtained by the multibiosensor. Thus, the results of the analysis of all considered water samples, obtained by traditional methods of toxicants determination, confirm the results obtained with a multibiosensor.

11.3.4 Quality Assessment of Wine and Wine Materials

Wine products as one of the elements of the food industry at all stages of production should undergo technical and chemical control aimed to determine the components of wine and must and their effect on the quality of final product (Saurina, 2010; Zeravik et al., 2009). Therefore, study on the conversion of various substances at must fermentation and wine formation is greatly important in wine processing.

Glycerol is the most important secondary product of alcoholic fermentation because it has a significant impact on the softness and viscosity of the wine, as well as on its taste characteristics, especially on the richness of taste, which gives wines oiliness and softness (Compagnone et al., 1998). Ethanol is the main product of alcoholic fermentation. It determines the toxic, addictive and caloric properties of wine and other alcoholic beverages. It was found that ethanol in moderate doses has antistress, cardioprotective and radioprotective effects. An analysis of glucose is essential considering that glucose, on the one hand, is a source of carbon for yeast, i.e., fermenting material, and, on the other hand, it is a substrate limiting yeast growth. Besides, the amount of fermented glucose determines the ethanol content in wine, and residual nonfermented glucose has a considerable effect contributing sweetness to the wine taste. The monitoring level of lactic acid, lactate, when producing high-quality beverages is also obligatory, since this parameter not only determines wine quality and flavor but is also an indicator of bacterial activity during fermentation (Avramescu et al., 2001). Therefore, reliable information about the lactate content in the must at different stages of wine production allows the control and regulation of the fermentation (Parra et al., 2006). Additionally, the stability of wines during storage also depends on the lactate concentration (Esti et al., 2004).

Traditionally, wine components are analyzed by chromatographic, enzymatic, spectrophotometric, refractometric, densitometric methods, capillary electrophoresis, etc. Biosensors providing express, highly sensitive, selective and cheap analysis can be a promising modern alternative. We developed enzyme amperometric biosensors for the determination of lactate, glucose, glycerol and ethanol; and designed their laboratory prototypes and specified the main analytical characteristics as well (Table 11.6) (Goriushkina et al., 2009; 2010; Shkotova et al., 2006; 2008; 2016).

An important task in the development of effective biosensors for use in winemaking is to test the developed laboratory prototypes under real conditions of analysis and to compare the results with the data obtained using traditional methods of wine analysis. Amperometric enzyme biosensors based on immobilized oxidases are a promising tool for monitoring key components of wine and must in the course of fermentation. However, it is a common practice that developed laboratory prototypes are quite suitable in

TABLE 11.6 ANALYTICAL CHARACTERISTICS OF DEVELOPED AMPEROMETRIC BIOSENSORS

Substrate (Enzyme)	Dynamic Range (mM)	Operational Mode	Time of Analysis (Minutes)	Stability	
				Operational (Hours)	Storage (Days)
Glucose (Glucose oxidase)	0.03–8	Steady-state	1–3	>20	90
Lactate (Lactate oxidase)	0.004–0.5	Steady-state	1–3	>20	25
Ethanol (Alcohol oxidase)	0.08–1	Steady-state	1–3	>120	33
Glycerol (Glycerol oxidase)	0.001–25	Steady-state	1–3	>30	40
Multisensor detection: Glucose, Lactate, Ethanol (Glucose Oxidase—Lactate Oxidase—Alcohol Oxidase)	Lactate 0.0005–0.64, Ethanol 0.1–6.4, Glucose 0.0005–1.6	Steady-state	1–5	>20	32

model solutions, whereas serious challenges appear once researchers attempt to utilize them for real samples (Dzyadevych & Soldatkin, 2008). Selectivity and stability of amperometric enzyme biosensors are key parameters when creating effective analytical systems intended for operation with real liquids. Additionally, an important task in designing efficient biosensors for wineries is testing developed laboratory prototypes under real conditions of analyses and comparison of the results obtained with the data of traditional methods of wine analysis.

11.3.4.1 Determination of Glycerol Concentration

To assess the possibility of using the developed biosensor for the analysis of real samples, the glycerol content in wine samples was determined by the method of standard additions. The samples of dry and semisweet table wines with an ethanol content of 10%–13% vol. (82–100 g/l) were used. Additionally, the glycerol content in these wines was determined by HPLC. The results of the analysis are presented in Table 11.7.

The results of biosensor analysis were approximately 1.5–3 times higher than the results of HPLC method. It is possible that, in addition to ethanol, GOD is sensitive to other components of wine. Anyway, the developed amperometric biosensor is able to detect glycerin even in such complex multicomponent mixtures as wine, although with low accuracy. We assume that optimization of the used glycerol oxidase preparation will improve the reliability of the results of analysis. So far, it is rather early to report a correlation between the results of glycerol determination using biosensor and HPLC analysis.

TABLE 11.7 ANALYSIS OF GLYCEROL IN WINE AND MUST

		Glycerol (g/l)	
Sample	**Type**	**Biosensor**	**HPLC**
Monte Blanc Koktebel	White, table, semisweet	23.70	6.61
Bear's Blood Koktebel	Red, table, semisweet	23.48	9.05
Bastardo Caliston	Red, table, semisweet	26.02	7.91
Rkatsiteli Koktebel	White, table, dry	21.48	7.69
Aligote Koktebel	White, table, dry	20.58	6.37
Fetyaska Golitsyn wines	White, table, dry	16.12	6.41
Merlot Koblevo	Red, table, dry	20.49	8.74
Cabernet Golitsyn wines	Red, table, dry	19.34	9.6
Pinot-Franc Koktebel	Red, table, dry	13.40	8.72

11.3.4.2 Determination of Ethanol, Lactate and Glucose Concentrations

The content of ethanol, lactate and glucose in 23 samples of wine of different types and in two samples of red and white must was determined using the developed enzyme amperometric biosensors based on oxidase. The data obtained were compared with the results of the ethanol analysis by the distillate densitometry (DD), whereas the HPLC method was used as control for lactate and glucose analyses. The results obtained during the experiments are presented in Table 11.8.

As can be seen from the table, we have a high degree of agreement between the results obtained using the developed amperometric biosensors for the determination of ethanol, with the results of the traditional method of alcohol distillation followed by densitometric determination ($R=0.98$). Moreover, the created biosensor proved its effectiveness also in the analysis of table wines with an ethanol content of about 10% by volume, and in the analysis of strong wines containing approximately twice as much alcohol. When determining lactate and glucose, a high correlation of results was shown with the data obtained using HPLC method (correlation coefficient for glucose $R=0.99$, for lactate $R=0.97$). Therefore, the developed and optimized amperometric biosensor based on immobilized oxidase is an effective and accurate instrument for analysis of wine and wine materials of different types with a wide range of content of both a target analyte and interfering substances.

11.3.4.3 Determination of Lactate and Glucose Concentrations Using a Multibiosensor

Along with the development of monobiosensors, there is a need for simultaneous determination of several analytes. The creation of multibiosensor combining a series of sensitive elements on a single chip is an actual challenge. Therefore, we developed the amperometric enzyme multibiosensor and designed its portable laboratory prototype for control of wine quality. It had the linear ranges of 0.005–0.8 mM lactate and 0.005–1 mM glucose, and sufficient storage stability. The developed system was tested in the analysis of real samples of wine products. Table 11.9 shows the data of the analysis of lactate and glucose in the studied wines.

Noteworthy good correlation of the results obtained by the multibiosensor with those of HPLC analysis (the correlation coefficient for glucose $R=0.998$, for lactate $R=0.718$). Thus, the developed enzyme multibiosensor can be used to control wine and wine materials with a high degree of reliability.

11.3.5 Determination of Sugars in Food

The organization of proper control of the content of mono- and disaccharides at different stages of production is an important task in various industries: food

TABLE 11.8 ANALYSIS OF ETHANOL, LACTATE AND GLUCOSE IN WINE AND MUST

Sample	Type	Ethanol (±0.1% vol.)		Glucose (±0.1 g/l)		Lactate (±0.1 g/l)	
		Biosensor	DD	Biosensor	HPLC	Biosensor	HPLC
Rkatsiteli Koktebel	White, table, dry	12.7	12.0	0.6	0.29	1.62	1.62
Aligote Koktebel	White, table, dry	11.6	11.2	0.47	0.12	2.2	1.98
Merlot Koblevo	Red, table, dry	12.1	12.0	0.84	0.51	1.21	1.27
Cabernet Koktebel	Red, table, dry	9.6	11.2	0.70	0.32	1.04	1.08
Monte Blanc Koktebel	White, table, semisweet	11.2	10.5	16.7	16.40	1.01	1.07
Porte wine 777 red	Red, strong	18.8	17.5	32.8	34.54	0.80	0.81
Porte 777 pink	Pink, strong	18.2	17.5	32.2	33.24	0.63	0.64
Porte white	White, strong	18.4	17.0	31.5	32.78	0.76	0.69
Madeira Massandra	White, strong	19.6	19.0	7.56	7.15	0.53	0.49
Kokur Koktebel	White, dessert	18.1	16.0	87.2	87.12	0.58	0.53
Kara-Dag Koktebel	Red, dessert	16.3	16.0	85.32	83.05	0.98	0.98
Cahors Ukrainian	Red, dessert	15.9	16.1	87.3	87.18	0.95	0.89
White wine wort	-	0.8	-	125.3	124.99	0	0
Red wine wort	-	0.3	-	131.	134.54	0	0

TABLE 11.9 COMPARATIVE ANALYSIS OF GLUCOSE AND LACTATE IN WINES

	Biosensor		HPLC	
Wine	**Glucose (±0.1 g/l)**	**Lactate (±0.1 g/l)**	**Glucose (±0.1 g/l)**	**Lactate (±0.1 g/l)**
Fetyaska	0.44	0.16	0.35	0.22
Cabernet	0.46	0.41	0.44	0.43
Madeira	8.7	0.18	11.27	0.68
Marsala	19.3	0.89	28.62	1.26
Pinot-Franc	2.72	0.55	2.79	0.33

industry, agriculture, and pharmacy, among others. In agriculture, the sucrose content of sugar beet is one of the important indicators that determine the efficiency of technological production of sugar at all stages—from beets growing and preservation to its complete processing. In dairy production, the content of lactose, one of the main components of milk, is an important indicator of the quality. Maltose is also of great importance in the food industry, because its content in maltose molasses determines the quality of the final product, in particular beer and kvass. Information on the presence and concentration of mono- and disaccharides in beverages and food is an important indicator of their quality (BeMiller, 2017; Nielsen, 2003). Control of carbohydrate content is also necessary to monitor the course of fermentation, etc.

We developed a number of electrochemical enzyme biosensors for the determination of basic natural carbohydrates, namely, glucose, maltose, sucrose, fructose and lactose (Dudchenko et al., 2016; Pyeshkova et al., 2015, 2021). Three specific hydrolases, mutarotase and glucose oxidase are used to determine sucrose, maltose and lactose, whereas only one, glucose oxidase, is used to determine glucose. In the course of reactions under the action of the enzymes invertase, β-galactosidase and α-glucosidase, the corresponding substrates—sucrose, lactose and maltose—are cleaved to α-D-glucose, which, under the action of mutarotase, is converted into β-D-glucose. The latter is cleaved by glucose oxidase to hydrogen peroxide and D-gluconolactone. Gluconolactone, in turn, spontaneously hydrolyzes to gluconic acid, which dissociates into an acid residue and a proton; as a result, the solution conductivity changes, which can be recorded using a conductometric transducer. Using sucrose and glucose biosensors, the sugar content was measured in different parts of the sugar beet roots and other samples of the sugar beet homogenate (Table 11.10).

Polarimetry with lead acetate as a control method was carried out at the Institute for sugar beets of the Ukrainian Academy of Agrarian Sciences (Kyiv).

TABLE 11.10 ANALYSIS OF SACCHARINITY OF SUGAR BEETS

Part of Sugar Beet Roots and Other Samples	Saccharinity (%)	
	Biosensor Method	**Control Method**
Top part	9.5±0.5	10.9±0.1
Middle part	14.7±0.6	14.2±0.1
Bottom part	12.9±0.4	13.0±0.1
Sample #1	10.1±0.6	12.4±0.1
Sample #2	18.3±0.7	17.2±0.2
Sample #3	13.3±0.8	15.2±0.2
Sample #4	16.6±0.6	18.8±0.2
Sample #5	18.7±0.5	17.7±0.2
Sample #6	17.6±0.7	16.6±0.2

TABLE 11.11 ANALYSIS OF GLUCOSE AND SUCROSE CONTENT IN JUICES

Samples of Drinks	Glucose (mM)		Sucrose (mM)	
	HPLC (Control)	**Biosensors**	**HPLC (Control)**	**Biosensors**
Pineapple juice	209.4±0.4	192.0±15.2	144.7±1.4	126.9±5.2
Grape juice	441.3±4.0	437.5±17.9	0.126±0.0003	0
Apple juice	170.2±0.6	166.0±10.3	29.07±0.1	32.5±0.9
Apricot and grape nectar	309.6±6.4	334.0±22.0	19.66±0.2	21.7±0.7

As can be seen from Table 11.11, the data obtained by the biosensor method are confirmed by the control method (R=0.914). Using sucrose and glucose biosensors, a number of studies were also carried out to determine sucrose and glucose in various drinks from "Dimes" (Turkey): apple, pineapple and grape juices, obtained from concentrates without sugar addition, and apricot–grape nectar made with the addition of sugar and fruit syrup. As a control, we used high-performance liquid chromatography (HPLC) (carried out at the Middle East Technical University, Ankara, Turkey). The measurement results are shown in Table 11.11.

The table shows good correlation between the results of the analysis of saccharides in juices, obtained by conductometric biosensor, with the data obtained by HPLC (R=0.996).

TABLE 11.12 ANALYSIS OF GLUCOSE AND SUCROSE CONTENT IN JUICES AND HONEY

	Sucrose Concentration (g/l)		Glucose Concentration (g/l)	
Sample	**Biosensor (n=5)**	**HPLC (n=3)**	**Biosensor (n=5)**	**HPLC (n=3)**
Grape and apple juice "Our juice"	2.76±1.27	5.1±0.13	50.9±3.1	53.7±1.35
Apple and carrot juice Truly juice	4.4±1.5	4.1±0.10	7.8±2.5	6±0.15
Apple juice Truly juice	7.9±1.44	10.7±0.27	22±1.5	22.3±0.56
Tomato juice	1.19±0.65	not found	10.4±1.2	11.3±0.26
Apple juice	12.42 ±1.34	12.5±0.31	22.7±2	22.2±0.55
Orange juice Sandora	33.66±3.9	37.8±0.95	19.3±1.8	21.6±0.53
Orange nectar Dooy	14.48±1.16	15.85±0.62	54.68±1.95	59.43±1.41
Orange juice Cappy	73.12±2.05	82.77±0.10	11.53±1.18	9.76±0.09
Apple juice Pinar	14.21±1.30	16.5±0.14	33.33±1.86	36.18±1.17
Orange nectar Göze	6.06±0.34	6.13±0.03	67.56±1.70	70.51±0.95
Flower honey (1)	8.4±3.28	10.5±0.26	26±1.4	27.1±0.68
Flower honey (2)	1.3±4.5	2.3±0.06	32.07±1.8	31.3±0.78

Ten samples of fruit and vegetable juices and nectars and two samples of honey were analyzed. The results of the analysis are shown in Table 11.12. For the analysis, we used commercially available beverage samples obtained by direct extraction of fruits and roots. The preparation of the samples containing fruit pulp consisted only of centrifugation for 5 min at 2,400 rpm. An aliquot of the tested sample was introduced into the measuring cell so that the final dilution was 1,000 times. This allowed working within the linear range of determination of sucrose and glucose, as well as avoiding the influence of the ionic strength of the solution introduced into the measuring cell. Each of the measurements was repeated at least five times. Thus, we received the responses to the introduction of each of the samples. Then, using the method of extrapolation on the calibration curve of biosensors and the developed method of calculation, the values of glucose and sucrose concentration in diluted beverage samples were found. HPLC with subsequent refractometric detection was chosen as a control, as it is known that this method is characterized by high accuracy, and the results obtained can be considered reliable. VarianProStar HPLC system with a column Microsorb

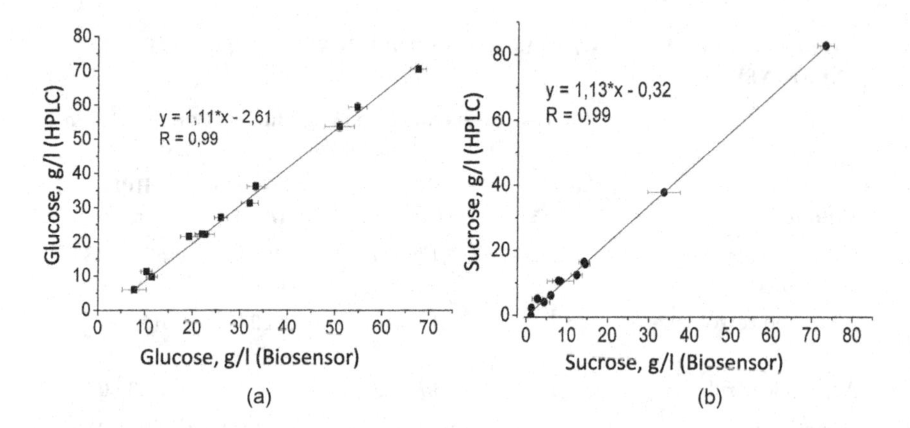

Figure 11.17 Correlation of the results of determination of glucose (A) and sucrose (B) obtained using a multi-enzyme biosensor and HPLC.

MV NH2 (4.6 mm × 250 mm, 5 mm) and liquid phase acetonitrile/H_2O (80:20) was used. The results of determination of the glucose and sucrose concentrations in beverages obtained by the biosensor method and HPLC were quite close (Table 11.12).

The data on the content of sucrose and glucose, obtained by the biosensor, were in good correlation with those of control method (R=0.99) (Figure 11.17). This indicates great prospects for the use of biosensors in the analysis of food products for the presence of sucrose and glucose in appropriate concentrations, which is an important indicator of nutritional value of the product and can sometimes even reveal falsification of the product (counterfeit juices, adulteration of honey by adding sugar).

Also, the samples of milk were analyzed using a three-enzyme lactose amperometric biosensor based on a platinum disk electrode modified by nanosized semipermeable poly (meta-phenylenediamine) film (Figure 11.18).

Lactose in the milk samples was determined using also HPLC, to compare with the lactose amperometric biosensor. For the HPLC determination (AGILENT HPLC 1260 SL), Microsorb-MV100 NH_2 (4.6 mm, 5 μm; R008670005, VARIAN) column was used at 25°C, with the mobile phase acetonitrile/H_2O (80:20) at 1.5 ml/min flow rate and AGILENTG1362A 1260 RID detector. Three milk samples were analyzed and quantified in 0.1–10 g/L lactose range. The results obtained by HPLC and the biosensor method were in good correlation. Therefore, such laboratory prototypes of conductometric multi- and mono-biosensors can be used in the future as a basis for creating a commercial device for simultaneous analysis of basic carbohydrates in the food industry, pharmaceutics and medicine.

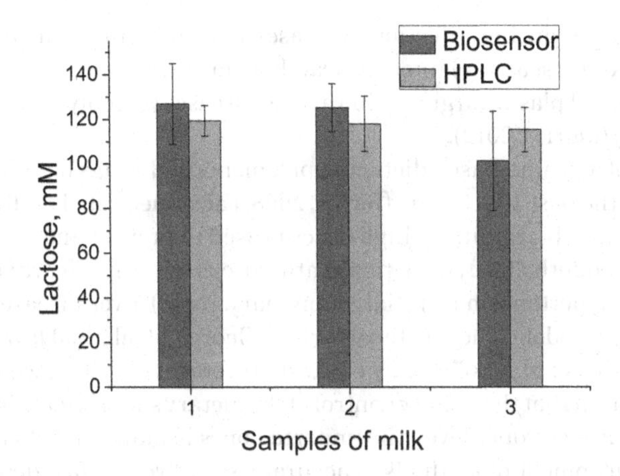

Figure 11.18 Results of determination of lactose concentration in milk samples obtained by amperometric enzyme biosensor and HPLC.

11.3.6 Determination of Arginine in Dietary Supplements and Drugs

Nowadays, treatment and prevention of deficiency of the key nutrients and biologically active substances in the human body are important for the restoration of body functions, and therefore are increasingly used in medicine for therapeutic and rehabilitation purposes. One of the compounds, which plays a critical role in the large number of synthetic processes in the body, is the conditionally essential amino acid arginine (α-amino-δ-guanidino-valeric acid). The presence of the guanidine group in the molecule of arginine makes it the most polar positively charged amino acid at a neutral pH value. Compared to other proteinogenic amino acids, arginine has the highest ratio of nitrogen to carbon in the molecule, which makes it a valuable source of organic nitrogen required in the large quantities for synthesis of nucleic acids and proteins (Winter et al., 2015).

In mammals, arginine is not only a component of peptides and proteins but also a precursor of a number of important individual compounds such as urea, proline, glutamate, creatine, agmatine, L-ornithine, L-citrulline, γ-aminobutyric acid and the only precursor of nitric oxide (NO). Noteworthy that arginine deficiency is a key link in the chain of mechanisms, which lead to the development of endothelial dysfunction and NO deficiency in a number of pathologies, including coronary heart disease and diabetes. Among the main factors, which cause a lack of arginine in the human body, are: (a) low content of arginine in the diet (relevant for sick, young and elderly people); (b) the endogenous arginine due to the high activity of arginase and inducible NO synthase in macrophages during infectious diseases and inflammations occurring in tissues and organs; and (c) intensive catabolism of arginine as a result of high activity of blood plasma arginase in various diseases (hemolytic anemia, sickle cell anemia) (Morris, 2012).

The use of arginine-based dietary supplements and drugs has become widespread over the past 10–15 years (Baylis, 2008; Facchinetti et al., 2007; Lucotti et al., 2009). Today, L-arginine in high doses is used to prevent and treat all manifestations of endothelial dysfunction of atherosclerotic origin (cardiac, cerebral, peripheral), hypertension (arterial, pulmonary, renal), liver diseases, diabetes, obesity, immunodeficiency, osteoarthritis (George et al., 2004; Moutaouakil et al., 2009; Oka et al., 2005; Ozsoy et al., 2011; Yoon et al., 2013), wounds, burns and other. Given that the quality control of the dietary supplements is not always carried out at the proper level (the imported ones usually are not controlled at all), the development of methods of accurate, selective and fast determination of the arginine concentration in such products is important for implementation of the facilitated quality control procedure for such category of goods.

For preparation of biosensors, we used arginase (EC 3.5.3.1) from bovine liver (activity 136 U/mg protein and 80 U/mg of protein), urease (EC 3.5.1.5) from *Canavalia ensiformis* (activity 66.3 U/mg protein) and a recombinant arginine deiminase (EC 3.5.3.6) from *Mycoplasma hominis* produced in cells of *E. coli* cells (volume activity 147 U/ml).

Quantification of arginine with the developed biosensors was performed in the dietary supplements (capsules) of different producers and the oral solution "Arginine Veyron" (Laboratoires Pierre Fabre Medicament, France). To measure the concentration of L-arginine in the oral solution "Arginine Veyron," two methods were used and compared: the classical biosensor method using a calibration curve and the standard addition method (Table 11.13).

TABLE 11.13 DETERMINATION OF ARGININE IN AMPOULES ARGININE VEYRON USING THE CONDUCTOMETRIC BIOSENSOR BASED ON ARGINASE AND UREASE

No. of Exp.	CC (mM)	CDF (mM)	SD/SE (mM)	SC (mM)
1	0.927	954	20.84/9.32	922.4
2	0.955	983		
3	0.955	983		
4	0.940	968		
5	0.908	934		

Notes: CC, individual concentration determined by the method of calibration curve for Arg-HCl; CDF, concentration of arginine taking into account the sample dilution factor (DF was 1029.25); SD, standard deviation; SE, standard error of the mean; SC, concentration stated by the producer.

Quantification of arginine in the dietary supplements was also performed by two control methods, namely, chromatographic and spectrophotometric methods. Chromatographic analysis was performed by the independent laboratory "LLC Expert Center of Diagnostics and Laboratory Support "Biolights" (Kyiv region, Ukraine). The spectrophotometric determination of arginine (enzymatic colorimetric assay) was performed using the commercial L-arginine assay kit (Sigma-Aldrich, cat. no. MAK370). The standard deviation between five repeated measurements (n=5) was found to be 20.84 mM, with a standard error of 9.32 mM (Table 11.13). Taking the reliability assessment as γ=0.95, the found arginine concentration can be given within the following confidence interval: ($946.13 \leq a \leq 982.67$) mM. The coefficient of variation of the obtained values was 2.16%. Comparing the stated value (922.4 mM) and the value obtained experimentally, we suppose that discrepancy between them might originate from the interference of the background (i.e., the presence of the additives in the sample matrix). Determination of the unknown concentrations of arginine in dietary supplements was performed using the method of standard additions. The results of the biosensor analysis are presented in Table 11.14. As found from the obtained data, the results of the biosensor analysis had high correlation with the data stated by the manufacturers (the coefficient of correlation was R^2=0.9992).

The results of the biosensor determination of arginine in dietary supplements in comparison with the results of control methods and data declared by the manufacturers are shown in Table 11.15.

Based on the calculated correlation coefficients, it was found that the results of the determination of L-arginine using the biosensor based on arginine deiminase correlated well with both the results of ion chromatography and spectrophotometric assay (the correlation coefficients were R=0.987 and R=0.997, respectively).

In the next stage, the effect of the procedure of the sample preparation on the analysis results was studied. The study was performed for one of the samples,

TABLE 11.14 DETERMINATION OF ARGININE IN DIETARY SUPPLEMENTS USING A CONDUCTOMETRIC BIOSENSOR BASED ON ARGINASE AND UREASE

Sample	CDF (mg/ml)	CV (%)	SC (mg/ml)
Tri-Amino (Now Foods, USA)	25.515	7.24	25.500
L-arginine 500 mg (Now Foods, USA)	52.298	0.56	50.000
L-arginine (Elit-Pharm, Ukraine)	36.376	3.94	35.000

Notes: CDF, concentration of arginine taking into account the sample dilution factor (DF was 2000); CV, coefficient of variation; SC, concentration stated by the producer.

TABLE 11.15 DETERMINATION OF ARGININE IN DIETARY SUPPLEMENTS USING A CONDUCTOMETRIC BIOSENSOR BASED ON ARGININE DEIMINASE

Sample	Concentration of L-arginine			
	Determined by Biosensor Method (mM)	Determined by Ion Chromatography (mM)	Determined by Spectrophotometric Method (mM)	Declared by the Manufacturer (mM)
L-arginine 500 m (Solgar Vitamin and Herb, USA)	143.86±10.95	150.642	153.4	149.7
L-arginine 500 mg (Now Foods, USA)	146.3±25.56	148.496	148.38	143.8
L-arginine (Elite-Pharm, Ukraine)	114.3±17.4	95.616	109.81	100

namely, for the dietary supplement "L-arginine 500 mg" (Solgar Vitamin and Herb, USA). We compared the results of the arginine quantification in the sample without pretreatment (only dissolving of the capsule content) with the results obtained after mild treatment of the sample to remove the compounds, which might affect the accuracy of the arginine determination. The additional procedures of the sample preparation were as follows: (a) dissolving the content of the capsules followed by filtering the resulting solution through 1 μm and 0.22 μm filters (cat. no. 729228, Chromafil Xtra, Macherey-Nagel, Germany and cat. no. CE 0459, Millex GP, Merck Millipore, Ireland, respectively) and (b) centrifugation of the unfiltered sample at 2,800 rpm for 10 min. The results of the arginine quantification in the sample by the biosensor after the additional preparations are presented in Table 11.16.

The obtained data allow us to conclude that filtering the sample before the analysis leads to the sample preconcentration, which causes the analysis to show a higher concentration of arginine in the sample than it was in the unfiltered sample. Comparing the data of the biosensor determination for the unfiltered sample and the results of the chromatographic analysis, we can assume that dissolving the sample without subsequent filtration is sufficient to obtain reliable results of the analysis. Noteworthy that centrifugation of the unfiltered

TABLE 11.16 THE EFFECT OF THE PROCEDURE OF THE SAMPLE PREPARATION ON THE RESULTS OF DETERMINATION OF L-ARGININE IN THE DIETARY SUPPLEMENT L-ARGININE 500 MG (SOLGAR VITAMIN AND HERB, USA) USING A CONDUCTOMETRIC BIOSENSOR BASED ON ARGININE DEIMINASE

	Procedure of Sample Preparation		
	Sample Filtering	Centrifugation of the Unfiltered Sample	Unfiltered Sample
Concentration of L-arginine, determined by biosensor (mM)	173.61±29.1	135.87±28.37	143.86±10.95

sample turned out to be also unnecessary, since it led to some underestimation of the measurement result, apparently due to the removal of a certain amount of L-arginine from the solution analyzed. The results of the study give grounds to state that the use of the developed biosensor for analysis of the dietary supplement does not require the sample pretreatment. This consideration supports a clear advantage of the biosensor method of analysis over traditional analytical methods.

11.3.7 Determination of Mycotoxins in Food Samples by Sensors Based on Molecularly Imprinted Polymers

Today, contamination of food and feed with mycotoxins is a global problem, as the metabolites of microscopic fungi are toxic to humans and animals in extremely low concentrations, and at favorable temperatures and humidity, they are found in all foods where fungal growth is possible. The most common mycotoxins are secondary metabolites of fungi of the genera *Fusarium, Aspergillus, Myrothecium, Stachybotrys, Trichoderma, Trichothecium, Penicillium*, which usually have neurotoxic, teratogenic, mutagenic, carcinogenic and immunosuppressive effects on animals and humans (Ostry et al., 2017; Pinotti et al., 2016). Consumption of contaminated food causes significant economic losses, both in terms of reduced productivity and associated medical costs. The most effective approaches to overcome this problem are the control of food contamination by fungal infections and the monitoring of the presence of mycotoxins at the stages of harvesting, storage and processing of crops. Due to high toxicity, the accurate, highly selective and rapid monitoring of mycotoxin accumulation is an extremely important task. None of the instrumental and immunochemical methods developed to date make it possible to monitor the presence of mycotoxins in the field/home, while there is an urgent need to analyze samples outside the laboratory. The development of biosensor devices for the detection of mycotoxins based on electrochemical and optoelectronic sensors and natural biomolecules for the detection of these substances is the most promising approach (Stepurska et al., 2015a, 2015b). However, the problem of stabilizing antibodies and receptors in biosensor systems during long-term storage remains unresolved. An alternative approach is the synthesis of artificial mimics of biological receptors by molecular imprinting (Wulff, 1995) and their use for the development of a new generation of sensor devices. This combines the affinity and selectivity of natural receptors with the stability of the physical and mechanical properties of synthetic polymers.

In terms of the practical application of sensor systems based on molecularly imprinted polymers (MIP), the most promising is the production of porous MIP membranes and thin films (Chen et al., 2021; Chrzanowska et al., 2019). Such selective elements of biosensor devices can be easily integrated into electrochemical biosensors as well as generate an optical sensor signal, which can be simply recorded using modern digital technologies, including smartphone cameras, and analyzed using the specialized image analysis software for either Windows or Android operating systems.

The effectiveness of the application of MIP membranes as a basis for easy-to-use, highly sensitive methods for determining small organic molecules in real time, which are suitable to be used outside the laboratory, was proven (Sergeyeva et al., 2017). MIP membranes containing the artificial receptor sites for mycotoxins' binding, in particular aflatoxin B1 and zearalenone, which enter cereal seeds, food and feed as a result of contamination with *Aspergillus* and *Fusarium* fungi, can be effectively used as selective elements of optical biosensor devices based on smartphones or portable fluorimeters (Sergeyeva et al., 2019, 2020, 2022). Such devices make it possible to monitor the presence of mycotoxins in samples of grain and food products in the field since they do not require stationary laboratory equipment and sophisticated sample preparation. MIP membranes function according to the "host-guest" principle and can be obtained using the principle of synthesis of interpenetrating polymer networks. The choice of a functional monomer responsible for the formation of artificial receptor binding sites with high affinity for target mycotoxins is usually optimized taking into account the data obtained by computer simulation methods (molecular dynamics, quantum chemical methods). The MIP membranes selective to aflatoxin B1 and zearalenone, as the basis of fluorescent sensor devices for detecting mycotoxins in real samples, operate according to the following principle. The analytes are selectively adsorbed by artificial receptor sites in the composition of polymers-biomimics (MIP membranes). Further irradiation of the latter with UV light initiates the natural fluorescence of the molecules of aflatoxin B1 and zearalenone, the intensity of which is proportional to their concentration in the analyzed extract obtained from cereals or some other type of foods (Figure 11.19).

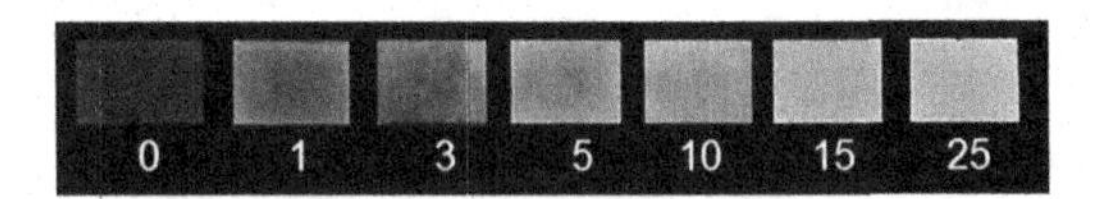

Figure 11.19 Dependence of the typical calibration graph of the biosensor system, based on MIP membranes with receptor properties and a smartphone, for determination of aflatoxin B1 concentration. Intensity of fluorescence of the MIP membranes with zearalenone, adsorbed on their surface, after UV irradiation (λ=320 nm) on zearalenone concentration in the analyzed sample.

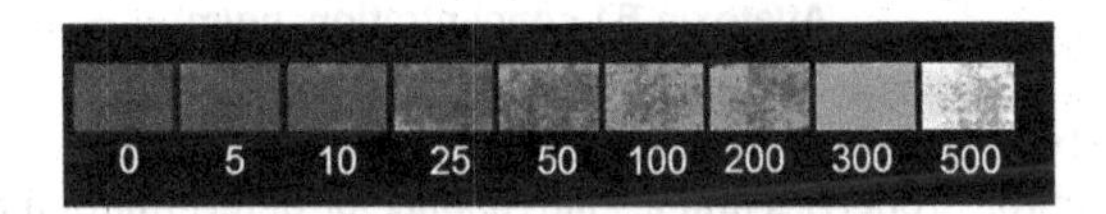

Figure 11.20 Dependence of the fluorescence intensity of MIP membranes with aflatoxin B1, adsorbed on their surface, after UV irradiation (λ=365 nm) on aflatoxin B1 concentration in the analyzed sample.

As can be seen in Figure 11.19, an increase in zearalenone concentration in the analyzed sample provided a proportional increase in the fluorescence intensity of MIP membranes, where zearalenone-selective receptor sites were formed, whereas for blank membranes significantly lower values of the sensor responses were observed.

The fluorescent sensor system for the aflatoxin B1 determination operates in a similar way. The measurements are carried out after irradiation of mycotoxin, adsorbed on the surface of the sensor-sensitive element, with UV light (λ=365 nm) (Figure 11.20).

If the analyte molecules are characterized by low-intensity fluorescence, their detection is based on a competitive variant of analysis using a highly fluorescent structural analog of the target analyte, which competes with it for binding to receptor sites in MIP (Sergeyeva et al., 2020); or using plasmon enhancement of fluorescence due to the formation in the MIP structure nanoparticles of precious metals along with supramolecular receptor sites. In particular, silver nanoparticles are capable of generating local surface plasmon resonance, and, as a consequence, significant enhancement of the fluorescent sensor response (Sergeyeva et al., 2022) after addition of aflatoxin B1 or zearalenone. MIP membranes containing supramolecular recognition sites of aflatoxin B1 and zearalenone, both in the form of polymer membranes with

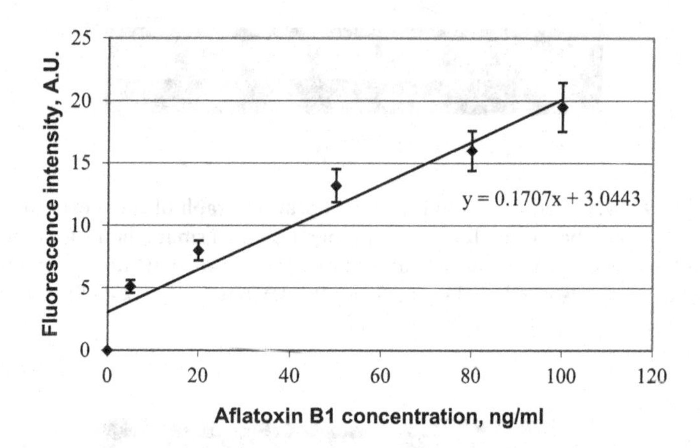

Figure 11.21 Typical calibration graph of the biosensor system, based on MIP membranes with receptor properties and a smartphone, for determination of aflatoxin B1 concentration.

stable physicochemical and mechanical properties and in the form of thin polymeric films immobilized on glass slide surfaces, usually show extremely high (up to 18 months) storage stability at room temperature. Sensors based on the MIP membranes provide highly selective and sensitive analysis of mycotoxins in the subnanomolar range. Additionally, all these systems are very effective for the analysis of the large number of real food samples contaminated with mycotoxins.

To determine the calibration graphs and working parameters of the MIP membrane-based biosensor systems, a set of measurements should be performed to detect aflatoxin B1 and zearalenone, which mostly differed in the conditions of the experiment (different days of the week, different operators, and different sets of laboratory glassware). Typical calibration graphs obtained for these sensors are shown in Figures 11.21 and 11.22.

The obtained calibration graphs of the biosensors for mycotoxins determination were used in measurements of the aflatoxin B1 and zearalenone concentrations in real samples, namely, food extracts.

To study the influence of the components of real samples on the accuracy of the developed method, we used flour samples that did not contain target mycotoxins (determined by classical instrumental and immunochemical methods) and added to them aflatoxin B1 and zearalenone in a known concentration. A number of flour samples of different origins (wheat, rye, corn) from different manufacturers were used in the experiment. In the later experiments, the presence of target mycotoxins in the extracts was determined using the developed sensor systems (Yarynka et al., 2021a, 2021b). The results are shown in Tables 11.17 and 11.18.

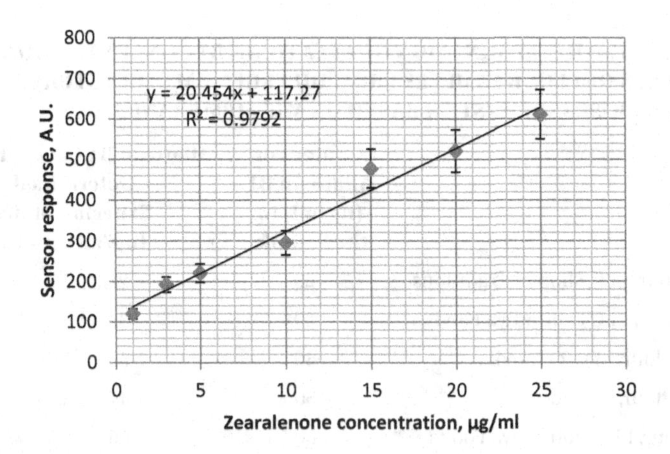

Figure 11.22 Typical calibration graph of a biosensor system, based on MIP membranes with receptor properties, for determination of zearalenone concentration.

Polymers-biomimics in the form of MIP membranes and thin films with receptor properties turned out to be effective for the determination of aflatoxin B1 and zearalenone in real samples (in particular, in flour samples from different manufacturers). The presence of other components in real samples had an insignificant effect on the determination accuracy of target analytes by the developed sensor system (Tables 11.17 and 11.18) (Yarynka et al., 2021a, 2021b). It has been proven that sensor systems based on MIP membranes can be successfully used to determine the presence and concentration of aflatoxin B1 and zearalenone in extracts of flour samples of different origins.

11.4 CONCLUSIONS

Interestingly, a number of electrochemical mono- and multibiosensors based on different enzymes, living cells and biomimics were developed, and their laboratory prototypes were fabricated and thoroughly investigated for real conditions of application (blood serum, blood dialysate, wine and wine must, natural fruit juices, environmental water and in some other samples). Noteworthy, electrochemical biosensors are adaptable to the technologies of large-scale production of miniaturized devices. Concerning the further wide application of the biosensors developed, the obtained results demonstrate the possibility to modulate their main characteristics to comply with the specific requirements for potential practical purposes. Furthermore, diverse biosensor modifications (with genetically modified enzymes and microorganisms, additional

TABLE 11.17 DETERMINATION OF AFLATOXIN B1 CONCENTRATION IN EXTRACTS OF FLOUR SAMPLES FROM DIFFERENT MANUFACTURERS USING A BIOSENSOR SYSTEM BASED ON MIP MEMBRANES

Sample	Content of Aflatoxin B1 (ng/ml) in the Sample	Aflatoxin B1 Concentration Determined Experimentally (ng/ml), n=15
Corn flour, LLC Shop of Traditions	20	26.3±3.7
Corn flour, LLC Dobrodiya Foods	30	32.9±5.1
Wheat flour, LLC Zernovita	40	38.5±1.9
Wheat flour, LLC Aro	50	47.9±6.9
Rrye flour, LLC Dobrodiya Foods	60	66.3±6.5
Corn flour, LLC Alta Vista	80	85.3±6.9
Wheat flour, LLC EuroMill	90	93.5±11.2
Wheat flour, LLC Khutorok	150	143.9±19.8
Wheat flour, LLC Zernari	250	259.2±21.3
Wheat flour, LLC KyivMlyn	300	316.5±22.1

Note: MIP, Molecularly imprinted polymer.

TABLE 11.18 DETERMINATION OF ZEARALENONE CONCENTRATION IN FLOUR SAMPLES FROM DIFFERENT MANUFACTURERS USING A BIOSENSOR SYSTEM BASED ON POLYMERS-BIOMIMICS WITH RECEPTOR PROPERTIES

Sample Number (Description)	Zearalenone Concentration (μg/ml)	
	In the Sample	Determined Experimentally (n=15)
1 (wheat flour, LLC "Zernari")	1	1.5±0.24
2 (wheat flour, LLC "Khutorok")	5	4.8±0.5
3 (wheat flour, LLC "EuroMill")	10	10.5±0.73
4 (corn flour, LLC "Shop of Traditions")	25	23.7±1.8
5 (corn flour, LLC "Alta-Vista")	50	50.1±0.53

membranes, different nanoscaled materials, etc.) can be elaborated as well. In brief, the potential of this technology has been shown here; its wide application is pending.

ACKNOWLEDGMENTS

The work was carried out with the financial support of the National Academy of Sciences of Ukraine within the framework of the target research program "Smart" sensors of the new generation based on modern materials and technologies" and the National Research Foundation of Ukraine in the framework of the competition of projects for research and development "Support of research of leading and young scientists" (project 2020.02/0097).

REFERENCES

Arkhipova, V.N., Dzyadevych, S.V., Soldatkin, A.P., El'skaya, A.V., Jaffrezic-Renault, N., Jaffresic, H., & Martlet, C. (2005). Multibiosensor based on enzyme inhibition analysis for determination of different toxic substances. *Talanta*, 55(5), 919–927. https://doi.org/10.1016/S0039-9140(01)00495-7

Avramescu, A., Noguer, T., Magearu, V., & Marty, J.L. (2001). Chronoamperometric determination of d-lactate using screen-printed enzyme electrodes. *Analytica Chimica Acta*, 433(1), 81–88. https://doi.org/10.1016/S0003-2670(00)01386-6

Baup, M. (1826). Extrait d'une lettre sur plusieurs nouvelles substances. *Annales des Chimie et des Physique*, 31, 108–109.

Baylis, C. (2007). Nitric oxide deficiency in chronic kidney disease. *American Journal of Physiology-Renal Physiology*, 294(1), F1–F9. https://doi.org/10.1152/ajprenal.00424.2007

BeMiller, J.N. (2017). Carbohydrate Analysis. In: Nielsen, S.S. (Ed.), *Food Analysis. Food Science Text Series*. Springer, Cham. https://doi.org/10.1007/978-3-319-45776-5_19

Bundesgesetzblatt. (1986). Verordnung über Trinkwasser und über Wasser für Lebensmittelbetriebe. *Bundesgesetzblatt Tail 1*, 22, 760–773. http://www.bgbl.de/xaver/bgbl/start.xav?startbk=Bundesanzeiger_BGBl&jumpTo=bgbl186s0760.pdf

Chen, Q., Liu, X., Yang, H., Zhang, S., Song, H., & Zhu, X. (2021). Preparation and evaluation of magnetic graphene oxide molecularly imprinted polymers (MIPs-GO-Fe_3O_4@SiO_2) for the analysis and separation of tripterine. *Reactive and Functional Polymers*, 169, 105055, https://doi.org/10.1016/j.reactfunctpolym.2021.105055

Chrzanowska, A.M., Díaz-Álvarez, M., Wieczorek, P.P., Poliwoda, A., & Martín-Esteban, A. (2019). The application of the supported liquid membrane and molecularly imprinted polymers as solid acceptor phase for selective extraction of biochanin A from urine. *Journal of Chromatography A*, 1599, 9–16. https://doi.org/10.1016/j.chroma.2019.04.005

Compagnone, D., Esti, M., Messia, M.C., Peluso, E., & Palleschi, G. (1998). Development of a biosensor for monitoring of glycerol during alcoholic fermentation. *Biosensors and Bioelectronics*, 13(7–8), 875–880. https://doi.org/10.1016/S0956-5663(98)00055-4

Cosnier, S. (2015). *Electrochemical Biosensors*. Pan Stanford Publishing, Singapore.

Dudchenko, O.Y., Pyeshkova, V.M., Soldatkin, O.O., Akata, B., Kasap, B.O., Soldatkin, A.P., & Dzyadevych, S.V. (2016). Development of silicalite/glucose oxidase-based biosensor and its application for glucose determination in juices and nectars. *Nanoscale Research Letters*, 11, 59. https://doi.org/10.1186/s11671-016-1275-2

Dzantiev, B.B., Yazynina, E.V., Zherdev, A.V., Plekhanova, Yu.V., Reshetilov, A.N., Chang, S.C., & McNeil, C.J. (2004). Determination of the herbicide chlorsulfuron by amperometric sensor based on separation-free bienzyme immunoassay. *Sensors and Actuators*, 98(2–3), 254–261. https://doi.org/10.1016/j.snb.2003.10.021

Dzyadevych, S.V. (2005). Conductometric enzyme biosensors theory, technology and application. *Biopolymers and Cell*, 21(2), 91–106. https://doi.org/10.7124/bc.0006E1

Dzyadevych, S.V., Arkhypova, V.N., Soldatkin, A.P., El'skaya, A.V., Martelet, C., & Jaffrezic-Renault, N. (2004). Enzyme biosensor for tomatine detection in tomatoes. *Analytical Letters*, 37(8), 1–14. https://doi.org/10.1081/AL-120037591

Dzyadevych, S.V., & Soldatkin, A.P. (2008). Solid-state electrochemical enzyme biosensors. IMBG, Kyiv, Ukraine.

Esti, M., Volpe, G., Micheli, L., Delibato, E., Compagnone, D., Moscone, D., & Palleschi, G. (2004). Electrochemical biosensors for monitoring malolactic fermentation in red wine using two strains of *Oenococcus oeni*. *Analytica Chimica Acta*, 513(1), 357–364. https://doi.org/10.1016/j.aca.2003.12.011

European Commission. (2014). Commission Regulation (EC) No 1881/2006 of 19 December 2006 setting maximum levels for certain contaminants in foodstuffs. Amended values. Retrieved from http://eur-lex.europa.eu/legal-content/EN/TXT/PDF/?uri=CELEX:02006R1881-20140701&from=FR

Facchinetti, F., Saade, G.R., Neri, I., Pizzi, C., Longo, M., & Volpe, A. (2007). L-arginine supplementation in patients with gestational hypertension: A pilot study. *Hypertension in Pregnancy*, 26(1), 121–130. https://doi.org/10.1080/10641950601147994

Friedman, M., & McDonald, G. (1997). Potato glycoalkaloids: Chemistry, analysis, safety, and plant physiology. *Critical Reviews in Plant Sciences*, 16(1), 55–132. https://doi.org/10.1080/07352689709701946

George, J., Shmuel, S.B., Roth, A., Herz, I., Izraelov, S., Deutsch, V., Keren, G., & Miller, H. (2004). L-arginine attenuates lymphocyte activation and anti-oxidized LDL antibody levels in patients undergoing angioplasty. *Atherosclerosis*, 174(2), 323–327. https://doi.org/10.1016/j.atherosclerosis.2004.01.025

Goriushkina, T.B., Shkotova, L.V., Gayda, G.Z., Klepach, H.M., Gonchar, M.V., Soldatkin, A.P., & Dzyadevych, S.V. (2010). Amperometric biosensor based on glycerol oxidase for glycerol determination. *Sensors and Actuators B Chemical*, 144(2), 361–367. https://doi.org/10.1016/j.snb.2008.11.051

Goriushkina, T.B., Soldatkin, A.P., & Dzyadevych, S.V. (2009). Application of amperometric biosensors for analysis of ethanol, glucose and lactate in wine. *Journal of Agricultural Chemistry*, 57(15), 6528–6535. https://doi.org/10.1021/jf9009087

Hellenäs, K.E., & Branzell, C. (1997). Liquid chromatographic determination of the glycoalkaloids alpha-solanine and alpha-chaconine in potato tubers: NMKL Interlaboratory Study. Nordic Committee on Food Analysis. *Journal of AOAC International*, 80(3), 549–554. https://doi.org/10.1093/jaoac/80.3.549

Johnson, B.L. (2000). A review of health-based comparative risk assessments in the United States. *Reviews on Environmental Health*, 15(3), 273–287. https://doi.org/10.1515/REVEH.2000.15.3.273

Johnson, H., & Hellenas, K.E. (1983). Glykoalkaloider i svensk potatis. (Glycoalkaloids in Swedish potatoes). *VårFöda*, 35, 299–314.

Kukla, A.L., Pavluchenko, A.S., Goltvianskyi, Y.V., Soldatkin, A.A., Arkhypova, V.M., Dzyadevych, S.V., & Soldatkin, A.P. (2008). Sensor arrays based on the differential ISFET elements for monitoring of toxic substances of natural and artificial origin, *Sensor Electronics and Microsystem Technologies*, 5(2), 69–73. https://doi.org/10.18524/1815-7459.2008.2.114446

Kumara, P., Kim, K., Bansal, V., Lazarides, T., & Kumar, N. (2017). Progress in the sensing techniques for heavy metal ions using nanomaterials. *Journal of Industrial and Engineering Chemistry*, 54, 30–43. https://doi.org/10.1016/j.jiec.2017.06.010

Lucotti, P., Monti, L., Setola, E., La Canna, G., Castiglioni, A., Rossodivita, A., Pala, M. G., Formica, F., Paolini, G., Catapano, A.L., Bosi, E., Alfieri, O., & Piatti, P. (2009). Oral L-arginine supplementation improves endothelial function and ameliorates insulin sensitivity and inflammation in cardiopathic nondiabetic patients after an aortocoronary bypass. *Metabolism*, 58(9), 1270–1276. https://doi.org/10.1016/j.metabol.2009.03.029

Marks, R. S., (Ed.), Lowe, C. R., (Ed.), Cullen, D. C. (Ed.), Weetall, H. H. (Ed.), & Karube, I. (2007). *Handbook of Biosensors and Biochips*. Wiley.

Morris, S.M. (2012). Arginases and arginine deficiency syndromes. *Current Opinion in Clinical Nutrition and Metabolic Care*, 15(1), 64–70. https://doi.org/10.1097/MCO.0b013e32834d1a08

Moutaouakil, F., El Otmani, H., Fadel, H., Sefrioui, F., & Slassi, I. (2009). L-arginine efficiency in MELAS syndrome. A case report. *Revue Neurologique*, 165(5), 482–485. https://doi.org/10.1016/j.neurol.2008.08.006

Nielsen, S.S. (2003). *Food Analysis*, 3rd edn. Kluwer Academic, New York.

Oka, R.K., Szuba, A., Giacomini, J.C., & Cooke, J.P. (2005). A pilot study of L-arginine supplementation on functional capacity in peripheral arterial disease. *Vascular Mediccine*, 10(4), 265–274. https://doi.org/10.1191/1358863x05vm637oa

Ostry, V., Malir, F., Toman, J., & Grosse, Y. (2017). Mycotoxins as human carcinogens—The IARC Monographs classification. *Mycotoxin Research*, 33(1), 65–73. https://doi.org/10.1007/s12550-016-0265-7

Ozsoy, Y., Ozsoy, M., Coskun, T., Namli, K., Var, A., & Ozyurt, B. (2011). The effects of L-arginine on liver damage in experimental acute cholestasis an immunohistochemical study. *HPB Surgery*, 2011, Article ID 306069. https://doi.org/10.1155/2011/306069

Parra, A., Casero, E., Vázquez, L., Pariente, F., & Lorenzo, E. (2006). Design and characterization of a lactate biosensor based on immobilized lactate oxidase onto gold surfaces. *Analytica Chimica Acta*, 555(2), 308–315. https://doi.org/10.1016/j.aca.2005.09.025

Pavluchenko, A.S., Kukla, A.L., Goltvianskyi, Yu.V., Soldatkin, O.O., Arkhypova, V.M., Dzyadevych, S.V., & Soldatkin, A.P. (2011). Investigation of stability of the pH-sensitive field-effect transistor characteristics. *Sensor Letters*, 9(6), 2392–2396. https://doi.org/10.1166/sl.2011.1797

Perumal, V., & Hashim, U. (2014). Advances in biosensors: Principle, architecture and applications. *Journal of Applied Biomedicine*, 12(1), 1–15. https://doi.org/10.1016/j.jab.2013.02.001

Pinotti, L., Ottoboni, M., Giromini, C., Dell'Orto, V., & Cheli, F. (2016). Mycotoxin Contamination in the EU Feed Supply Chain: A Focus on Cereal Byproducts. *Toxins (Basel)*, 8(2), 45–69. https://doi.org/10.3390/toxins8020045

Pyeshkova, V.M., Dudchenko, O.Y., Soldatkin, O.O., Alekseev, S.A., Seker, T., Akata Kurç, B., & Dzyadevych, S.V. (2022). Development of three-enzyme lactose amperometric biosensor modified by nanosized poly (meta-phenylenediamine) film. *Applied Nanoscience*, 12, 1267–1274. https://doi.org/10.1007/s13204-021-01859-8

Pyeshkova, V.M., Dudchenko, O.Y., Soldatkin, O.O., Kasap, B.O., Lagarde, F., AkataKurç, B., & Dzyadevych, S.V. (2015). Application of silicalite-modified electrode for the development of sucrose biosensor with improved characteristics. *Nanoscale Research Letters*, 10, 149. https://doi.org/10.1186/s11671-015-0853-z

Rotariu, L., Lagarde, F., Jaffrezic-Renault, N., & Bala, C. (2016). Electrochemical biosensors for fast detection of food contaminants – Trends and perspective. *TrAC Trends in Analytical Chemistry*, 79, 80–87. https://doi.org/10.1016/j.trac.2015.12.017

Saurina, J. (2010). Characterisation of wines using compositional profiles and chemometrics. *Trends in Analytical Chemistry*, 29(3), 234–245. https://doi.org/10.1016/j.trac.2009.11.008

Schuman, S.H., & Simpson, W.M. (1997). A clinical historical overview of pesticide health issues. *Occupational Medicine (Philadelphia, Pa.)*, 12(2), 203–207.

Sergeyeva, T., Yarynka, D., Dubey, L., Dubey, I., Piletska, E., Linnik, R., Antonyuk, M., Ternovska, T., Brovko, O., Piletsky, S., & El'skaya, A. (2020). Sensor based on molecularly imprinted polymer membranes and smartphone for detection of *Fusarium* contamination in cereals. *Sensors*, 20(15), 4304–4324. https://doi.org/10.3390/s20154304

Sergeyeva, T., Yarynka, D., Lytvyn, V., Demydov, P., Lopatynskyi, A., Stepanenko, Y., Brovko, O., Pinchuk, A., & Chegel, V. (2022). Highly-selective and sensitive plasmon-enhanced fluorescence sensor of aflatoxins. *Analyst*, 147(6), 1135–1143. https://doi.org/10.1039/D1AN02173G

Sergeyeva, T., Yarynka, D., Piletska, E., Lynnik, R., Zaporozhets, O., Brovko, O., Piletsky, S., & El'skaya, A. (2017). Fluorescent sensor systems based on nanostructured polymeric membranes for selective recognition of Aflatoxin B1. *Talanta*, 175, 101–107. https://doi.org/10.1016/j.talanta.2017.07.030

Sergeyeva, T., Yarynka, D., Piletska, E., Lynnik, R., Zaporozhets, O., Brovko, O., Piletsky, S., & El'skaya, A. (2019). Development of a smartphone-based biomimetic sensor for aflatoxin B1 detection using molecularly imprinted polymer membranes. *Talanta*, 201, 204–210. https://doi.org/10.1016/j.talanta.2019.04.016

Shkotova, L.V., Goriushkina, T.B., Tran-Minh, C., Soldatkin, A.P., & Dzyadevych, S.V. (2008). Amperometric biosensor for lactate analysis in wine and must during fermentation. *Materials Science and Engineering C*, 28(5–6), 943–948. https://doi.org/10.1016/j.msec.2007.10.038

Shkotova, L.V., Piechniakova, N.Y., Kukla, O.L., & Dzyadevych, S.V. (2016). Thin-film amperometric multibiosensor for simultaneous determination of lactate and glucose in wine. *Food Chemistry*, 197(Part A), 972–978. https://doi.org/10.1016/j.foodchem.2015.11.066

Shkotova, L.V., Soldatkin, A.P., Gonchar, M.V., Schuhmann, W., & Dzyadevych, S.V. (2006). Amperometric biosensor for ethanol detection based on alcohol oxidase immobilised within electrochemically deposited Resydrol film. *Materials Science and Engineering C*, 26(2–3), 411–414. https://doi.org/10.1016/j.msec.2005.10.031

Sin, M.L., Mach, K.E., Wong, P.K., & Liao, J.C. (2014). Advances and challenges in biosensor-based diagnosis of infectious diseases. *Expert Review of Molecular Diagnostics*, 14(2), 225–244. https://doi.org/10.1586/14737159.2014.888313

Soldatkin, O.O, Pavluchenko, O.S., Kukla, O.L, Arkhypova, V.M., Dzyadevych, S.V., Soldatkin, A.P., & El'skaya, A.V. (2008). Optimization of multibiosensor usage for inhibitory analysis of toxins. *Biopolymers and Cell*, 24(6), 494–502. https://doi.org/10.7124/bc.0007C2

Soldatkin, O.O., Peshkova, V.M., Saiapina, O.Y., Kucherenko, I.S., Dudchenko, O.Y., Melnyk, V.G., Vasylenko, O.D., Semenycheva, L.M., Soldatkin, A.P., & Dzyadevych, S.V. (2013). Development of conductometric biosensor array for simultaneous determination of maltose, lactose, sucrose and glucose. *Talanta*, 115, 200–207. https://doi.org/10.1016/j.talanta.2013.04.065

Stepurska, K.V., Soldatkin, O.O., Arkhypova, V.M., Soldatkin, A.P., Lagarde, F., Jaffrezic-Renault, N., & Dzyadevych, S.V. (2015a). Development of novel enzyme potentiometric biosensor based on pH-sensitive field-effect transistors for aflatoxin B1 analysis in real samples. *Talanta*, 144, 1079–1084. https://doi.org/10.1016/j.talanta.2015.07.068

Stepurska, K.V., Soldatkin, O.O., Kucherenko, I.S., Arkhypova, V.M., Dzyadevych, S.V., & Soldatkin, A.P. (2015b). Feasibility of application of conductometric biosensor based on acetylcholinesterase for the inhibitory analysis of toxic compounds of different nature. *Analytica Chimica Acta*, 854, 161–168. https://doi.org/10.1016/j.aca.2014.11.027

Thévenot, D.R., Toth, K., Durst, R.A., & Wilson, G.S. (1999). Electrochemical biosensors: Recommended definitions and classification (Technical report). *Pure Applied Chemistry*, 71(12), 2333–2348. https://doi.org/10.1351/pac199971122333

Verscheuren, K. (1983). *Handbook of Environmental Data on Organic Chemicals*. Van Norstrand Reinhold, New York.

Winter, G., Todd, C.D., Trovato, M., Forlani, G., & Funck, D. (2015). Physiological implications of arginine metabolism in plants. *Frontiers in Plant Science*, 6, 534. https://doi.org/10.3389/fpls.2015.00534

Wulff, G. (1995). Molecular imprinting in cross-linked materials with the aid of molecular templates – A way towards artificial antibodies. *Angewandete Chemie International Edition*, 34(17), 1812–1832. https://doi.org/10.1002/anie.199518121

Yarynka, D., Sergeyeva, T., Piletska, E., Linnik, R., Antonyuk, M., Brovko, O., Piletsky, S., & El'skaya, A. (2021). Validation of aflatoxin B1 MI P membrane-based smartphone sensor system for real sample applications. *Biopolymers and Cell*, 37(5), 346–356. https://doi.org/10.7124/bc.000A60

Yarynka, D.V., Sergeyeva, T.A., Piletska, E.V., Stepanenko, Y., Brovko, O.O., Piletsky, S.A., El'skaya, A.V. (2021). Zearalenone-selective biomimetic-based sensor system and its validation for real samples' analysis. *Biopolymers and Cell*, 37(6), 438–446. https://doi.org/10.7124/bc.000A69

Yatsenko, V. (1996). Determining the characteristics of water pollutants by neural sensors and pattern recognition methods. *Journal of Chromatography A*, 722(1–2), 233–243. https://doi.org/10.1016/0021-9673(95)00571-4

Yoon, J., Frankel, A., Feun, L., Ekmekcioglu, S., & Kim, K. (2013). Arginine deprivation therapy for malignant melanoma. *Clinical Pharmacology*, 5(1), 11–19. https://doi.org/10.2147/CPAA.S37350

Zeravik, J., Hlavacek, A., Lacina, K., & Skladal, P. (2009). State of the art in the field of electronic and bioelectronic tongues–towards the analysis of wines. *Electroanalysis*, 21(23), 2509–2520. https://doi.org/10.1002/elan.200900285

Index

Note: **Bold** page numbers refer to tables and *italic* page numbers refer to figures.

Printed in the United States
by Baker & Taylor Publisher Services

Printed in the United States
by Baker & Taylor Publisher Services